AF749782

Organosilanes in Radical Chemistry

Organosilanes in Radical Chemistry

Chryssostomos Chatgilialoglu
Consiglio Nazionale delle Ricerche,
Bologna, Italy

John Wiley & Sons, Ltd

Copyright © 2004 John Wiley & Sons Ltd, The Atrium, Southern Gate,
Chichester, West Sussex PO19 8SQ, England
Telephone (+44) 1243 779777
Email (for orders and customer service enquiries): cs-books@wiley.co.uk
Visit our Home Page on www.wileyeurope.com or www.wiley.com

All Rights Reserved. No part of this publication may be reproduced, stored in a retrieval system or transmitted in any form or by any means, electronic, mechanical, photocopying, recording, scanning or otherwise, except under the terms of the Copyright, Designs and Patents Act 1988 or under the terms of a licence issued by the Copyright Licensing Agency Ltd, 90 Tottenham Court Road, London W1T 4LP, UK, without the permission in writing of the Publisher. Requests to the Publisher should be addressed to the Permissions Department, John Wiley & Sons Ltd, The Atrium, Southern Gate, Chichester, West Sussex PO19 8SQ, England, or emailed to permreq@wiley.co.uk, or faxed to (+44) 1243 770620.

This publication is designed to provide accurate and authoritative information in regard to the subject matter covered. It is sold on the understanding that the Publisher is not engaged in rendering professional services. If professional advice or other expert assistance is required, the services of a competent professional should be sought.

Other Wiley Editorial Offices

John Wiley & Sons Inc., 111 River Street,
Hoboken, NJ 07030, USA

Jossey-Bass, 989 Market Street,
San Francisco, CA 94103-1741, USA

Wiley-VCH Verlag GmbH, Boschstr. 12, D-69469
Weinheim, Germany

John Wiley & Sons Australia Ltd, 33 Park Road,
Milton, Queensland 4064, Australia

John Wiley & Sons (Asia) Pte Ltd, 2 Clementi Loop #02-01,
Jin Xing Distripark, Singapore 129809

John Wiley & Sons Canada Ltd, 22 Worcester Road,
Etobicoke, Ontario, Canada M9W 1L1

Library of Congress Cataloging-in-Publication Data

Chatgilialoglu, Chryssostomos.
Organosilanes in radical chemistry/Chryssostomos Chatgilialoglu.
p. cm.
Includes bibliographical references and index.
ISBN 0-471-49870-X (cloth : alk. paper)

1. Organosilicon compounds. 2. Free radicals (Chemistry) I. Title.
QD305. S54C43 2004
547′.08–dc22

2003016573

British Library Cataloguing in Publication Data

A catalogue record for this book is available from the British Library

ISBN 0 471 49870 X

Typeset in 10/12 pt Times by Kolam Information Services Pvt Ltd, Pondicherry, India.
Printed and bound in Great Britain by TJ International, Padstow, Cornwall
This book is printed on acid-free paper responsibly manufactured from sustainable forestry in which at least two trees are planted for each one used for paper production.

DEDICATION

To my parents Χρήστος and Πηνελόπη

CONTENTS

PREFACE

A large number of papers dealing with silyl radicals, dating back to the late 1940s, have been published. To my knowledge, there are no books on this subject. Several reviews and book chapters on silyl radicals have appeared from time to time on specific aspects. This book focuses on the recent literature of silyl radicals in the liquid phase. However, some related gas-phase data of pivotal species such as the $H_3Si\bullet$ and $Me_3Si\bullet$ radicals are taken into consideration when necessary. In the last decade, silyl radicals have thoroughly penetrated areas as diverse as organic synthesis and material sciences, and the eight chapters in this book survey the most exciting aspects of their chemistry.

Fundamental aspects of silyl radicals such as methods of formation, structural characteristics and thermodynamic data are discussed in Chapters 1 and 2. We will see that α-substituents have a profound influence on the geometry of silyl radicals as well as on the homolytic bond dissociation energies of silicon–silicon and silicon–heteroatom bonds. Gas-phase data are essential in order to understand the thermochemistry of organosilanes. Chapter 3 considers the elementary reaction steps, which play an essential role in the majority of radical chain reactions involving organosilanes. Research over the last two decades has indeed revealed the factors governing the reactivity of silicon hydrides towards a variety of radicals. In Chapters 4, 5 and 7, the concepts and guidelines for using silicon hydrides as radical-based reducing agents and as mediators for consecutive radical reactions will be illustrated. Nowadays radical chain reactions are of considerable importance in the development of synthetic methodologies and have allowed the synthesis of complicated polyfunctional molecules to be afforded in recent years. The art of synthesizing complex molecules from relatively simple starting materials in one-pot reactions driven by the radical reactivity is really impressive and will be illustrated by numerous examples. In Chapter 6 the various unimolecular reactions involving silyl radicals are considered, which have enabled synthetic organic chemists to explore reactivities and strategies incorporating these processes. In Chapter 8 silyl radicals in polymers and materials are contemplated. A unified mechanism for understanding the oxidation of poly(hydrosilane)s and hydrogen-terminated silicon surfaces has been proposed. In Chapter 8, a general discussion of how silicon surfaces are used to obtain monolayers is also presented. As mentioned above, it is not my purpose to consider the entire chemistry of silyl radicals or to discuss their applications. For example, I have taken into consideration the radical chemistry dealing with the monolayer formation of the silicon surfaces, but I have not entered into the field of silicon-containing ceramics obtained by chemical vapour deposition techniques, although gaseous silyl radicals are thought to be essential.

Since this book mainly deals with the literature on silyl radicals after the 1980s, the references quoted for the early work are not always the seminal ones but the available reviews. I hope that early experts in the field will forgive me if they find their pet paper uncited. I have tried to maintain an essential simplicity and readability of the text, and hope that I have succeeded so that the book is easily consulted also by nonexperts. I also hope that this book serves as an important link between the various areas of chemistry.

Chryssostomos Chatgilialoglu
Bologna, July 2003

ACKNOWLEDGEMENTS

I thank Keith U. Ingold for having introduced me to this subject. When I arrived in Ottawa at the National Research Council of Canada in 1979 for three years' postdoctoral work with him, very little was known on the reactivity of silyl radicals. At that time, several papers dealing with kinetics of silyl radicals were published, which allowed the reactivity of silyl radical to be translated into a quantitative base. Special thanks go to David Griller for his collaboration on the initial work on hydrogen donor abilities of silicon hydrides during the late 1980s.

Many thanks to Fluka Chemie AG for the Prize of 'Reagent of the Year 1990'. The discovery of tris(trimethylsilyl)silane as a good radical-based reducing agent stimulated our research during the 1990s. I am grateful to the colleagues who have worked with me over these years on this subject for the privilege of their collaboration and friendship. I am especially grateful to Carla Ferreri for her longstanding collaboration during these years as well as her continuing support and encouragement for completing this book.

Finally, I thank Hanns Fischer, Philippe Renaud, Vitaliy I. Timokhin and Andreas A. Zavitsas, for having critically read some of the chapters and for their valuable suggestions.

1 Formation and Structures of Silyl Radicals

1.1 METHODS OF GENERATION OF SILYL RADICALS

The reaction of atoms, radicals or excited triplet states of some molecules with silicon hydrides is the most important way for generating silyl radicals [1,2]. Indeed, Reaction (1.1) in solution has been used for different applications. Usually radicals X• are centred at carbon, nitrogen, oxygen, or sulfur atoms depending on the objective.

$$R_3SiH + X\bullet \longrightarrow R_3Si\bullet + XH \quad (1.1)$$

For example, photochemically produced *t*-BuO• radicals have been mainly used for the generation of silyl radicals to be studied by spectroscopic techniques (see Chapters 1 and 2). Carbon-centred X• radicals are of great importance in chemical transformations under reducing conditions, where an appropriate silane is either the reducing agent or the mediator for the formation of new bonds (see Chapters 4, 5 and 7). Chapter 3 is entirely dedicated to the hydrogen donor abilities of silicon hydrides towards a variety of radicals. In particular, a large number of available kinetic data are collected and analysed in terms of the substituent influence on the Si—H moiety and on the attacking radical.

Several methods for generating of silyl radicals exist using direct interaction of silanes with light (Reaction 1.2). However, none of them is of general applicability, being limited to some specific application [3].

$$R_3Si{-}Y \xrightarrow{h\nu} R_3Si\bullet + Y\bullet \quad (1.2)$$

The best example is the photochemistry of aryldisilanes, which undergo essentially three principal photoprocesses [4–6]. These include the silylene extru-

Organosilanes in Radical Chemistry C. Chatgilialoglu
© 2004 John Wiley & Sons, Ltd ISBN: 0-471-49870-X

sion, 1,3-Si shift to the *ortho* position of the aryl group to afford silatrienes and homolytic cleavage of Si—Si bond to give silyl radicals. Silenic products are derived from the lowest excited singlet state and are the major products in nonpolar solvents, while silyl radicals are derived from the lowest excited triplet state and are the major products in polar solvents such as acetonitrile [5]. The homolytic cleavage can also be promoted when the 1,3-Si migration is sterically hindered as shown in Reaction (1.3) [7]. Regarding the alkyl substituted oligo- and polysilanes, the silylene extrusion is the principal photoprocess in the far-UV photochemistry whereas reductive elimination of silylsilylene and homolytic Si—Si scission is also detected [8,9].

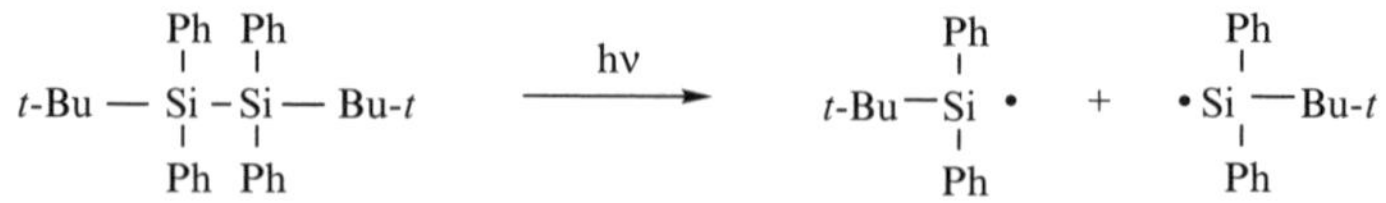

(1.3)

Organosiliconboranes having bulky substituents on the boron, e.g. $R_3SiB[N(CHMe)_2]_2$, exhibit UV absorption at wavelengths longer than 300 nm. Photolysis of this band afforded a pair of silyl and boryl radicals that can be trapped quantitatively by nitroxide (TEMPO) as shown in Reaction (1.4) [10].

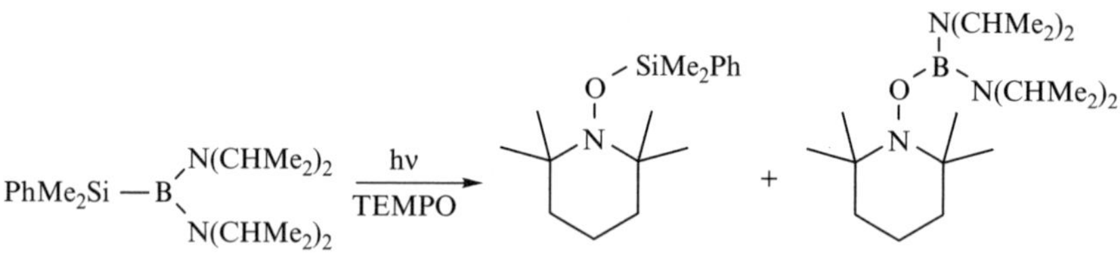

(1.4)

Silyl radicals have been produced by one-electron oxidation of silyl metals [11]. This is found to be the method of choice for the generation of persistent silyl radicals and allowed the preparation of the first isolable silyl radical (see later in this chapter). Reactions (1.5) and (1.6) show two sterically hindered silyl anions with Na^+ as the counter-cation, and their oxidation by the nitrosyl cation [12] and the complex $GeCl_2$/dioxane [13], respectively.

$$(t\text{-BuMe}_2\text{Si})_3\text{SiNa} \xrightarrow[n\text{-heptane}]{\text{NO}^+\text{BF}_4^-} (t\text{-BuMe}_2\text{Si})_3\text{Si}\bullet \qquad (1.5)$$

$$(t\text{-Bu}_2\text{MeSi})_3\text{SiNa} \xrightarrow[\text{Et}_2\text{O}]{\text{GeCl}_2/\text{dioxane}} (t\text{-Bu}_2\text{MeSi})_3\text{Si}\bullet \qquad (1.6)$$

Silyl radicals are also involved as the reactive intermediates during one-electron reduction of bromosilanes. As an example, Reaction (1.7) shows the reduction by sodium of a silyl bromide to produce a persistent radical, which has been characterized by EPR spectroscopy [12].

$$(t\text{-BuMe}_2\text{Si})_3\text{SiBr} \xrightarrow[n\text{-heptane}]{\text{Na}} (t\text{-BuMe}_2\text{Si})_3\text{Si}\bullet \quad (1.7)$$

Processes involving photoinduced electron transfer of organosilanes [3,14,15] have not been covered in this book with the exception of the following method that was successfully applied to various radical reactions, such as cyclizations, intermolecular additions and tandem annulations (see Chapters 4, 5 and 7). Silyl radicals have been obtained by a complex but efficient method using $PhSeSiR_3$ as the reagent. The strategy is based on the mesolysis of $PhSeSiR_3^{-\bullet}$ to give $R_3Si\bullet$ radical and $PhSe^-$ [16–18]. Indeed, the selective formation of $PhSeSiR_3^{-\bullet}$ is accomplished by visible-light irradiation (410 nm) of solutions containing $PhSeSiR_3$, 9,10-dimethoxyanthracene (DMA) as the electron donor, and ascorbic acid (H_2A) as the co-oxidant. Scheme 1.1 shows the photoinduced electron transfer with the formation of $PhSeSiR_3^{-\bullet}$ and $DMA^{+\bullet}$, together with the regeneration of DMA at the expense of ascorbic acid. The choice of the substituents is limited by their stabilities. Trialkyl substituted derivatives are highly sensitive to air and prone to hydrolysis, whereas the t-$BuPh_2Si$ derivative was found to be the most stable.

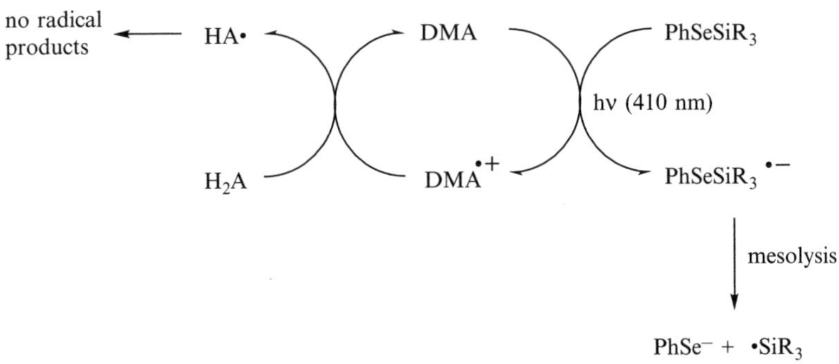

Scheme 1.1 Generation of silyl radicals by a photoinduced electron transfer method

$PhSeSiR_3$ reacts with Bu_3SnH under free radical conditions and affords the corresponding silicon hydride (Reaction 1.8) [19,20]. This method of generating $R_3Si\bullet$ radicals has been successfully applied to hydrosilylation of carbonyl groups, which is generally a sluggish reaction (see Chapter 5).

$$\text{PhSeSiEt}_3 + \text{Bu}_3\text{SnH} \xrightarrow[\text{benzene}]{\text{AIBN, 80 °C}} \text{PhSeSnBu}_3 + \text{Et}_3\text{SiH} \quad (1.8)$$

Although a detailed mechanistic study is still lacking, it is reasonable to assume that the formation of $R_3Si\bullet$ radicals occurs by means of the mesolysis of reactive intermediate $PhSeSiR_3^{-\bullet}$, by analogy with the mechanistic information reported above. Indeed, an electron transfer between the initially formed stannyl radical and the silyl selenide is more plausible (Reaction 1.9), than a bimolecular homolytic substitution at the seleno moiety.

$$PhSeSiEt_3 + Bu_3Sn\bullet \longrightarrow PhSeSiEt_3^{\bullet -} + Bu_3Si^+ \quad (1.9)$$

1.2 STRUCTURAL PROPERTIES OF SILYL RADICALS

Trisubstituted carbon-centred radicals chemically appear planar as depicted in the π-type structure **1**. However, spectroscopic studies have shown that planarity holds only for methyl, which has a very shallow well for inversion with a planar energy minimum, and for delocalized radical centres like allyl or benzyl. Ethyl, isopropyl, *tert*-butyl and all the like have double minima for inversion but the barrier is only about 300–500 cal, so that inversion is very fast even at low temperatures. Moreover, carbon-centred radicals with electronegative substituents like alkoxyl or fluorine reinforce the non-planarity, the effect being accumulative for multi-substitutions. This is ascribed to $n\sigma^*$ bonds between n electrons on the heteroatom and the bond to another substituent. The degree of bending is also increased by ring strain like in cyclopropyl and oxiranyl radicals, whereas the disubstituted carbon-centred species like vinyl or acyl are 'bent' σ radicals [21].

1

2

For a long time, this knowledge on carbon-centred radicals has driven the analysis of spectroscopic data obtained for silicon-centred (or silyl) radicals, often erroneously. The principal difference between carbon-centred and silyl radicals arises from the fact that the former can use only 2s and 2p atomic orbitals to accommodate the valence electrons, whereas silyl radicals can use 3s, 3p and 3d. The topic of this section deals mainly with the shape of silyl radicals, which are normally considered to be strongly bent out of the plane (σ-type structure **2**) [1]. In recent years, it has been shown that α-substituents have had a profound influence on the geometry of silyl radicals and the rationalization of the experimental data is not at all an extrapolation of the knowledge on alkyl radicals. Structural information may be deduced by using chemical, physical or theoretical methods. For better comprehension, this section is divided in subsections describing the results of these methods.

1.2.1 CHEMICAL STUDIES

The pyramidal structure of triorganosilyl radicals ($R_3Si\bullet$) was first indicated by chirality studies on optically active compounds containing asymmetric silicon.

For example, the α-naphthylphenylmethylsilyl radical (**3**) generated by hydrogen abstraction from the corresponding chiral silane reacts with CCl_4 to give optically active chlorosilane that has retained, at least in part, the configuration of the starting material [22]. Thus, the silyl radical is chiral and exists in a pyramidal form with considerable configurational stability, and it abstracts a chlorine atom from CCl_4 faster than its inversion (Reaction 1.10). Moreover, it was observed that the α-naphthylphenylmethylsilyl radical gave varying degrees of optical purity in the products as the concentration of CCl_4 was progressively diluted with benzene or cyclohexane. Analysis of these results by using a Stern–Volmer type of approach, yielded $k_{inv}/k = 1.30$ M at 80 °C, where k_{inv} is the rate constant for inversion at the silicon centre (Reaction 1.10) and k is the rate constant for the reaction of silyl radical with CCl_4 [23]. From these data, $k_{inv} = 6.8 \times 10^9\ s^{-1}$ at 80 °C is obtained which corresponds to an activation barrier of ca 23.4 kJ/mol if a normal preexponential factor of inversion is assumed, i.e., $\log(A/s^{-1}) = 13.3$. A number of other optically active organosilanes behave similarly, when the α-naphthyl group in α-NpSi*(Ph)(Me)H, is replaced by neo-C_5H_{11}, C_6F_5or Ph_2CH [22]. Under the same conditions, however, $Ph_3SiSi^*(Ph)(Me)H$ gave a chloride that was racemic indicating either that the inversion rate of the disilyl radical is much faster than its rate of reaction with CCl_4, or that the radical centre is planar.

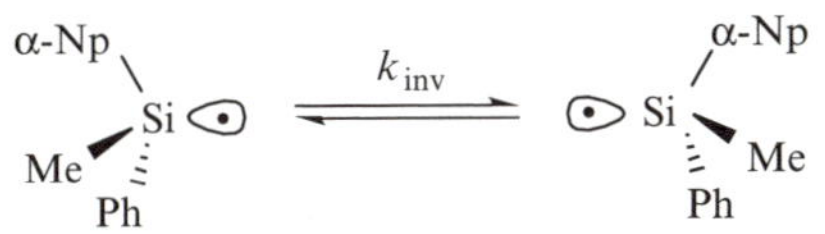

3

(1.10)

Analogous competitive kinetic studies have been reported for the inversion of silyl radicals **4** and **5** generated from corresponding silanes (Reaction 1.11) [24]. Rate constants for the interconversion of the two isomer radicals were estimated to be $k_1 \approx 9 \times 10^9\ s^{-1}$ and $k_{-1} \approx 4 \times 10^9\ s^{-1}$ at 0°C [23]. The equilibrium is slightly shifted to the right ($K \approx 2.3$) which suggests that radical **5** is a few hundred calories more stable than radical **4**. Activation energies for the forward and reverse inversion processes can be estimated to be ca 17–21 kJ/mol by assuming $\log(A/s^{-1}) = 13.3$.

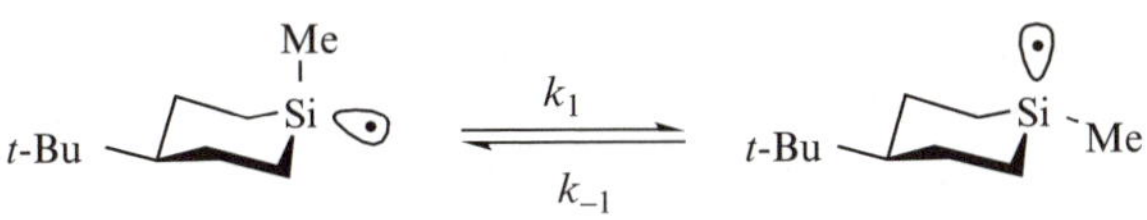

4 **5**

(1.11)

For comparison, it is worth mentioning that in the gas phase $H_3Si\bullet$ is bent out of the plane by $(16.0 \pm 2.0)^\circ$, corresponding to an H—Si—H bond angle of $(112.5 \pm 2.0)^\circ$ and with an inversion barrier of 22.6 kJ/mol [25].

Structural information on silyl radicals has also been obtained from the isomerization of 9,10-dihydro-9,10-disilaanthracene derivatives **6** and **7** [26,27].

Indeed, irradiation of a pentane solution of either the *cis* isomer **6** or the corresponding *trans* isomer **7** in the presence of di-*tert*-butyl peroxide as radical initiator affords the same *cis*/*trans* mixture. For R = Me or Ph, a ratio of 47/53 is observed whereas for the more sterically hindered R = *t*-Bu a ratio of 81/19 is obtained. It was proposed that the radicals **9** and **10** generated by hydrogen abstraction from **6** and **7**, respectively, undergo inversion of the radical centre (Reaction 1.12) followed by hydrogen abstraction from the parent silanes (an identity reaction, see Chapter 3) [27]. Interestingly, the analogous 9-silaanthracene derivative **8** does not isomerize under identical conditions [8], suggesting that the disilaanthracene skeleton plays an important role either in lowering the activation energy of the identity reaction or fastening the inversion of silyl radical in Reaction (1.12).

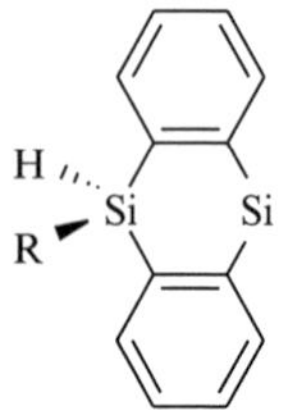

6

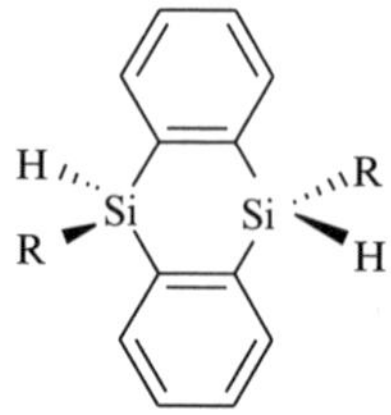

7

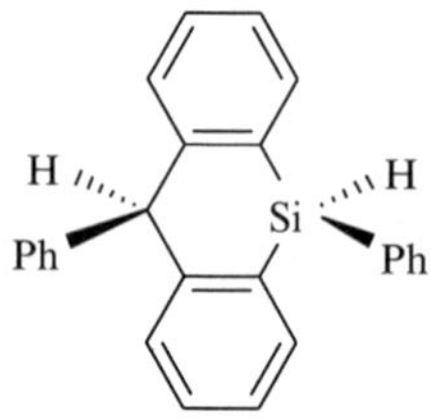

8

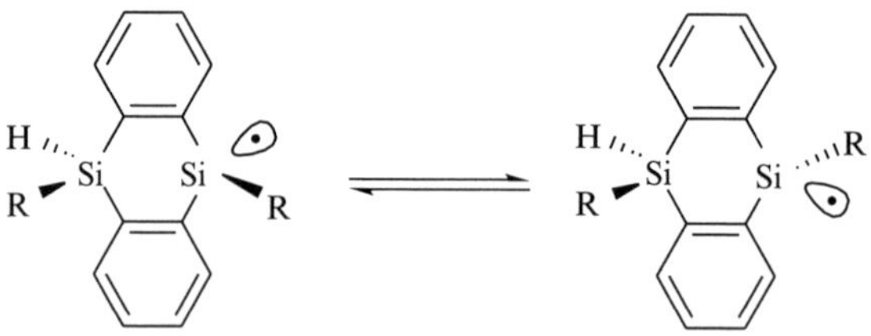

9 **10** (1.12)

1.2.2 ELECTRON PARAMAGNETIC RESONANCE (EPR) SPECTRA

EPR spectroscopy is the most important method for determining the structures of transient radicals. Information obtained from the EPR spectra of organic radicals in solution are: (i) the centre position of the spectra associated with *g* factors, (ii) the number and spacing of the spectral lines related to hyperfine splitting (hfs) constants, (iii) the total absorption intensity which corresponds to the radical concentration, and (iv) the line widths which can offer kinetic information such as rotational or conformational barriers. The basic principles as well as extensive treatments of EPR spectroscopy have been described in a number of books and reviews and the reader is referred to this literature for a general discussion [28–30].

Generally, the EPR spectra of silyl radicals show a central set of lines due to ^{1}H hfs constants and weaker satellites due to the coupling with $^{29}Si(I = 1/2, 4.7\ \%)$. The data for silyl radicals, presented in Table 1.1, have

Table 1.1 EPR data for a variety of α-substituted silyl radicals[a]

Silyl radical	$a(^{29}Si)^b$(G)	a(others) (G)	g factor
$H_3Si\bullet$	189	7.96 (3 H)[c]	2.0032
$H_2MeSi\bullet$	181	11.82 (2 H)[c] 7.98 (3 H)	2.0032
$HMe_2Si\bullet$	183	16.99 (1 H)[c] 7.19 (6 H)	2.0031
$Me_3Si\bullet$	181	6.28 (9 H)	2.0031
$Et_3Si\bullet$	170	5.69 (6 H) 0.16 (9 H)	2.0030
t-$Bu_3Si\bullet$	163	0.43 ($3\,^{13}C$)	
$Ph_3Si\bullet$[d]	150		
$Mes_3Si\bullet$[e]	135	0.70 (33 H)	2.0027
$(MeO)_3Si\bullet$	339		2.0012
$(t$-$BuO)_3Si\bullet$	331	0.23 (27 H)	2.0014
$F_3Si\bullet$	498	136.6 (3 F)	2.0003
$MeCl_2Si\bullet$	295	10.5 ($2\,^{35}Cl$)	2.0035
$Cl_3Si\bullet$	416	12.4 ($3\,^{35}Cl$)	2.0035
$(Me_3Si)Me_2Si\bullet$	137	8.21 (6 H) 0.47 (9 H)	2.0037
$(Me_3Si)_2MeSi\bullet$	90	9.28 (3 H) 0.44 (18 H)	2.0045
$(Me_3Si)_3Si\bullet$	64	7.1 ($3\,^{29}Si$) 0.43 (27 H)	2.0053

[a] See Reference [1] for the original citations.
[b] Because the magnetogyric ratio of ^{29}Si is negative, the signs of $a(^{29}Si)$ will also be negative.
[c] The sign is found to be positive by *ab initio* calculations [34].
[d] Phenyls are perdeuterated.
[e] Mes=2,4,6-trimethylphenyl.
Reprinted with permission from Reference [1]. Copyright 1995 American Chemical Society.

been chosen in order to include a variety of different substituents. In addition, isotropic hyperfine splitting and g factors are reported and most were obtained directly from solution spectra, although a few were taken from solid-state experiments. As an example, Figure 1.1 shows the EPR spectrum of $(Me_3Si)_2Si(\bullet)Me$ radical obtained at −40 °C by reaction of photogenerated t-BuO• radical with the parent silane [31]. The central quartet of relative intensity 1:3:3:1 with $a_H = 9.28$ G is caused by hyperfine coupling with the α-methyl protons. Each of these lines exhibits an additional hyperfine structure from 18 equivalent protons (six β-methyl groups) with $a_H = 0.44$ G (inset). The ^{29}Si-satellite regions were recorded with a 10-fold increase of the gain and are associated with $a(\alpha\text{-}^{29}Si) = 90.3$ G.

Table 1.1 shows that the nature of the α-substituent in the radical centre enormously influences the ^{29}Si hfs constants. These constants, which can be used as a guide to the distribution of unpaired electron density, were initially correlated to changes in geometry at the radical centre by analogy with ^{13}C hfs constants of α-substituted alkyl radicals. Indeed, it was suggested that by

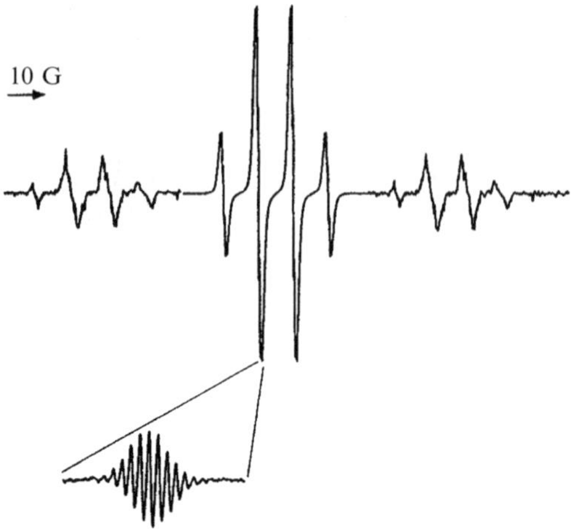

Figure 1.1 EPR spectrum of $(Me_3Si)_2Si(\bullet)Me$ recorded at 223 K. The satellite regions were recorded with a 10-fold increase of the gain. The inset shows an enlargement of the second spectral line recorded at lower modulation amplitude revealing hyperfine structure from 18 equivalent protons. Reprinted with permission from Reference [31]. Copyright 1992 American Chemical Society.

increasing the electronegativity of the α-substituents, the pyramidality of the silyl radical would increase, which would also mean a higher percentage of 3s character in the single occupied molecular orbital (SOMO), and therefore an increase in the ^{29}Si hfs, as well [32]. However, a theoretical study at the UMP2/DZP level reported that for a variety of α-substituted silyl radicals ($X_3Si\bullet$, where X = H, CH_3 NH_2, OH, F, SiH_3, PH_2, SH, Cl) the arrangement of atoms around silicon is essentially tetrahedral except for X = SiH_3 and that the large variation of the ^{29}Si hfs constants are due to the different distribution of the spin population at the Si center among 3s, 3p and 3d orbitals rather than to a change of geometry at the radical centre (see Section 1.2.5) [33,34]. The *g* factor of silyl radicals decreases along the series of substituents alkyl > alkoxyl > fluorine and silyl > chlorine (Table 1.1) while the spin–orbit coupling constant increases along the series C < O < F and Si < Cl [28]. Generally the *g* factor is larger than the free electron value of 2.00229 if spin–orbit coupling mixes the SOMO with low lying LUMOs and smaller if the mixing is with high lying doubly occupied orbitals. Moreover, the extent of the odd electron delocalization onto the atoms or groups attached to silicon is also expected to have an important influence on the *g* factor trend. Another factor affecting the magnitude of the *g* value is the geometry of the radical centre. Readers should refer to a general text on EPR for a more detailed discussion on the interpretation of hfs constants and *g* factors [29,30].

α-Aryl-substituted silyl radicals have been a subject of attraction in order to evaluate the extent to which a silicon centre radical can conjugate with an adjacent aromatic system. However, the high reactivity of the silyl radical towards aromatic substitution (see Section 5.1.1), limited the detection of this type of transients by EPR spectroscopy. For example, $PhH_2Si•$, $Ph_2HSi•$ and $Ph_3Si•$ radicals have not been observed in solution whereas the corresponding perdeuterated silyl radicals have been detected in a solid matrix [35]. Two sterically hindered analogous radicals, trimesitylsilyl and tris(3,5-di-*tert*-butylphenyl)silyl have been observed by EPR in solution and appear to be partially delocalized species according to the ring proton hfs constants [36,37]. Similar considerations and analogous experiments have been extended to α-vinyl substituted silyl radicals and the results are in line with the α-phenyl substituted case [38]. The spectra of Me_3Si-substituted silyl radicals are of particular interest. Thus, when Me_3Si groups progressively replace methyl groups, the ^{29}Si hfs constants decrease from 181 G in the $Me_3Si•$ radical to 64 G in the $(Me_3Si)_3Si•$ radical (Table 1.1). This trend is due mainly to the spin delocalization onto the Si—C β-bond and in part to the decrease in the degree of pyramidalization at the radical centre caused by the electron-releasing Me_3Si group [39].

Kinetic information from the line width alterations of EPR spectra by changing the temperature has been obtained for a number of silacycloalkyl radicals [40,41]. For example, silacyclopentyl radical exists at low temperature (−119 °C) in two equivalent twist conformations (**11** and **12**), which interconvert at higher temperature (15 °C). The Arrhenius parameters for such interconversion are $\log A/s^{-1} = 12.0$ and $E_a = 21.3$ kJ/mol.

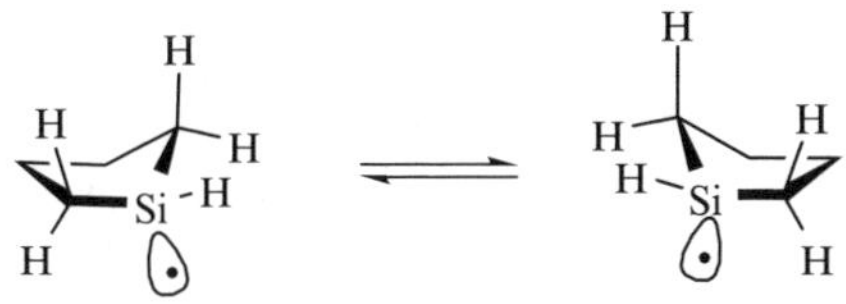

(1.13)

11 **12**

Persistent and stable silyl radicals have attracted considerable attention [42]. Bulky aryl or alkyl groups that generally make carbon-centred radicals persistent [43,44] have a much weaker effect on the silyl radicals. The high reactivity of the $Ph_3Si•$ radical contrary to the stable $Ph_3C•$ radical is mentioned above. The decay of the trimesitylsilyl radical at −63°C follows a first-order kinetics with a half-life of 20 s [37]. Tri-*tert*-butylsilyl radical is also not markedly persistent showing the modest tendency of *tert*-butyl groups to decrease pyramidalization [45]. The most persistent trialkyl-substituted silyl radical is $[(Me_3Si)_2CH]_3Si•$, which at 20 °C follows a first-order decay with a half-life of 480 s [36]. An exceptionally stable diradical was isolated by reaction of 1,1-dilithio-2,3,4,5-tetraphenylsilole with 1,1-dichloro-2,3-diphenylcyclopropene, for which the structure **13** was suggested on the basis of EPR data and theoretical calculations [46]. The remarkable unreactivity of this diradical has been explained by steric hindrance, as well as delocalization of the unpaired electrons over the silole ring.

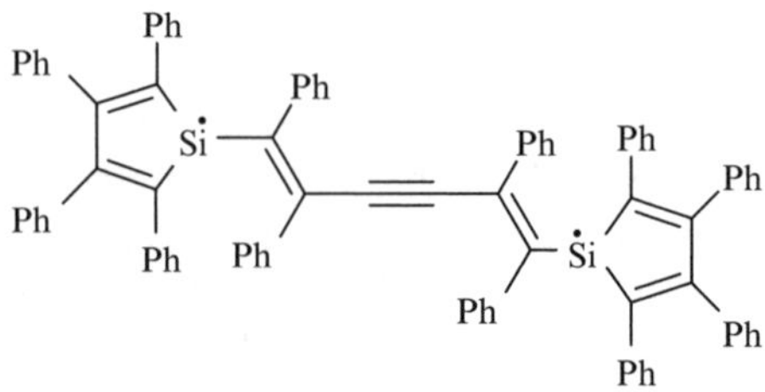

13

On the other hand, bulky trialkylsilyl substituents have a profound effect on the structure of silyl radicals. Indeed, by increasing the steric effect with more crowded trialkylsilyl substituents the persistency of silyl radicals increases substantially. Table 1.2 reports the EPR data for a variety of tris(trialkylsilyl)silyl radicals in comparison with the prototype $(Me_3Si)_3Si\bullet$. Inspection of the data shows the α-^{29}Si hfs constants tend to decrease and the β-^{29}Si hfs slightly increase when the methyl group is progressively replaced by a bulkier group, the effect being cumulative, e.g., along the series $(Me_3Si)_3Si\bullet$, $(Et_3Si)_3Si\bullet$, $(i-Pr_3Si)_3Si\bullet$ and $(Me_3Si)_3Si\bullet$, $(Et_2MeSi)_3Si\bullet$, $(t\text{-}Bu_2MeSi)_3Si\bullet$. These trends have been associated with an increase of the polysilane skeleton flattening through the series [12,13,48–50]. Indeed, the half-lives of the radicals increase within the series and the $(t\text{-}Bu_2MeSi)_3Si\bullet$ radical is found to be stable and isolable in a crystal form. Therefore, the radicals $(Et_3Si)_3Si\bullet$, $(i\text{-}Pr_3Si)_3Si\bullet$, $(t\text{-}BuMe_2Si)_3Si\bullet$ and $(t\text{-}Bu_2MeSi)_3Si\bullet$ have a practically planar structure due to the steric repulsions among the bulky silyl substituents. The small differences of their α-^{29}Si hfs constants are presumably due to different degrees of spin delocalization onto the Si—C β-bond, as a consequence of conformational effects in order to minimize the steric hindrance. Persistent silyl radicals have also been formed upon

Table 1.2 EPR data for a variety of tris(trialkylsilyl)silyl radicals

Silyl radical	a(α-^{29}Si)(G)	a(β-^{29}Si)(G)	a(others) (G)	g factor	Reference
$(Me_3Si)_3Si\bullet$	63.8	7.1	0.43 (27 H)	2.0053	[47]
$(EtMe_2Si)_3Si\bullet$	62.8	7.1	0.37 (18 H) 0.14 (6 H)	2.0060	[48]
$(Et_2MeSi)_3Si\bullet$	60.3	7.3	0.27 (12 H) 0.15 (9 H) 3.2 (3 ^{13}C)	2.0060	[48]
$(Et_3Si)_3Si\bullet$	57.2	7.9	0.12 (18 H) 3.0 (3 ^{13}C)	2.0063	[48]
$(i\text{-}Pr_3Si)_3Si\bullet$	55.6	8.1	2.2 (3 ^{13}C)	2.0061	[49]
$(t\text{-}BuMe_2Si)_3Si\bullet$	57.1	8.1	0.33 (27 H) 0.11 (18 H)	2.0055	[12]
$(Me_3SiMe_2Si)_3Si\bullet$	59.9	7.4		2.0065	[50]
$(t\text{-}Bu_2MeSi)_3Si\bullet$	58.0	7.9		2.0056	[13]

photolysis of poly(di-*n*-alkylsilanes) in solution via a complex reaction mechanism [8]. Radical **14** (Hx = *n*-hexyl) with $g = 2.0047$, $a(\alpha\text{-}^{29}\mathrm{Si}) = 75\,\mathrm{G}$ and $a(\beta\text{-}^{29}\mathrm{Si}) = 5.8\,\mathrm{G}$, showed line-broadening effects as the temperature was lowered. This observation has been correlated to the restricted rotational motion about the C—Si• bond and, in particular, to a rocking interchange of the two α-hydrogens. Isolation of 'allylic-type' silyl radical **15** has also been achieved [51]. The EPR spectrum consists of a broad singlet ($g = 2.0058$) with three doublet satellite signals due to coupling with ^{29}Si of 40.7, 37.4 and 15.5 G. The two doublets with 40.7 and 37.4 G broaden upon raising the temperature and coalesce at 97 °C due to the rotation of the *t*-BuMe$_2$Si group. The magnitude of ^{29}Si hfs constants is consistent with the delocalization of the unpaired electron over the three silicon atoms in the ring, but it is noteworthy that the coupling constants of the outer Si atoms are not equal. This is explained below.

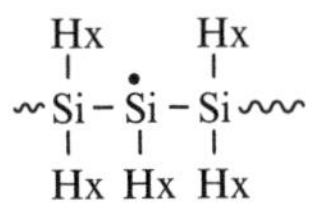

14

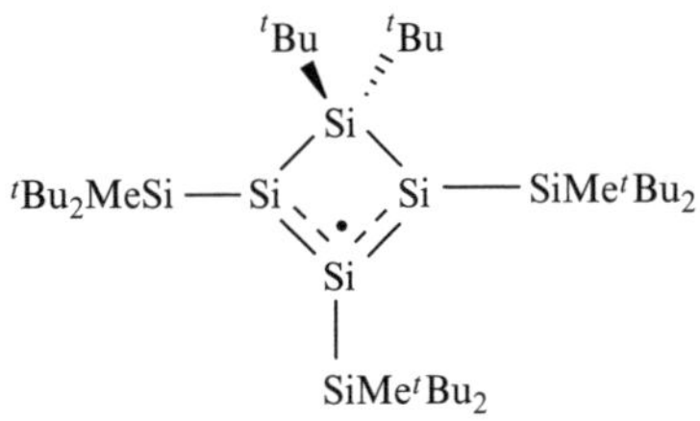

15

1.2.3 CRYSTAL STRUCTURES

The crystal structures of two isolable silyl radicals have recently been reported. The bulky substituted (t-Bu$_2$MeSi)$_3$Si• radical was isolated as air-sensitive yellow needles [13], whereas the conjugated and bulky substituted cyclotetrasilenyl radical **15** was obtained as red–purple crystals [51].

Figure 1.2 shows a completely planar geometry around the Si1 atom of (t-Bu$_2$MeSi)$_3$Si• radical. Indeed, the bond angles Si2—Si1—Si3, Si2—Si1—Si4 and Si3—Si1—Si4 are 119.49°, 120.08° and 120.43°, respectively, their sum being exactly 360°. The Si—Si bonds are larger (2.42 ± 0.01 Å) than normal. Interestingly, all the methyl substituents at the α-Si atoms (i.e., C1, C4 and C7) are located in the plane of the polysilane skeleton in order to minimize steric hindrance. As reported in the previous section, the planarity of this radical is retained in solution.

Figure 1.3 shows the ORTEP drawing of the conjugated radical **15**. The four-membered ring is nearly planar with the dihedral angle between the radical part Si1—Si2—Si3 and Si1—Si4—Si3 being 4.7°. The Si1 and Si2 atoms have planar geometry (the sums of the bond angles around them are 360.0° and 359.1°, respectively) whereas the Si3 atom is slightly bent (356.2°). This small asymmetry of the moiety where the radical is delocalized is also observed in the

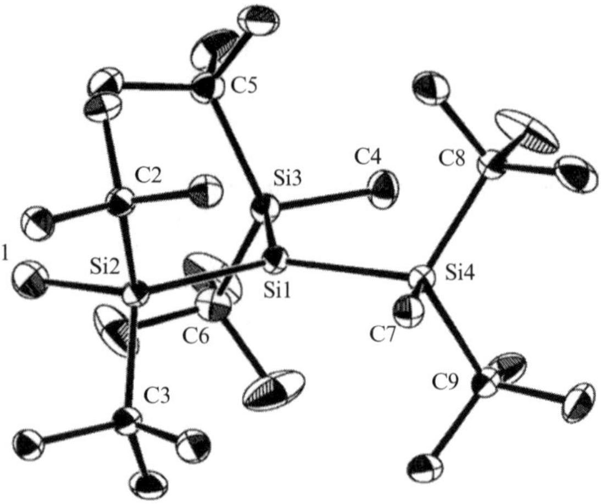

Figure 1.2 Molecular structure of (t-$Bu_2MeSi)_3Si$• radical with thermal ellipsoids drawn at the 30 % level (hydrogen atoms are omitted for clarity). Reprinted with permission from Reference [13]. Copyright 2002 American Chemical Society.

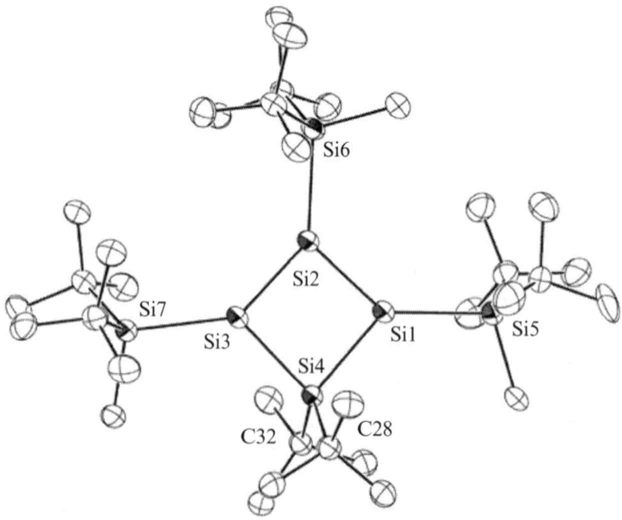

Figure 1.3 ORTEP Drawing of cyclotetrasilenyl radical **15**. Hydrogen atoms are omitted for clarity. Reprinted with permission from Reference [51]. Copyright 2001 American Chemical Society.

Si—Si bond length, the Si1—Si2 being slightly shorter than Si2—Si3 (2.226 vs 2.263 Å), and explains the magnetic inequivalence of Si1 and Si3 noted above.

The reaction of (t-$Bu_2MeSi)_3Si$• radical with lithium in hexane at room temperature afforded the silyllithium **16** for which the crystal structure shows the central anionic silicon atom to be almost planar (119.7° for Si—Si—Si bond angles) and the Si—Si bond lengths significantly shorter (2.36 Å) than in the radical (2.42 Å) [52]. Similarly, the cyclotetrasilenyl radical **15** reacted with lithium to give the corresponding lithiated derivative, which has a π-type structure with coordination of a lithium cation to a trisilaallyl moiety [53]. It is also worth mentioning that the crystal structure of $[(i\text{-Pr})_3Si]_3SiH$ shows a nearly planar structure of the polysilane skeleton [49]. In fact, the Si—Si—Si bond angles are 118.1° and the sum of the three angles around the central silicon atom is 354.3°. The Si—Si—H bond angle is 98.0°. Therefore, the introduction of bulky silyl groups induces a significant flattening of the silicon skeleton by large steric repulsion even in silicon hydride. Such steric hindrances should play an even more important role in the planarization of the corresponding silyl radicals.

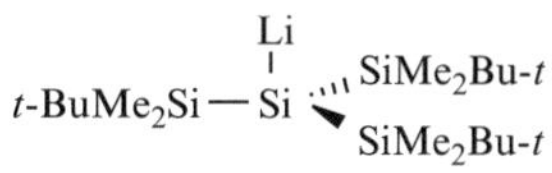

16

1.2.4 UV–VISIBLE SPECTRA

The electronic absorption spectra of few trialkylsilyl radicals have been recorded in both gas and liquid phases. Radicals $R_3Si•$ (R = Me, Et, *n*-Pr) generated in the gas phase exhibit a strong band in the region of 220–300 nm with a maximum at ca 260 nm ($\varepsilon_{max} = 7500\,M^{-1}\,cm^{-1}$ for $Me_3Si•$) [54,55]. Figure 1.4 shows the UV–visible spectrum of $Et_3Si•$ in liquid isooctane, which exhibits a continuously increasing absorption below 340 nm, with no maximum above 280 nm ($\varepsilon_{308} = 1100\,M^{-1}\,cm^{-1}$), and a weak symmetric band between 350 and 450 nm, with a maximum at 390 nm [56].

The transient absorption spectra of silyl radicals with the Me group of $Me_3Si•$ progressively replaced by Ph or Me_3Si groups were also studied. $PhMe_2Si•$, $Ph_2MeSi•$, and $Ph_3Si•$ exhibit a strong band in the range of 290–360 nm attributed to the electronic transition involving the aromatic rings and a weak absorption between 360 and 550 nm (see Figure 1.4 for $Ph_3Si•$) [56]. In the series of Me_3Si-substituted silyl radicals, $Me_3SiSi(•)Me_2$ exhibits a band between 280 and 450 nm with a maximum at ca 310 nm and a shoulder at longer wavelengths [57], whereas the spectrum of $(Me_3Si)_3Si•$ radical shows a continuously increasing absorption below ca 350 nm and no maximum above 280 nm [47].

The absorption spectra of the $(RS)_3Si•$ radicals (R = Me, *i*-Pr) exhibit a strong band at 300–310 nm. In addition, the absorption envelopes extend well out into the visible region of the spectrum to about 500 nm and show a shoulder at ca 425 nm (see Figure 1.4 for $(MeS)_3Si•$) [58].

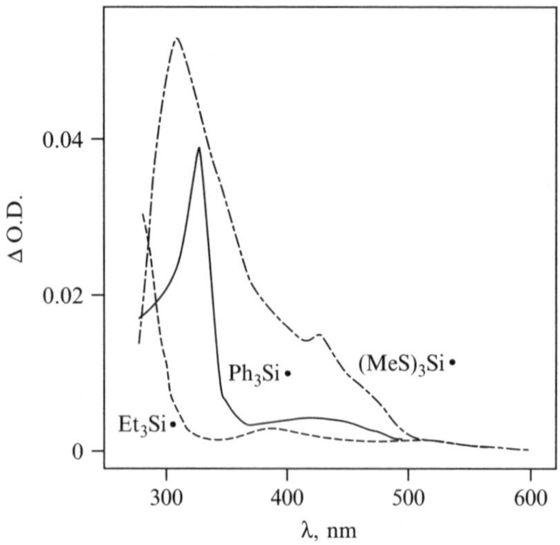

Figure 1.4 Transient spectra for some R_3Si• radicals generated by reaction of *t*-BuO• with the corresponding silanes under similar experimental conditions. Reprinted in part with permission from Reference [56]. Copyright 1983 American Chemical Society.

1.2.5 THEORETICAL STUDIES

There has been a number of theoretical studies on a variety of silyl radicals at various levels of *ab initio* theory. The structural parameters for a variety of halogenated silyl radicals, i.e., $F_{3-n}Si(\bullet)H_n$, $Cl_{3-n}Si(\bullet)H_n$, and $Cl_{3-n}Si(\bullet)F_n$, (with *n* having values from 0 to 3) have been examined with the 6-31 ++ G* basis set, with optimization at the UHF level and single point calculations at the UMP2 level [59]. All radicals have bond angles close to the ideal tetrahedral angle. Both vertex inversion (transition state **17**) and edge inversion (transition state **18**) mechanisms were taken into consideration. For the H_3Si• radical, the calculated barriers for the **17** and **18** transition states are 20.5 and 277.4 kJ/mol, respectively. Similarly, FH_2Si•, ClH_2Si• and Cl_2HSi• all invert by the vertex mechanism. However, for the F_2HSi• radical the calculated barriers for the two mechanisms are almost identical, and increased halogenation results in a change of mechanism. Thus all $Cl_{3-n}Si(\bullet)F_n$ radicals invert by the edge mechanism.

17

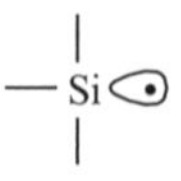

18

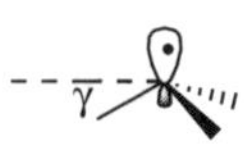

19

The optimized structural parameters of the α-trisubstituted silyl radical (X_3Si•, where X = H, CH_3, NH_2OH, F, SiH_3, PH_2, SH and Cl) were performed at the UMP2/DZP level of theory [33]. As expected, the bond lengths decrease according to electronegativity. The calculated angles γ (see structure **19**) for the silyl radicals with different substituents (in parentheses) are: 17.73° (H), 18.69° (CH_3), 21.53 ° (NH_2), 20.76 ° (OH), 20.77 ° (F), 13.40 ° (SiH_3), 22.68 ° (PH_2), 20.51 ° (SH), and 19.43 ° (Cl). Therefore, for all these trisubstituted radicals, the arrangement of atoms around the silicon is found to be essentially tetrahedral with the exception of the $(H_3Si)_3Si\bullet$ radical which is much less bent. The magnitude and the trend of the ^{29}Si hfs constants from EPR spectra are well reproduced by these calculations and are due to more 3s character of the unpaired electron orbital at the Si-center rather than to a general change of geometry at radical centre. The calculations show that in the SOMO the delocalization of the unpaired electron onto the α-substituent increases from second to third row elements, whereas the population on Si-3s increases linearly with the increasing electronegativity of the α-substituent. For example, the calculated distribution of the unpaired electron density for $Me_3Si\bullet$ is 81 % on silicon (14.3 % in 3s, 64.6 % in 3p, and 2.1 % in 3d) and 19 % on methyls; for $F_3Si\bullet$ it is 84.4 % on silicon (41.8 % in 3s, 32.6 % in 3p, and 6.4 % in 3d) and 15.6 % on fluorines; for $Cl_3Si\bullet$ it is 57.3 % on silicon (21.5 % in 3s, 32.6 % in 3p, and 3.2 % in 3d) and 42.7 % on chlorines. UMP2/DZP/TZP calculations have been extended to the series $H_{3-n}Si(\bullet)Me_n (n = 0\text{–}2)$ addressing the early controversy about the signs of α-^1H hfs constants [34]. The sign was found to be positive for all these radicals, which have a nearly tetrahedral geometry at silicon. The same level of theory has been used to calculate the α-^{29}Si hfs constants of series $Me_{3-n}Si(\bullet)Cl_n$ and $Me_{3-n}Si(\bullet)(SiMe_3)_n (n = 0\text{–}3)$ and to analyse observed trends [60]. The large increase of the ^{29}Si hfs when Me is successively replaced by Cl is mainly due to the change of the Si orbital populations rather than to structural changes, whereas when Me is replaced by $SiMe_3$ the considerable decrease is due to the increased spin delocalization and the flatter geometry.

The structural parameters of $(HS)_3Si\bullet$ radicals were computed at the HF/6-31G* level for C_3 symmetry [58]. The radical centre at silicon is pyramidal. Two minima have been found along the energy surface generated by the synchronous rotation of the SH groups. In the most stable conformation **20**, the hydrogens adopt a gauche conformation ($\omega = 50°$) with respect to the SOMO, which is mainly the sp_3 atomic orbital (AO) of Si. In the other minimum **21**, which is 18.4 kJ/mol higher in energy, the hydrogens are nearly anti ($\omega = 150°$) with respect to the SOMO.

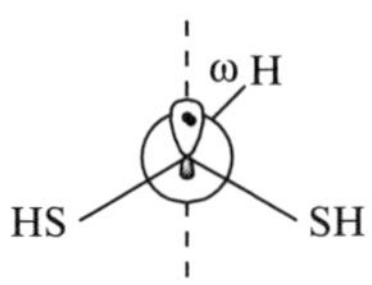

20

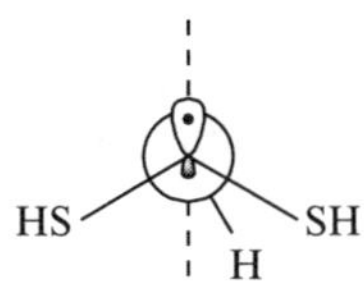

21

Multiple scattering Xα (MSXα) method was applied to assign the optical absorption spectra of $Me_3Si\bullet$ and $(MeS)_3Si\bullet$ radicals. The strong band observed for $(alkyl)_3Si\bullet$ radicals at ca 260 nm has been attributed to the superimposition of the valence transition from the MO localized at the Si—C bond to the SOMO and of the transition from the SOMO to the 4p Rydberg orbital [61]. Furthermore the observed weak band between 350 and 450 nm for $Et_3Si\bullet$ radical (Figure 1.4) has been assigned to the transition from SOMO to the 4s Rydberg orbital. For the $(MeS)_3Si\bullet$ radical [58], the strong band has been attributed to transitions from the SOMO (localized mainly at the 3p AO of Si) to the antibonding σ^*_{Si-S}MOs (a_1 and e symmetry). A contribution to the intensity of this band could also derive from the valence transition from the $\sigma_{Si-S}(a_1)$ MO to the SOMO and from the Rydberg transition from the SOMO to the $4p(a_1)$ orbital. The weak band/shoulder at ca 425 nm has been assigned to the valence excitation from the MO localized at the Si—S bond to the SOMO. Transitions from sulfur lone pairs to the SOMO have much lower oscillator strengths and are predicted to occur in the near-infrared region.

1.3 REFERENCES

1. Chatgilialoglu, C., *Chem. Rev.*, 1995, **95**, 1229.
2. Chatgilialoglu, C., and Newcomb, M., *Adv. Organomet. Chem.*, 1999, **44**, 67.
3. Steinmetz, M.G., *Chem. Rev.*, 1995, **95**, 1527.
4. Leigh, W.J., and Sluggett, G.W., *J. Am. Chem. Soc.*, 1993, **115**, 7531.
5. Leigh, W.J., and Sluggett, G.W., *Organometallics*, 1994, **13**, 269.
6. Sluggett, G.W., and Leigh, W.J., *Organometallics*, 1994, **13**, 1005.
7. Sluggett, G.W., and Leigh, W.J., *Organometallics*, 1992, **11**, 3731.
8. McKinley, A.J., Karatsu, T., Wallraff, G.M., Thompson, D.P., Miller, R.D., and Michl, J., *J. Am. Chem. Soc.*, 1991, **113**, 2003.
9. Davidson, I.M.T., Michl, J., and Simpson, T., *Organometallics*, 1991, **10**, 842.
10. Matsumoto, A., and Ito, Y., *J. Org. Chem.*, 2000, **65**, 5707.
11. Tamao, K., and Kawachi, A., *Adv. Organomet. Chem.*, 1995, **38**, 1.
12. Kira, M., Obata, T., Kon, I., Hashimoto, H., Ichinohe, M., Sakurai, H., Kyushin, S., and Matsumoto, H., *Chem Lett.*, 1998, 1097.
13. Sekiguchi, A., Fukawa, T., Nakamoto, M., Lee, V. Ya., and Ichinohe, M., *J. Am. Chem. Soc.*, 2002, **124**, 9865.
14. Shizuka, H., and Hiratsuka, H., *Res. Chem. Intermed.*, 1992, **18**, 131.
15. Kako, M., and Nakadaira, Y., *Coord. Chem. Rev.*, 1998, **176**, 87.
16. Pandey, G., and Rao, K.S.S.P., *Angew. Chem. Int. Ed. Engl.*, 1995, **34**, 2669.
17. Pandey, G., Rao, K.S.S.P., Palit, D.K., and Mittal, J.P., *J. Org. Chem.*, 1998, **63**, 3968.
18. Pandey, G., Rao, K.S.S.P., and Rao, K.V.N., *J. Org. Chem.*, 2000, **65**, 4309.
19. Nishiyama, Y., Kajimoto, H., Kotani, K., and Sonoda, N., *Org. Lett.*, 2001, **3**, 3087.
20. Nishiyama, Y., Kajimoto, H., Kotani, K., Nishida, T., and Sonoda, N., *J. Org. Chem.*, 2002, **67**, 5696.
21. Symons, M.C.R., *Chemical and Biochemical Aspects of Electron-Spin Resonance Spectroscopy*, Van Nostrand Reinhold, Melbourne, 1978.
22. Sommer, L.H., and Ulland, L.A., *J. Org. Chem.*, 1972, **37**, 3878.
23. Chatgilialoglu, C., Ingold, K.U., and Scaiano, J.C., *J. Am. Chem. Soc.*, 1982, **104**, 5123.
24. Sakurai, H., and Murakami, M., *Bull. Chem. Soc. Jpn.*, 1977, **50**, 3384.

25. Nimlos, M.R., and Ellison, G.B., *J. Am. Chem. Soc.*, 1986, **108**, 6522.
26. Kyushin, S., Shinnai, T., Kubota, T., and Matsumoto, H., *Organometallics*, 1997, **16**, 3800.
27. Nishiyama, K., Oba, M., Takagi, H., Saito, T., Imai, Y., Motoyama, I., Ikuta, S., and Hiratsuka, H., *J. Organomet. Chem.*, 2001, **626**, 32.
28. Fischer, H. In *Free Radicals*, Vol. II, J.K. Kochi (Ed.), Wiley, New York, 1973, p. 435.
29. Atherton, N.M., *Principles of Electron Spin Resonance*, Ellis Horwood, Chichester, 1993.
30. Weil, J.A., Bolton, J.R., and Wertz, J.E., *Electron Paramagnetic Resonance: Elementary Theory and Practical Applications*, Wiley, New York, 1994.
31. Chatgilialoglu, C., Guerrini, A., and Lucarini, M., *J. Org. Chem.*, 1992, **57**, 3405.
32. Alberti, A., and Pedulli, G.F., *Rev. Chem. Intermed.*, 1987, **8**, 207.
33. Guerra, M., *J. Am. Chem. Soc.*, 1993, **115**, 11926.
34. Guerra, M., *Chem. Phys. Lett.*, 1995, **246**, 251.
35. Lim, W-L., and Rhodes, C.J., *J. Chem. Soc., Chem. Commun.*, 1991, 1228.
36. Gynane, M.J.S., Lappert, M.F., Riley, P.I., Rivière, P., and Rivière-Baudet, M., *J. Organomet. Chem.*, 1980, **202**, 5.
37. Sakurai, H., Umino, K., and Sagiyama, H., *J. Am. Chem. Soc.*, 1980, **102**, 6837.
38. Jackson, R.A., and Zarkadis, A.K., *Tetrahedron Lett.*, 1988, **29**, 3493.
39. Guerra, M., *J. Chem. Soc., Perkin Trans. 2*, 1995, 1817.
40. Jackson, R.A., and Zarkadis, A.K., *J. Chem. Soc., Perkin Trans. 2*, 1990, 1139.
41. Jackson, R.A., and Zarkadis, A.K., *J. Chem. Soc., Faraday Trans.*, 1990, **86**, 3229.
42. Power, P.P., *Chem. Rev.*, 2003, **103**, 789.
43. Griller, D., and Ingold, K.U., *Acc. Chem. Res.*, 1976, **8**, 13.
44. Griller, D., and Ingold, K.U., *Acc. Chem. Res.*, 1980, **13**, 193.
45. Jackson, R.A., and Weston, H., *J. Organomet. Chem.*, 1984, **277**, 13.
46. Toulokhonova, I.S., Stringfellow, T.C., Ivanov, S.A., Masunov, A., and West, R., *J. Am. Chem. Soc.*, 2003, **125**, 5767.
47. Chatgilialoglu, C., and Rossini, S., *Bull. Soc. Chim. Fr.*, 1988, 298.
48. Kyushin, S., Sakurai, H., Betsuyaku, T., and Matsumoto, H., *Organometallics*, 1997, **16**, 5386.
49. Kyushin, S., Sakurai, H., and Matsumoto, H., *Chem. Lett.*, 1998, 107.
50. Apeloig, Y., Bravo-Zhivotovskii, D., Yuzefovich, M., Bendikov, M., and Shames, A.I., *Appl. Magn.Reson.*, 2000, **18**, 425.
51. Sekiguchi, A., Matsuno, T., and Ichinohe M., *J. Am. Chem. Soc.*, 2001, **123**, 12436.
52. Nakamoto, M., Fukawa, T., Lee, V. Ya., and Sekiguchi, A., *J. Am. Chem. Soc.*, 2002, **124**, 15160.
53. Matsuno, T., Ichinohe M., and Sekiguchi, A., *Angew. Chem. Int. Ed.*, 2002, **41**, 1575.
54. Shimo, N., Nakashima, N., and Yoshihara, K., *Chem. Phys. Lett.*, 1986, **125**, 303.
55. Brix, Th., Paul, U., Potzinger, P., and Reimann, B., *J. Photochem. Photobiol., A: Chem.*, 1990, **54**, 19.
56. Chatgilialoglu, C., Ingold, K.U., Lusztyk, J., Nazran, A.S., and Scaiano, J.C., *Organometallics,* 1983, **2**, 1332.
57. Lusztyk, J., Maillard, B., and Ingold, K.U., *J. Org. Chem.*, 1986, **51**, 2457.
58. Chatgilialoglu, C., Guerra, M., Guerrini, A., Seconi, G., Clark, K.B., Griller, D., Kanabus-Kaminska, J., and Martinho-Simões, J. A., *J. Org. Chem.*, 1992, **57**, 2427.
59. Hopkinson, A.C., Rodriquez, C.F., and Lien, M.H., *Can. J. Chem.*, 1990, **68**, 1309.
60. Guerra, M., *J. Chem. Soc., Perkin Trans. 2*, 1995, 1817.
61. Chatgilialoglu, C., and Guerra, M., *J. Am. Chem. Soc.*, 1990, **112**, 2854.

2 Thermochemistry

2.1 GENERAL CONSIDERATIONS

In the gas phase, homolytic bond dissociation enthalpies (D*H*) relate the thermochemical properties of molecules to those of radicals while ionization potentials (IP) and electron affinities (EA) tie the thermochemistry of neutral species to those of their corresponding ions. For example, Scheme 2.1 represents the relationships between R_3SiH and its related radicals, ions, and radical ions. This representation does not define thermodynamic cycles (the H fragment is not explicitly considered) but it is rather a thermochemical mnemonic that affords a simple way of establishing the experimental data required to obtain a chosen thermochemical property.

In Scheme 2.1 the horizontal arrows represent the IPs and EAs while the vertical arrows define homolytic cleavages. The diagonal arrows that ascend from left to right imply the formation of H^+ while for those that ascend from right to left it is the H^- that results. Seven pieces of experimental data are required to define the eleven thermodynamic properties of Scheme 2.1.

In the liquid phase, the equivalents of IPs and EAs are the electrochemical oxidation and reduction potentials and analogous thermochemical cycles have been used in the literature to calculate p*K* values. However, oxidation and reduction potentials of $R_3Si\bullet$ radicals are not yet established experimentally and, therefore, the solution thermochemical cycles suffer from these limitations.

In this chapter, we have collected and discussed the available data in both gas and liquid phases related to Scheme 2.1. Emphasis will be given to homolytic bond dissociation enthalpies of silanes. Generally, D*H* values are extrapolated from the gas phase to solution without concerning solvent effects (particularly in the

Organosilanes in Radical Chemistry C. Chatgilialoglu
© 2004 John Wiley & Sons, Ltd ISBN: 0-471-49870-X

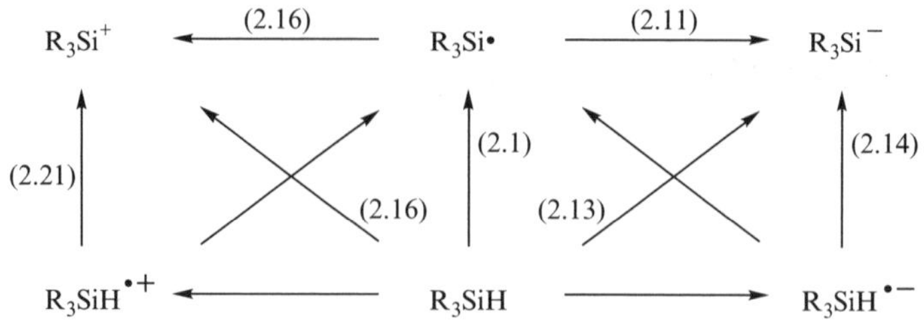

Scheme 2.1 Relationships between R_3SiH and its related radicals, ions, and radical ions (The numbers refer to reaction numbers as they appear in the text)

absence of hydroxylic or acidic bonds). This is true also in organosilanes since good agreement between the gas-phase and the liquid-phase D*H*s has been observed.

2.2 BOND DISSOCIATION ENTHALPIES

Knowledge of bond dissociation enthalpies (D*H*) has always been considered fundamental for understanding kinetics and mechanisms of free radicals. D*H*s offer an interesting window through which to view stability of radicals. Indeed, based on Reaction (2.1) the bond dissociation enthalpy of silanes $DH(R_3Si—H)$ is related to enthalpy of formation of silyl radicals, $\Delta H_f^\circ(R_3Si\bullet)$, by Equation (2.2).

$$R_3Si—H \longrightarrow R_3Si\bullet + H\bullet \tag{2.1}$$

$$DH(R_3Si—H) = \Delta H_f^\circ(R_3Si\bullet) + \Delta H_f^\circ(H\bullet) - \Delta H_f^\circ(R_3SiH) \tag{2.2}$$

Several techniques have been used to determine silane bond enthalpies: radical kinetics, gas-phase acidity cycles and photoacoustic calorimetry. However, the following sections concentrate on the results rather than the details of experimental techniques. D*H*s of silanes have changed substantially over time. For example, the $DH(Me_3Si—H)$ has changed significantly over the last 60 years. Indeed, values of 314, 339, 377 and 397 kJ/mol were reported in 1942, 1971, 1982, 1994, respectively [1]. Consequently, many bond dissociation enthalpies that were derived using these reference values have changed as well. Nevertheless, it is worth recalling that today the majority of data are consistent within the different experimental approaches. For consistency, all the D*H* and ΔH_f° values are normally extrapolated to standard state conditions (i.e., 25 °C in the gas phase).

2.2.1 RADICAL KINETICS

Studies of the kinetics of chemical equilibrium (Reactions 2.3/–2.3) have provided very accurate thermodynamic information from the series

Table 2.1 Bond dissociation enthalpies and standard enthalpies of formation of silanes, and enthalpies of associated silyl radicals (kJ/mol)[a]

Silane (R_3SiH)	$DH(R_3Si—H)$[b]	$\Delta H_f°(R_3SiH)$[c]	$\Delta H_f°(R_3Si•)$[d]
$H_3Si—H$	384.1 ± 2	34.3 ± 2	200.5 ± 2.5
$MeSiH_2—H$	388 ± 5	-29.1 ± 4	141 ± 6
$Me_2SiH—H$	392 ± 5	-94.7 ± 4	79 ± 6
$Me_3Si—H$	397.4 ± 2	-163.4 ± 4	16 ± 6

[a] At 25 °C
[b] From References [2–4]
[c] From Reference [5]
[d] Calculated from Equation (2.2) using $\Delta H_f°(H•) = 218.0$ kJ/mol; rounded to the nearest 0.5 kJ/mol.

$Me_{3-n}SiH_{n+1}$ (with n having values from 0 to 3) [2–4]. In particular, the rate constants $k_{2.3}$ and $k_{-2.3}$ obtained by time-resolved experiments allow the determination of the reaction enthalpy (ΔH_r) by either second or third law methods. $DH(R_3Si—H)$ is obtained by Equation (2.4) and then $\Delta H_f^\circ(R_3Si•)$ from Equation (2.2). The values are collected in Table 2.1.

$$R_3SiH + Br• \underset{k_{-2.3}}{\overset{k_{2.3}}{\rightleftharpoons}} R_3Si• + HBr \qquad (2.3/-2.3)$$

$$DH(R_3Si—H) = -\Delta H_r + DH(H—Br) \qquad (2.4)$$

A similar value has been obtained for $\Delta H_f^\circ(Me_3Si•)$ from an independent kinetic study on the very low pressure pyrolysis of hexamethyldisilane (Reaction 2.5) [6]. A bond dissociation enthalpy $DH(Me_3Si—SiMe_3) = 332 \pm 12$ kJ/mol was obtained which is related to $\Delta H_f^\circ(Me_3Si•)$ by Equation (2.6).

$$Me_3Si—SiMe_3 \longrightarrow 2Me_3Si• \qquad (2.5)$$

$$DH(Me_3Si—SiMe_3) = 2\,\Delta H_f^\circ(Me_3Si•) - \Delta H_f^\circ(Me_3SiSiMe_3) \qquad (2.6)$$

From Table 2.1 it emerges that, when the Me group progressively replaces the H atom, the Si—H bond strength in the silane significantly increases (ca 4 kJ/mol), the effect being cumulative. It is worth mentioning that this effect is opposite to that for hydrocarbons, where $DH(R_3C—H)$ is 438.5, 423.0, 412.5 and 404.0 kJ/mol for $H_3C—H$, $MeCH_2—H$, $Me_2CH—H$ and $Me_3C—H$, respectively [7]. A rationalization is based on the fact that C is more electronegative than Si. It is suggested that the electron-deficient central C atom is stabilized by electron donation from the methyl groups, whereas the central Si atom is destabilized by electron withdrawal by the methyl groups (inductive effect). Table 2.1 also brings to light that $\Delta H_f^\circ(R_3Si•)$ decreases by a fixed amount, which is approximately of 60 kJ/mol, by replacements of H atoms with methyl groups.

2.2.2 PHOTOACOUSTIC CALORIMETRY

Photoacoustic calorimetry is a thermodynamic method to determine a bond strength in solution [8]. Indeed, this technique has been used to quantify the enthalpy change occurring in a photoinduced reaction of di-*tert*-butyl peroxide with a silane (Reaction 2.7). Few bond dissociation enthalpies of silanes have been measured by this technique. In the original reports [9,10], the absolute values of D*H*s were underestimated by ca 20 kJ/mol due mainly to the reaction volume changes and the change in solvation enthalpies [1,8]. Therefore, relative data by this technique should be reliable since solvation correction is not necessary. Table 2.2 reports the relative bond dissociation enthalpies (DH_{rel}) for a few silanes. The data demonstrate that silicon–hydrogen bonds can be dramatically weakened by successive substitution of the Me_3Si group at the Si—H functionality. A substantial decrease in bond strength is also observed by replacing alkyl with methylthio groups. It is worth mentioning that in the analogous experiments with other group 14 hydrides, the bond strengths decrease by 27 and 69 kJ/mol, going from Et_3Si—H to Bu_3Ge—H and to Bu_3Sn—H, respectively [11,12].

$$t\text{-BuOOBu-}t + 2\ R_3SiH \xrightarrow{h\nu} 2\ t\text{-BuOH} + 2\ R_3Si\bullet \qquad (2.7)$$

The relative bond enthalpies from the photoacoustic calorimetry studies can be placed on an absolute scale by assuming that the value for $DH(Et_3Si—H)$ is similar to $DH(Me_3Si—H)$. In Table 2.2 we have converted the DH_{rel} values to absolute D*H* values (third column). On the basis of thermodynamic data, an approximate value of $DH(Me_3SiSiMe_2—H) = 378$ kJ/mol can be calculated that it is identical to that in Table 2.2 [1]. A recent advancement of photoacoustic calorimetry provides the solvent correction factor for a particular solvent and allows the revision of bond dissociation enthalpies and conversion to an absolute scale, by taking into consideration reaction volume effects and heat of solvation [8]. In the last column of Table 2.2 these values are reported and it is gratifying to see the similarities of the two sets of data.

Table 2.2 Relative and absolute bond dissociation enthalpies (kJ/mol)[a]

Silane (R_3SiH)	DH_{rel}[b]	$DH(R_3Si—H)$[c]	$DH(R_3Si—H)$[d]
Et_3Si—H	0	398	402
$Me_3SiSiMe_2$—H	−20	378	381
$(MeS)_3Si$—H	−32	366	364
$(Me_3Si)_3Si$—H	−46.5	351.5	351.5

[a] At 25 °C.
[b] From References [9,10].
[c] Estimated based on the DH_{rel} (see text).
[d] Revised values taking into consideration solvent correction factors [8].

2.2.3 THEORETICAL DATA

The continuous development and implementation of molecular orbital theory *ab initio* methods have enlarged the applications to this area too. Indeed, the impact of theoretical calculations in thermochemistry is substantial. Experimental groups often use calculations as a supplement to the interpretation of their results. In this section we will mention a few recent and representative studies that are directly associated with the bond dissociation energies of silanes. Early theoretical investigations of the Si—H bond strength in silanes have been summarized [13].

The most reliable calculations so far are relative bond dissociation energy studies by means of the isodesmic Reaction (2.8). $\Delta H_{2.8}$ can be reliably calculated even at modest levels of theory because errors arising from deficient basis sets and incomplete corrections for electron correlation largely are canceled. The $\Delta H_{2.8}$ for R = Me and $n = 0$ is calculated as 13.5 kJ/mol at the MP3/6–31G* level in excellent agreement with the experimental finding [14]. MP4SDTQ/6–31G* level of theory was used to study the effects of substituents on $DH(XSiH_2—H)$ by means of the isogyric reaction shown in Reaction (2.9) [15]. The results indicated that electropositive substituents with low-lying empty orbitals (Li, BeH, and BH_2) decrease the Si—H bond strengths by 30–50 kJ/mol. A 12 kJ/mol decrease in bond strength from $H_3Si—H$ to $H_3SiSiH_2—H$ and a difference of 34 kJ/mol between $H_3Si—H$ and $(H_3Si)_3Si—H$ were also computed.

$$R_{3-n}SiH_{1+n} + H_3Si\bullet \longrightarrow R_{3-n}SiH_n\bullet + H_3SiH \tag{2.8}$$

$$XH_2Si\bullet + H_4Si \longrightarrow XSiH_3 + H_3Si\bullet \tag{2.9}$$

Phenyl substitution was calculated to decrease the Si—H bond strength by 6 kJ/mol using the isogyric Reaction (2.8) for R = Ph and $n = 2$ at the HF/STO-3G* and MP2/STO-3G* levels [16]. These calculations further indicated that the presence of a second phenyl group, i.e. Reaction (2.8) with R = Ph and $n = 1$, has no additional effect. UB3LYP/6–31G and ROMP2/6–311++G(d,2p) methods were used to calculate the Si—H bond strengths of 15 *para*-substituted silanes p-Z—$C_6H_4SiH_2$—H and no significant substituent effect was found in DH(Si—H), while DH(Si—X) in the same series for X = Cl, F, Li showed such effects [17].

Calculations on the enthalpy change for Reaction (2.1) were also reported. DFT methods substantially underestimate the absolute bond dissociation energies, whereas the relative ones are reliable enough and allow the rationalization of substituent effects. Indeed, the substituent effect on the Si—H bond strength was addressed at the BLYP/6–31G* level of theory and indicated that successive Me substitutions strengthen the bonds, while successive SMe and SiH_3 substitutions weaken the bond in excellent accord with experimental data [18]. Good absolute $DH(XSiH_2—H)$ were obtained for relatively small molecules, using the G3(MP2) method for calculating the enthalpy change of Reaction (2.1) [17]. The calculated DH values for the silanes with the following substituents (in parentheses) are 383.3 (H) 387.8 (CH_3), 384.4 (Cl), 394.5 (F), 379.3 (NH_2), 388.5 (OH) and 376.1 (SH) kJ/mol. It is worth noting that the replacement of H by CH_3 or OH increases the bond strength of ca 5 kJ/mol, whereas the replacement of H with NH_2 or SH decreases the bond strength by 4 and 7 kJ/mol, respectively.

2.2.4 DERIVED BOND DISSOCIATION ENTHALPIES

Due to the importance of homolytic bond dissociation enthalpies for understanding radical chemistry, a set of Me_3Si—X bond dissociation enthalpies was derived via the relationship

$$DH(Me_3Si{-}X) = \Delta H_f^\circ(Me_3Si\bullet) + \Delta H_f^\circ(X\bullet) - \Delta H_f^\circ(Me_3SiX) \quad (2.10)$$

Table 2.3 shows the ΔH_f° values for a variety of radicals and their corresponding Me_3Si derivatives, together with the calculated like DH(Me_3Si—X) from Equation (2.10).

The DH(Me_3Si—X) varies enormously through the series of compounds in Table 2.3 and strictly depends on the electronegativity of the X group. In general, the trends of DH(Me_3Si—X) are the following. (i) For a particular column of the periodic table, the bond strength decreases going from top to bottom, i.e.,

Table 2.3 Derived Me_3Si—X bond dissociation enthalpies (kJ/mol)

X•	$\Delta H_f^\circ(X\bullet)$[a]	$\Delta H_f^\circ(Me_3SiX)$[c,e]	DH(Me_3Si—X)[g]
$H_3C\bullet$	146.5 ± 0.5	-233.2 ± 3.2	396
$H_3Si\bullet$	200.5 ± 2	-112.5	329
$Me_3Si\bullet$	16 ± 6[b]	-303.7 ± 5.5	336[h]
$H_2N\bullet$	188.7 ± 1.5	-291[f]	496
$Me_2N\bullet$	145[c]	-248 ± 4	409
HO•	39.3 ± 0.2	-500 ± 3	555
MeO•	17.2 ± 4	-480 ± 8	513
HS•	143.0 ± 3	-273[f]	432
BuS•	59.5[d]	-381 ± 3	457
F•	79.4 ± 0.3	-568[f]	663
Cl•	121.3 ± 0.1	-354 ± 3	491
Br•	111.9 ± 0.1	-298 ± 4	426
I•	106.8 ± 0.1	-222 ± 4	345

[a] From Reference [7], unless otherwise mentioned.
[b] From Table 2.1.
[c] From Reference [5].
[d] Calculated assuming DH(BuS—H) equal to DH((MeS—H) = 365.6 kJ/mol [7].
[e] Experimental data, unless otherwise mentioned.
[f] Obtained from enthalpy/electronegativity correlation.
[g] Rounded to the nearest 1 kJ/mol; Uncertainties ±10 kJ/mol.
[h] Direct measurement: 332 ± 12 kJ/mol [6].

Table 2.4 Recommended R—H bond dissociation enthalpies for some selected organic molecules (kJ/mol)[a]

X—H	D*H*(X—H)	X—H	D*H*(X—H)
H_3C—H	439 ± 0.5	H_2N—H	452.5 ± 1.5
$MeCH_2$—H	423 ± 1.5	MeNH—H	419[b]
Me_2CH—H	412.5 ± 1.5	PhNH—H	368[b]
Me_3C—H	404 ± 1.5	HO—H	499.15 ± 0.20
CH_2=CH—H	465 ± 3.5	MeO—H	436 ± 4
C_6H_5—H	465 ± 3.5	PhO—H	371.3 ± 2.3[c]
CH_2=$CHCH_2$—H	369 ± 9	HS—H	381.5 ± 3
$PhCH_2$—H	370 ± 6	MeS—H	365.5 ± 2
HC(O)CH_2—H	394.5 ± 9	PhS—H	349.4 ± 4.5[d]
N≡CCH_2—H	396.5 ± 9	H_3Ge—H	349 ± 8
$HOCH_2$—H	402 ± 0.5	Bu_3Ge—H	368[e]
$HSCH_2$—H	393 ± 8	Ph_3Ge—H	356[e]
MeC(O)—H	374 ± 1.5	Bu_3Sn—H	326[e]

[a] From Reference [7], unless otherwise mentioned.
[b] The D*H*(X—H) in $MeNH_3^+$ is 447 kJ/mol (i.e., 28 kJ/mol stronger) and in $PhNH_3^+$ is 307.5 kJ/mol (i.e., 60.5 kJ/mol weaker); see Reference [19].
[c] From Reference [20].
[d] From Reference [21].
[e] Data in solution [8].

Si—F > Si—Cl > Si—Br > Si—I, or Si—O > Si—S, or Si—C > Si—Si. (ii) For a particular row, the bond strength increases going from left to right, i.e., Si—C < Si—N < Si—O < Si—F, or Si—Si < Si—S < Si—Cl.

In Table 2.4, we have collected background information for discussion in the following chapters. Recommended C—H bond dissociation enthalpies of selected organic compounds are reported in the first two columns, followed by a variety of heteroatom–hydrogen bond strengths including N—H, O—H, S—H, Ge—H, and Sn—H bonds.

2.3 ION THERMOCHEMISTRY

Thermochemical information about neutral species can also be obtained from measurements of ions. Indeed, accurate bond dissociation energies for neutral molecules have been obtained from gas-phase ion chemistry techniques. In this section, we will summarize both the negative-ion and hydride-affinity cycles involving silicon hydrides (R_3SiH) which are connected to electron affinity (EA) and ionization potential (IP) of silyl radicals, respectively [22–24].

2.3.1 NEGATIVE-ION CYCLES

Thermodynamic properties related to R_3SiH can be obtained from negative-ion gas-phase studies. The following thermochemical cycle (cf. Scheme 2.1):

$$R_3Si{-}H \longrightarrow R_3Si\bullet + H\bullet \quad DH(R_3Si{-}H) \tag{2.1}$$

$$R_3Si\bullet + e^- \longrightarrow R_3Si^- \quad -EA(R_3Si\bullet) \tag{2.11}$$

$$H\bullet \longrightarrow H^+ + e^- \quad IP(H\bullet) \tag{2.12}$$

$$R_3Si{-}H \longrightarrow R_3Si^- + H^+ \quad \Delta H_{acid} = DH(R_3Si{-}H) - EA(R_3Si\bullet) + IP(H\bullet) \tag{2.13}$$

links the gas-phase acidity of silane (ΔH_{acid}) to the bond dissociation enthalpy of Si—H, the electron affinity of the silyl radical and the ionization potential of a hydrogen atom. Equation (2.13) can be used to obtain different parameters depending on what information is already known [22,23].

The EAs of a variety of silyl radicals have been measured by means of electron photodetachment experiments (Table 2.5), in which the anion population was monitored as a function of the wavelength of irradiating light [26,27]. It is worth mentioning that the EAs reported in Table 2.5 indicate that (i) replacement of a hydrogen by a methyl group decreases EA, the effect being cumulative, (ii) substitution of a phenyl for hydrogen has essentially no effect on EA, and (iii) replacement of a hydrogen by a Me_3Si group increases EA, the effect being cumulative. EAs of $H_3Si\bullet$ (1.39 eV), $H_2FSi\bullet$ (1.53) and $H_2ClSi\bullet$ (1.44) have been calculated at the MP4SDTQ/6–311++G(2df,p) level [28].

Gas-phase acidities of a few R_3SiH have been measured by the equilibrium method and are expected to be quite accurate (see Table 2.5, third column in roman). The bond dissociation enthalpies in Table 2.5 (last column in italic) were calculated using Equation (2.13) and the appropriate ΔH_{acid} and EA data together with the IP(H•) = 13.6 eV. These $DH(R_3Si{-}H)$ values, although associated with large errors (any errors in ΔH_{acid} or EA propagate into DH), are in good agreement with those available from equilibrium kinetics studies (cf. Table 2.1) and indicate that replacement of a hydrogen by a phenyl group decreases the Si—H bond strength by 6–7 kJ/mol. Analogously, Equation (2.13) has been used to estimate ΔH_{acid} values when EA and $DH(R_3Si{-}H)$ are available (see Table 2.5, third column in italic). Since gas-phase acidity reactions are always endothermic, larger values of ΔH_{acid} correspond to weaker acids. Therefore, H_4Si is ~54 kJ/mol stronger acid than Me_3SiH and ~92 kJ/mol weaker acid than $(Me_3Si)_3SiH$ whereas both H_4Si and $PhSiH_3$ have about the same acidity. For a comparison with carbon analogues, Me_3CH is 17 kJ/mol stronger acid than H_4C whereas H_4C is 151 kJ/mol stronger acid than $PhCH_3$ [25].

Negative-ion thermodynamic cycles similar to the one mentioned above can be easily constructed in order to estimate the energetics of the two fragmentation paths associated with the radical anion $R_3SiH^{\bullet -}$ (Scheme 2.1). However, EAs of silanes in Table 2.5 are unknown. Considering the Si—H bond dissociation of $Me_3SiH^{\bullet -}$ as an example (Reaction 2.14), the thermochemical cycle in Equation (2.15) links ΔH_{acid} to the $DH(Me_3SiH^{\bullet -})$. Assuming

Table 2.5 Electron affinities of silyl radicals together with gas-phase acidities and bond dissociation enthalpies of silanes[a]

Silane (R_3SiH)	EA($R_3Si\bullet$) (eV)	ΔH_{acid}(R_3Si—H)[b] (kJ/mol)	D*H*(R_3Si—H)[b] (kJ/mol)
H_3Si—H	1.41 ± 0.03	1560 ± 8	*383 ± 8*
$MeSiH_2$—H	1.19 ± 0.03	1583 ± 8	*386 ± 12*
Me_2SiH—H	1.07 ± 0.02	*1601*	392[c]
Me_3Si—H	0.97 ± 0.03	*1616*	397.4[c]
$PhSiH_2$—H	1.44 ± 0.01	1551 ± 8	*377 ± 8*
PhMeSiH—H	1.33 ± 0.04	1566 ± 8	*382 ± 12*
$Me_2Si(H)SiMe_2$—H	1.40 ± 0.03	*1555*	378[d,e]
$(Me_3Si)_2SiH$—H	1.94 ± 0.08		
$(Me_3Si)_3Si$—H	2.03 ± 0.09	*1468*	351.5[d]

[a] Taken from References [25,26], unless otherwise mentioned.
[b] Experimental data in roman and calculated values from Equation (2.13) in italic.
[c] From Table 2.1.
[d] From Table 2.2.
[e] For $Me_3SiSiMe_2$—H.

that EA(Me_3SiH) ≈ EA(Me_4Si) = 3.8 eV [29], a bond strength of 670 kJ/mol is calculated.

$$Me_3SiH^{\bullet -} \longrightarrow Me_3Si^- + H\bullet \quad (2.14)$$

$$DH(Me_3SiH^{\bullet -}) = \Delta H_{acid}(Me_3SiH) - IP(H\bullet) + EA(Me_3SiH) \quad (2.15)$$

2.3.2 HYDRIDE-AFFINITY CYCLES

Thermodynamic properties related to R_3SiH can also be obtained from the hydride-affinities of R_3Si^+. The following thermochemical cycle (cf. Scheme 2.1):

$$R_3Si\text{—}H \longrightarrow R_3Si\bullet + H\bullet \qquad DH(R_3Si\text{—}H) \quad (2.1)$$

$$R_3Si\bullet \longrightarrow R_3Si^+ + e^- \qquad IP(R_3Si\bullet) \quad (2.16)$$

$$H\bullet + e^- \longrightarrow H^- \qquad -EA(H\bullet) \quad (2.17)$$

$$R_3Si\text{—}H \longrightarrow R_3Si^+ + H^- \qquad \Delta H_{base} = DH(R_3Si\text{—}H) + IP(R_3Si\bullet) - EA(H\bullet) \quad (2.18)$$

links the gas-phase basicity of silane (ΔH_{base}) to the bond dissociation enthalpy of Si—H, the ionization potential of the silyl radical and the electron affinity of a hydrogen atom. Equation (2.18) can be used to obtain different parameters depending on what information is already known [22,23].

Table 2.6 Gas-phase basicities of silanes and adiabatic ionization potential of silyl radicals[a]

Silane (R_3SiH)	$\Delta H_{base}(R_3Si—H)$[b] (kJ/mol)	$IP(R_3Si\bullet)$[c] (eV)
$H_3Si—H$	1094	8.11[d]
$MeSiH_2—H$	1029	7.40
$Me_2SiH—H$	963	6.67
$Me_3Si—H$	923	6.20
$PhSiH_2—H$	962	6.81

[a] From References [30,31].
[b] Estimated errors ±4 kJ/mol.
[c] Calculated using Equation (2.16).
[d] A value of 8.14 ± 0.01 eV is obtained from the photoelectron spectrum [32].

Ion cyclotron resonance (ICR) spectroscopy has been used to determine the reaction enthalphy (ΔH_r) of hydride-transfer reaction of silanes with various hydrocarbons having known hydride affinities (Reaction 2.19). The hydride affinities of R_3Si^+, $DH(X_3Si^+—H^-) = \Delta H_{base}$, were obtained from Equation (2.20) and are summarized in Table 2.6 [30,31].

$$R_3SiH + X_3C^+ \rightleftharpoons R_3Si^+ + X_3CH \tag{2.19}$$

$$DH(R_3Si^+—H^-) = DH(X_3C^+—H^-) + \Delta H_r \tag{2.20}$$

The adiabatic ionization potentials of substituted silyl radicals, shown in Table 2.6, were recalculated from Equation (2.18) by using the appropriate ΔH_{base} and $DH(R_3Si—H)$ values together with the $EA(H\bullet) = 0.754$ eV. Replacement of hydrogen by a methyl or phenyl group decreases the IP by ~0.7 and ~1.3 eV, respectively. Moreover, the first and second methyl substitutions decrease the IP by an equal amount, the third being significantly less.

Other positive-ion cycles can afford complementary data related to the two dissociation paths of radical cation $R_3Si^{+\bullet}$ (Scheme 2.1). Considering the Si—H bond dissociation of $PhSiH_3^{+\bullet}$ as an example (Reaction 2.21), the thermochemical cycle in Equation (2.22) links the gas-phase basicity of silane (ΔH_{base}) to the $DH(PhSiH_3^{+\bullet})$. Taking $IP(PhSiH_3) = 9.09$ eV [31], a bond strength of 159 kJ/mol is calculated.

$$PhSiH_3^{+\bullet} \longrightarrow PhSiH_2^+ + H\bullet \tag{2.21}$$

$$DH(PhSiH_3^{+\bullet}) = \Delta H_{base}(PhSiH_3) + EA(H\bullet) - IP(PhSiH_3) \tag{2.22}$$

2.4 REFERENCES

1. Chatgilialoglu, C., *Chem. Rev.*, 1995, **95**, 1229.
2. Seetula, J.A., Feng, Y., Gutman, D., Seakins, P.W., and Pilling, M.J., *J. Chem. Phys.*, 1991, **95**, 1658.

3. Goumri, A., Yuan, W.-J., and Marshall, P., *J. Am. Chem. Soc.*, 1993, **115**, 2539. Kalinovski, I.J., Gutman, D., Krasnoperov, L.N., Goumri, A., Yuan, W.-J., Marshall, P., *J. Chem. Phys.*, 1994, **98**, 9551.
4. Ding, L., and Marshall, P., *J. Chem. Soc., Faraday Trans.*, 1993, **89**, 419.
5. Becerra, R., and Walsh, R. In *The Chemistry of Organic Silicon Compounds*, Z. Rappoport, and Y. Apeloig, (Eds), Wiley, Chichester, 1998, Chapter 4, pp. 153–180.
6. Bullock, W.J., Walsh, R., and King, K.D., *J. Phys. Chem.*, 1994, **98**, 2595.
7. Berkowitz, J., Ellison, G.B., and Gutman, D., *J. Phys. Chem.*, 1994, **98**, 2744.
8. Laarhoven, L.J.J., Mulder, P., and Wayner, D.D.M., *Acc. Chem. Res.*, 1999, **32**, 342.
9. Kanabus-Kaminska, J., Hawari, J.A., Griller, D., and Chatgilialoglu, C., *J. Am. Chem. Soc.*, 1987, **109**, 5267.
10. Chatgilialoglu, C., Guerra, M., Guerrini, A., Seconi, G., Clark, K.B., Griller, D., Kanabus-Kaminska, J., and Martinho-Simões, J.A., *J. Org. Chem.*, 1992, **57**, 2427.
11. Clark, K.B., and Griller, D., *Organometallics*, 1991, **10**, 746.
12. Burkey, T., Majewski, M., and Griller, D., *J. Am. Chem. Soc.*, 1986, **108**, 2218.
13. Apeloig, Y. In *The Chemistry of Organosilicon Compounds*, S. Patai, and Z. Rappoport (Eds), Wiley, Chichester, 1989, Chapter 2, pp. 57–225.
14. Ding, L., and Marshall, P., *J. Am. Chem. Soc.*, 1992, **114**, 5754.
15. Coolidge, M.B., and Borden, W.T., *J. Am. Chem. Soc.*, 1988, **110**, 2298. Coolidge, M.B., Hrovat, D.A., and Borden, W.T., *J. Am. Chem. Soc.*, 1992, **114**, 2354.
16. Ballestri, M., Chatgilialoglu, C., Guerra, M., Guerrini, A., Lucarini, M., and Seconi, G., *J. Chem. Soc., Perkin Trans. 2,* 1993, 421.
17. Cheng, Y.-H., Zhao, X., Song, K.-S., Liu, L., and Guo, Q.-X., *J. Org. Chem.*, 2002, **67**, 6638.
18. Wu, Y.-D., and Wong, C.-L., *J. Org. Chem.*, 1995, **60**, 821.
19. Liu, W.-Z., and Bordwell, F.G., *J. Org. Chem.*, 1996, **61**, 4778.
20. Borges dos Santos, R.M., and Martinho Simões, J.A., *J. Phys. Chem. Ref. Data*, 1998, **27**, 707.
21. Borges dos Santos, R.M., Muralha, V. S. F., Correia, C.F., Guedes, R.C., Costa Cobral, B.J., and Martinho Simões, J. A., *J. Phys. Chem. A*, 2002, **106**, 9883.
22. Sablier, M., and Fujii, T., *Chem. Rev.*, 2002, **102**, 2855.
23. Ervin, K.M., *Chem. Rev.*, 2001, **101**, 391.
24. Rienstra-Kiracofe, J.C., Tschumper, G.S., Schaefer III, H.F., Nandi, S., and Ellison, G.B., *Chem. Rev.*, 2002, **102**, 231.
25. Wetzel, D.H., Salomon, K.E., Berger, S., and Brauman, J.I., *J. Am. Chem. Soc.*, 1989, **111**, 3835.
26. Brinkman, E.A., Berger, S., and Brauman, J.I., *J. Am. Chem. Soc.*, 1994, **116**, 8304.
27. Damrauer, R., and Hankin, J.A., *Chem. Rev.*, 1995, **95**, 1137.
28. Rodriquez, C.F., and Hopkinson, A.C., *Can. J. Chem.*, 1992, **70**, 2234.
29. Modelli, A., Jones, D., Favaretto, L., and Distefano, G., *Organometallics*, 1996, **15**, 380.
30. Shin, S.K., and Beauchamp, J.L., *J. Am. Chem. Soc.*, 1989, **111**, 900.
31. Nagano, Y., Murthy, S., and Beauchamp, J.L., *J. Am. Chem. Soc.*, 1993, **115**, 10805.
32. Dyke, J.M., Jonathan, N., Morris, A., Ridha, A., and Winter, M.J., *Chem. Phys.*, 1983, **81**, 481.

3 Hydrogen Donor Abilities of Silicon Hydrides

The hydrogen abstraction from the Si—H moiety of silanes is fundamentally important not only because it is the method of choice for studying spectroscopically the silyl radicals but also because it is associated with the reduction of organic molecules, process stabilizers and organic modification of silicon surfaces.

This chapter focuses on the kinetics of reactions of the silicon hydrides with radicals. Kinetic studies have been performed with many types of silicon hydrides and with a large variety of radicals. The data can be interpreted in terms of the electronic properties of the silanes imparted by substituents for each attacking radical. Therefore the carbon-, nitrogen-, oxgygen-, and sulfur-centred radicals will be treated separately followed by a section on ketone triplets. It is worth mentioning that the reactivity of atoms and small organic radicals with silanes in the gas phase has been studied extensively. Although we deal in this chapter mainly with organic radicals in solution, the reactions of hydrogen atom with some silanes will be given as examples of gas-phase chemistry. Theoretical models for calculating activation energies will also be considered.

The kinetic data reported in this chapter have been determined either by direct measurements, using for example kinetic EPR spectroscopy and laser flash photolysis techniques or by competitive kinetics like the radical clock methodology (see below). The method for each given rate constant will be indicated as well as the solvent used. An extensive compilation of the kinetics of reaction of Group 14 hydrides (R_3SiH, R_3GeH and R_3SnH) with radicals is available [1].

Organosilanes in Radical Chemistry C. Chatgilialoglu
© 2004 John Wiley & Sons, Ltd ISBN: 0-471-49870-X

3.1 CARBON-CENTRED RADICALS

3.1.1 PRIMARY ALKYL RADICALS AND FREE-RADICAL CLOCK METHODOLOGY

The kinetic data for the reaction of primary alkyl radicals ($RCH_2\bullet$) with a variety of silanes are numerous and were obtained by applying the free-radical clock methodology. The term *free-radical clock* or *timing device* is used to describe a unimolecular radical reaction in a competitive study [2–4]. Three types of unimolecular reactions are used as clocks for the determination of rate constants for this class of reactions. The neophyl radical rearrangement (Reaction 3.1) has been used for the majority of the kinetic data, but the ring expansion rearrangement (Reaction 3.2) and the cyclization of 5-hexenyl radical (Reaction 3.3) have also been employed.

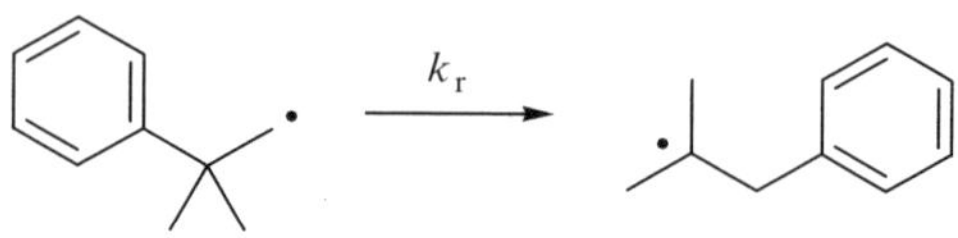

1

(3.1)

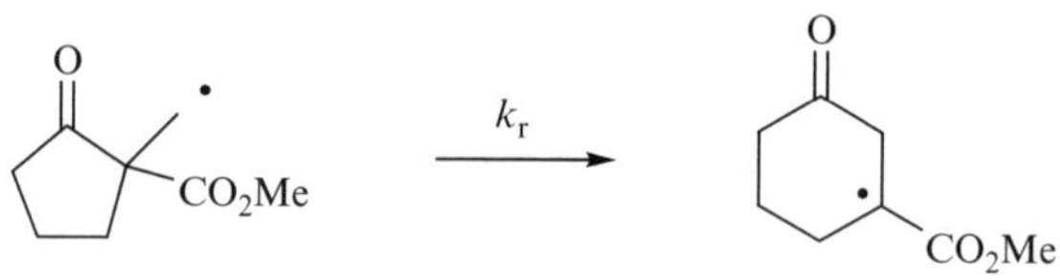

2

(3.2)

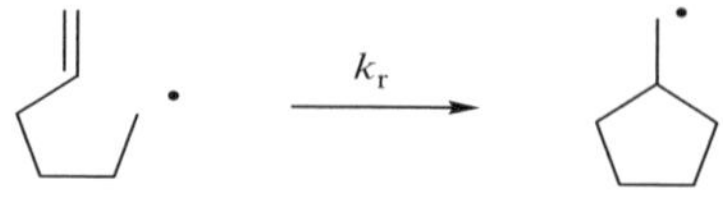

3

(3.3)

The concept of this method is illustrated in Scheme 3.1, where the clock reaction ($U\bullet \rightarrow R\bullet$) is the unimolecular radical rearrangement with a known rate constant (k_r). The rate constant for the H atom abstraction from R_3SiH by a primary alkyl radical U• can be obtained, provided that conditions are found in which the unrearranged radical U• is partitioned between the two reaction channels, i.e., the reaction with R_3SiH and the rearrangement to R•. At the end of the reaction, the yields of unrearranged (UH) and rearranged (RH) products can be determined by GC or NMR analysis. Under pseudo-first-order conditions of silane concentration, the following relation holds: $UH/RH = (k_H/k_r)[R_3SiH]$. A number of reviews describe the radical clock approach in detail [3,4].

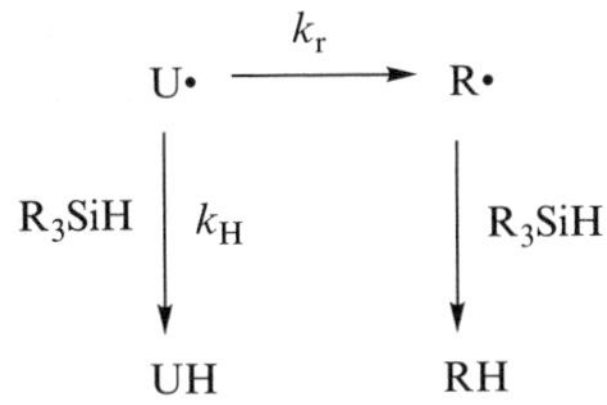

Scheme 3.1 Free-radical clock methodology

In Table 3.1 are collected the rate constants with various silanes and silanthrane derivatives **4–7** and the available Arrhenius parameters for some of them.

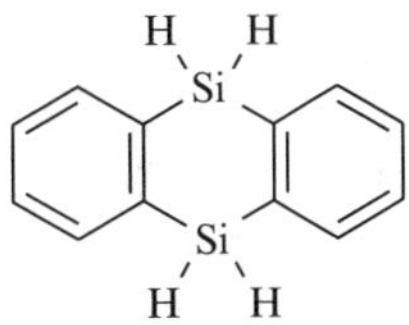

4

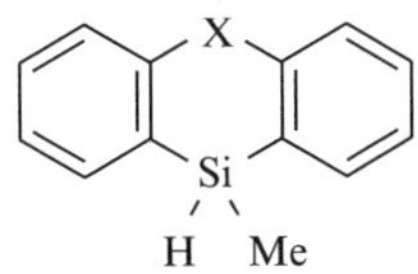

5 X = O
6 X = $SiMe_2$
7 X = S

Table 3.1 Rate constants (at 80 °C)[a] and Arrhenius parameters for the reactions of primary alkyl radicals with silicon hydrides[b]

Silane	Radical	$k/M^{-1}\,s^{-1}$	$\log A/M^{-1}\,s^{-1}$	E_a/kJ mol^{-1}	Reference
Et_3SiH	**1**	5.2×10^3	8.7	33.5	[5]
$PhSiH_3$	**1**	2.9×10^{4c}			[6]
Ph_2SiH_2	**1**	5.6×10^{4c}			[6]
Ph_3SiH	**1**	4.6×10^{4c}			[6]
5	**1**	2.2×10^4			[7]
6	**1**	4.5×10^4			[7]
7	**1**	8.7×10^4			[7]
4	**2**	2.1×10^6			[8]
$(Me_3Si)Me_2SiH$	**1**	2.5×10^4	9.0	31.4	[6]
RMe_2SiH[d]	**1**	3.6×10^4	9.0	30.1	[6]
$(Me_3Si)_2MeSiH$	**1**	1.5×10^5	8.9	25.1	[9]
$(Me_3Si)_3SiH$	**3**	1.2×10^6	8.9	18.8	[10]
$(MeS)_3SiH$	**2**	3.9×10^5			[8]
$(i\text{-}PrS)_3SiH$	**2**	3.7×10^5			[8]

[a] At 80 °C, if not otherwise mentioned.
[b] Benzene, *tert*-butylbenzene or 1,3-di-*tert*-butylbenzene were used as the solvent; for errors see the original references.
[c] At 110 °C.
[d] R = $SiMe_2SiMe_3$

The rate constants increase along the series $Et_3SiH < Ph_3SiH < (MeS)_3SiH < (Me_3Si)_3SiH$ with the expected intermediate values for silanes having mixed substituents. Mechanistic studies have shown that the attack of primary alkyl radicals on Et_3SiH occurs in about 60% of the cases at the SiH moiety and in 40% at the ethyl groups at 130 °C. The deuterium kinetic isotope effect (k_H/k_D) for the attack on the Si—D bond in Et_3SiD have been found to be 2.2 (at 130 °C) [4]. In the case of $(Me_3Si)_3SiH$ the H atom abstraction from the Si—H moiety amounted to about 95% of the total reactions [1]. The preexponential factors all lie in the expected range, and the activation energy is clearly the major factor in determining the radical–silane reactivity.

Phenyl substitution has only a small effect on the rate constant in contrast with that on the carbon analogues. The rate constants increase along the series $PhSiH_3 < Ph_3SiH < Ph_2SiH_2$ for neophyl radical; however, by taking into account the statistical number of hydrogens abstracted, the order changes as expected: $PhSiH_3 < Ph_2SiH_2 < Ph_3SiH$. For the discussion on the lack of resonance stabilization by phenyl groups see Section 1.2. The silanthrane **4** is one order of magnitude more reactive than Ph_2SiH_2 towards primary alkyl radicals (taking into account the statistical number of hydrogen abstracted). The reason for this enhancement in the reactivity is probably due to the stabilization of the silyl radical induced either by a transannular interaction of the vicinal Si substituent or by the quasi planar arrangement of the radical center [8].

Table 3.1 shows that the rate constant increases substantially with successive substitution of alkyl or phenyl by thiyl or silyl groups at the Si—H function. In particular, by replacing a methyl with a Me_3Si group, the rate increases by an order of magnitude, this effect being cumulative. These results are in good agreement with the thermodynamic data for these silanes, which show the weakening of the Si—H bond strength on replacing the alkyl substituent with thiyl or silyl groups (see Section 2.2).

Intramolecular hydrogen abstraction by primary alkyl radicals from the Si—H moiety has been reported as a key step in several unimolecular chain transfer reactions [11]. In particular, the 1,5-hydrogen transfer of radicals **8–11** (Reaction 3.4), generated from the corresponding iodides, was studied in competition with the addition of primary alkyl radicals to the allyltributylstannane and approximate rate constants for the hydrogen transfer have been obtained. Values at 80 °C are in the range of $(0.4–2) \times 10^4\ s^{-1}$, which correspond to effective molarities of about 1–2 M.

The competitive kinetics of Scheme 3.1 can also be applied to calibrate the unimolecular radical reactions provided that k_H is a known rate constant. In particular the reaction of primary alkyl radicals with $(Me_3Si)_3SiH$ has been used to obtain kinetic data for some important unimolecular reactions such as the β-elimination of octanethiyl radical from **12** (Reaction 3.5) [12], the ring expansion of radical **13** (Reaction 3.6) [8] and the 5-*endo-trig* cyclization of radical **14** (Reaction 3.7) [13]. The relative Arrhenius expressions shown below for the

(3.4)

8 R = *t*-Bu, X = O, Y = H
9 R = *t*-Bu, X = O, Y = Ph
10 R = *t*-Bu, X = CH_2, Y = H
11 R = Ph, X = CH_2, Y = H

Reactions (3.5)–(3.7), where $\theta = 2.3RT$ kJ/mol, combined with the Arrhenius parameters for the reaction of primary alkyl radicals with $(Me_3Si)_3SiH$ afford the temperature-dependent function for the unimolecular reactions. It is worth mentioning that the calibration of these radical clocks is based on the assumption that the radicals **12, 13** and **14** will react with $(Me_3Si)_3SiH$ with a rate constants equal to that of a simple model of primary alkyl radical **3**. This is reasonable for the radical **12** and **13**, whereas the influence of α-amide group in radical **14** is unknown, although it is expected to be similar based on the knowledge of the reactivity of Group 14 hydrides [1].

$$RSCH_2CH_2^\bullet \xrightarrow{k_f} RS^\bullet + H_2C = CH_2$$

$$\log k_H/k_f(M^{-1}) = -5.4 + 25.9/\theta \qquad (3.5)$$
$$\log k_f/s^{-1} = 14.2 - 44.4/\theta$$

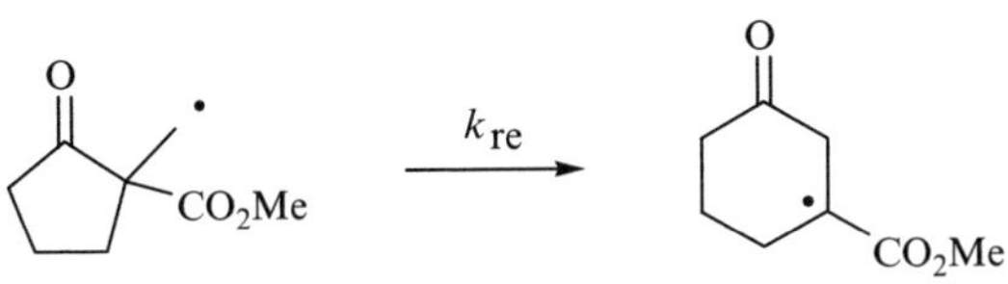

(3.6)

13 $\log k_H/k_{re}(M^{-1}) = -1.7 + 14.6/\theta$
$\log k_{re}/s^{-1} = 10.6 - 33.5/\theta$

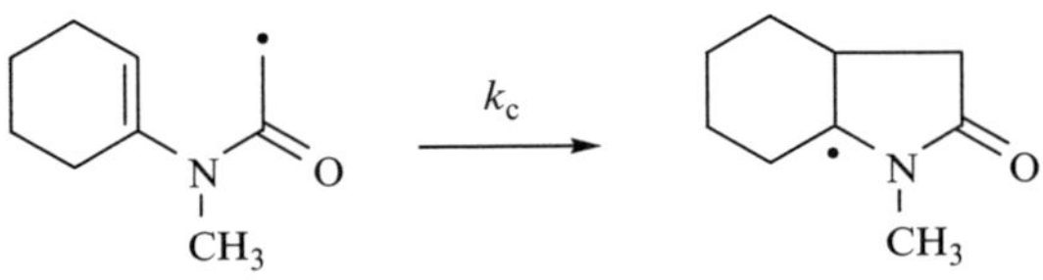

(3.7)

14 $\log k_H/k_c(M^{-1}) = -1.9 + 18.4/\theta$
$\log k_c/s^{-1} = 10.8 - 37.2/\theta$

3.1.2 OTHER TYPES OF CARBON-CENTRED RADICALS

In Table 3.2 are collected the rate constants of some silicon hydrides with a variety of radicals like the π-type secondary or tertiary alkyl radicals, the σ-type phenyl or acyl radicals, and the halogenated carbon-centred radicals. A few Arrhenius parameters are also available and reported here below.

The number of available rate constants or secondary and tertiary alkyl radicals is limited with respect to the primary ones. Their reactions with $(Me_3Si)_3SiH$ have been studied in detail by free-radical clock methods analogous to Reaction (3.3). The Arrhenius parameters are $\log A/M^{-1}\,s^{-1} = 8.3$ and $E_a = 18.0$ kJ/mol for secondary and $\log A/M^{-1}\,s^{-1} = 7.9$ and $E_a = 14.2$ kJ/mol for tertiary alkyl radicals [10]. It is worth mentioning that the rate constants for the reaction of primary, secondary, and tertiary alkyl radicals with $(Me_3Si)_3SiH$ are very similar in the range of temperatures that are useful for chemical transformation in the liquid phase. This is due to compensation of entropic and enthalpic effects through this series of alkyl radicals.

The Arrhenius expression for the reaction of the *o*-(allyloxy)phenyl radical (**15**) with $(Me_3Si)_3SiH$ relative to this unimolecular rearrangement (Reaction 3.8) has been established [10]. A rate constant $k_c = 8 \times 10^9\ s^{-1}$ has been estimated for

Table 3.2 Rate constants for the reactions of radicals with silicon hydrides

Radical	Silane	Solvent	T (°C)	$k/M^{-1}\,s^{-1}$	Reference
$RCH_2\dot{C}HCH_2$	$(Me_3Si)_3SiH$[a]	*n*-tetradecane	27	1.4×10^5	[10]
$RCH_2{-}C^{\bullet}(CH_3)_2$	Et_3SiH	benzene	50	3×10^{3}[b]	[14]
	Ph_3SiH	benzene	50	9×10^{3}[b]	[14]
	$(Me_3Si)_3SiH$[a]	*n*-tetradecane	27	2.6×10^5	[10]
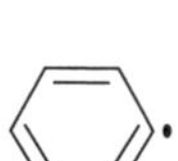	Ph_2SiH_2	CCl_4	60	3.4×10^7	[1]
	Ph_3SiH	CCl_4	60	2.1×10^7	[1]
	$(Me_3Si)_3SiH$	*n*-octane	20	3×10^8	[c]
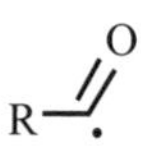	$(Me_3Si)_3SiH$[a]	toluene	27	1.8×10^4	[15]
$R_fCF_2CF_2\bullet$	Et_3SiH	Et_3SiH	30	7.5×10^5	[16]
	t-$BuMe_2SiH$	1,3-$(CF_3)_2$-C_6H_4	25	4.9×10^7	[17]
	$(Me_3Si)Me_2SiH$	1,3-$(CF_3)_2$-C_6H_4	25	3.1×10^6	[17]
	$(Me_3Si)_2MeSiH$	benzene-d_6	30	1.6×10^7	[16]
	$(Me_3Si)_3SiH$	benzene-d_6	30	5.1×10^7	[16]
$(CF_3)_2CF\bullet$	Et_3SiH	Et_3SiH	26	3.6×10^6	[18]
$(CF_3)_3C\bullet$	Et_3SiH	Et_3SiH	26	2.4×10^8	[18]
$Cl_3C\bullet$	Et_3SiH[a]	CCl_4/c-C_6H_{12}	27	5.0×10^2	[19]

[a] For Arrhenius parameters, see text.
[b] Approximate values.
[c] See text.

the cyclization reaction at 25 °C [4], which allows the calculation of a value of $k_H = 3 \times 10^8\ M^{-1}\ s^{-1}$ at this temperature. Hydrogen abstractions from Ph_2SiH_2 or Ph_3SiH by phenyl radical are more than one order of magnitude slower and obtained by competing bimolecular processes (Cl atom abstraction from CCl_4).

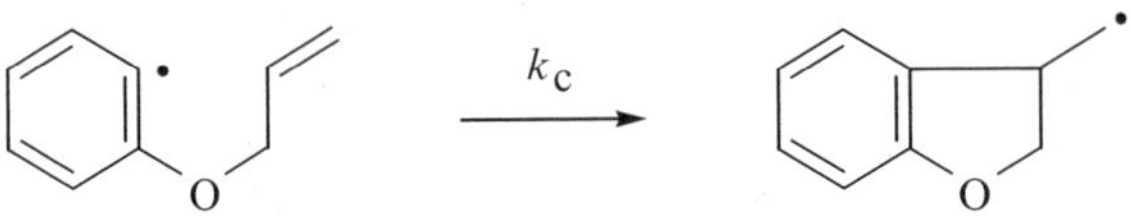

(3.8)

15 $\log k_H/k_c(M^{-1}) = -2.6 + 6.7/\theta$

Rate constants for hydrogen atom abstraction from $((Me_3Si)_3SiH$ by acyl radicals were measured by using competing decarbonylation reactions as the radical clock (Reaction 3.9), and the reference data being α-cleavage of the propanoyl radical in the gas phase [15]. It is worth mentioning that the hydrogen donation toward RCO• radicals is essentially independent of the nature of the alkyl substituent R and the Arrhenius parameters are $\log A/M^{-1}\ s^{-1} = 8.2$ and $E_a = 29.3$ kJ/mol.

(3.9)

The kinetic data for halogenated carbon-centred radicals with silicon hydrides are also numerous and a few examples are shown in Table 3.2. The kinetic data for perfluoroalkyl radicals were obtained by competition of the appropriate silane with the addition to an olefin [16–18]. The kinetic deuterium isotope effects (k_H/k_D) on the attack of n-C_4F_9• on the Si—D bond of t-$BuMe_2SiD$ and $Me_3SiSi(D)Me_2$ have been found to be 3.3 and 2.4, respectively, at 25 °C [17]. The rate constant for the reaction of Cl_3C• radical with Et_3SiH and its temperature dependence were obtained relative to the reaction of H atom abstraction from cyclohexane. The Arrhenius parameters are $\log A/M^{-1}\ s^{-1} = 8.6$ and $E_a = 33.9$ kJ/mol [19].

The trend in reactivity observed for the primary alkyl radical seems also to hold for the other carbon-centred radicals (Table 3.2), i.e., the rate constants increase along the series $Et_3SiH < Ph_3SiH < (Me_3Si)_3SiH$ with the expected intermediate values for silanes having mixed substituents. For triethylsilane, the rate constants decrease along the series $(CF_3)_3C$• > $(CF_3)_2CF$• > $R_fCF_2CF_2$• > RCH_2• > R_3C• > Cl_3C• and cover almost six orders of magnitude. For tris(trimethylsilyl)silane, the rate constants decrease along the series C_6H_5• > $R_fCF_2CF_2$• > RCH_2• ≈ R_2CH• ≈ R_3C• > RC(•)O and cover more than four orders of magnitude. The trends outlined above can be entirely attributed to the interplay of entropic, enthalpic and polar effects, the last effect manifested in the triplet repulsion of the silicon and carbon atoms of the reacting three-electron system (see Section 3.7).

The free-radical clock methodology has been also applied to calibrate unimolecular radical reactions based on the k_H values of Table 3.2 and, when available, on their Arrhenius parameters. The data for the reaction of secondary alkyl radicals with $(Me_3Si)_3SiH$ have been used for the determination of Arrhenius parameters related to the ring expansion of radical **16** (Reaction 3.10, where $\theta = 2.3RT$ kJ/mol) [20]. Trapping of the acyl radical **17** by $(Me_3Si)_3SiH$ has been used as a reference reaction for obtaining the temperature-dependent function of the acyl radical cyclization (Reaction 3.11) [20]. Reaction (3.12) is an example of perfluoroalkyl radical clocks, for which Et_3SiH is generally found to be the most appropriate hydrogen donor. Indeed, a $k_H/k_c = 2.0\,M^{-1}$ was obtained at 30 °C which corresponds to a $k_c = 3.8 \times 10^5\,s^{-1}$ [16].

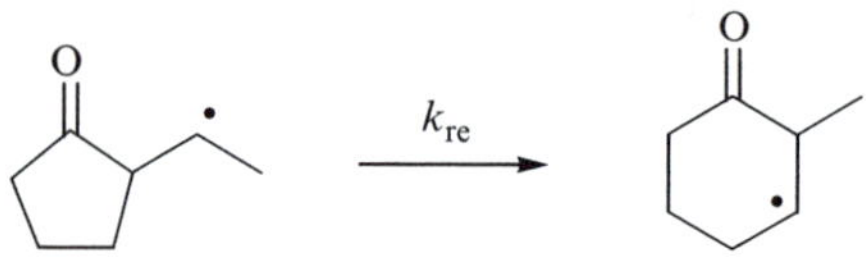

(3.10)

16

$$\log k_H/k_{re}(M^{-1}) = -2.2 + 21.3/\theta$$
$$\log k_{re}/s^{-1} = 0.5 - 39.3/\theta$$

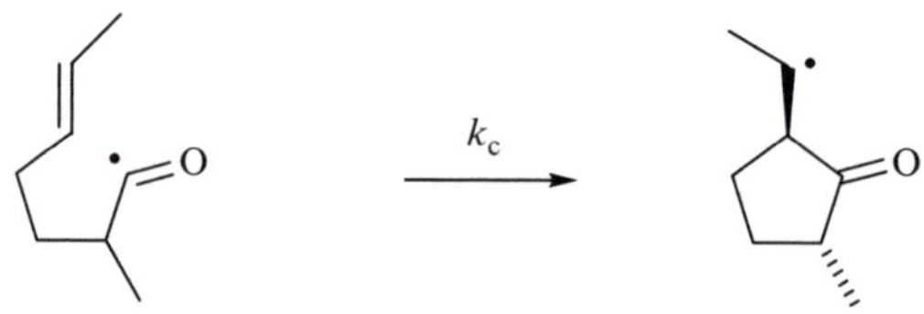

(3.11)

17

$$\log k_H/k_c(M^{-1}) = -1.4 + 10.5/\theta$$
$$\log k_c/s^{-1} = 9.6 - 25.1/\theta$$

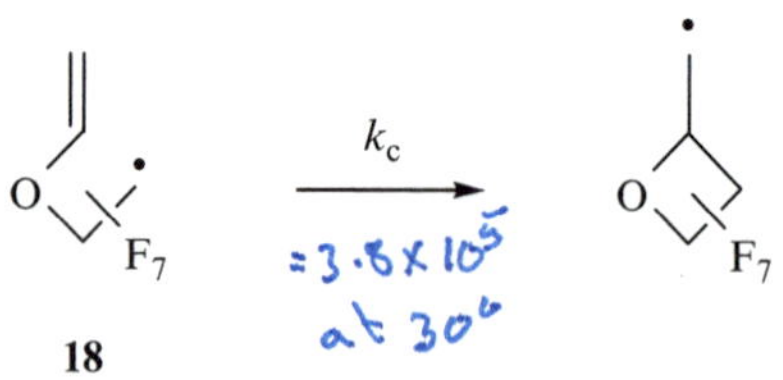

(3.12)

18

3.2 NITROGEN-CENTRED RADICALS

The kinetic data of the reactions of dialkylaminyl radicals with silicon hydrides are limited to the hindered 2,2,6,6,-tetramethylpiperidinyl radical (**19**) and are obtained by using EPR spectroscopy [21]. In Table 3.3 are collected the activation parameters together with the calculated rate constants at 25 °C. Two remarkable features are the low preexponential factors and the order of reactivity of phenyl substituted derivatives, where the rate constants increase along the series: $Ph_3SiH < Ph_2SiH_2 < PhSiH_3$ (taking into account the statistical number of abstractable hydrogen atoms). This behaviour, which is the opposite to the analogous reactions of alkyl and alkoxyl radicals, was explained in terms of steric effects in the transition state that demands a very specific orientation between the abstracting aminyl radical and the hydrogen–silicon bond in the silane [21].

Rate constants for the hydrogen abstraction from $(Me_3Si)_3SiH$ by diaryl aminyl radicals **20–24** were determined by using a method in which the corresponding amines catalyse the reaction of 2,4,6-tri-*tert*-butylphenoxyl radical with $(Me_3Si)_3SiH$ (see below). At 91 °C, rate constants are 33.5, 7.27, 5.63, 2.45 and 45.8 M^{-1} s^{-1} for radicals **20**, **21**, **22**, **23** and **24**, respectively [22]. Based

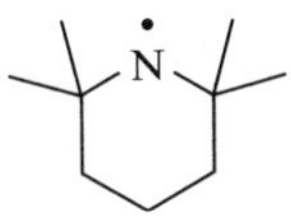

19

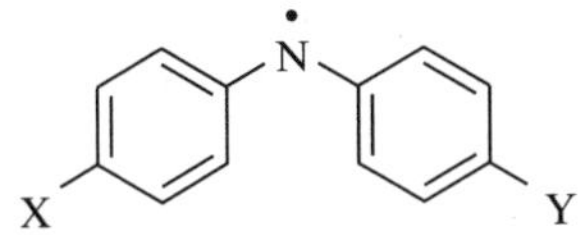

20 X = Y = H
21 X = Y = CH_3
22 X = H, Y = OCH_3
23 X = Y = OCH_3
24 X = Y = Br

on the analysis of the data, it was suggested that polarized transition states should be unimportant and the reactivity trend simply reflects the differences in the bond dissociation enthalpies of amine, D*H*(Ar_2N—H). The Arrhenius parameters for the reaction of Ph_2N• radical with $(Me_3Si)_3SiH$ have been measured (Table 3.3) and fit very well with a reaction in which both substrates are sterically hindered.

Table 3.3 Rate constants (25 °C) and Arrhenius parameters for the reactions of aminyl radicals with silicon hydrides [21,22][a]

Silane	Radical	$k/M^{-1}\ s^{-1}$	$\log A/M^{-1}\ s^{-1}$	E_a/kJ mol^{-1}
Et_3SiH	**19**	2.7	5.2	27.2
$PhSiH_3$	**19**	119.2	5.7	28.0
Ph_2SiH_2	**19**	32.1	5.8	24.3
Ph_3SiH	**19**	5.3	6.9	27.6
$(Me_3Si)Me_2SiH$	**19**	10.0	4.8	21.8
$(Me_3Si)_3SiH$	**19**	34.4	4.8	18.4
$(Me_3Si)_3SiH$[b]	Ph_2N•	4.4	5.2	25.9

[a] ESR spectroscopy and *tert*-butylbenzene as the solvent if not otherwise stated.
[b] Quasi-equilibrium method (see text) and chlorobenzene as the solvent.

3.3 OXYGEN-CENTRED RADICALS

3.3.1 ALKOXYL RADICALS

The reaction of thermally and photochemically generated *tert*-butoxyl radicals with silicon hydrides (Reactions 3.13 and 3.14) has been extensively used for the generation of silyl radicals in EPR studies, time-resolved optical techniques, and organic synthesis.

$$t\text{-BuOOBu-}t \xrightarrow{h\nu \text{ or } \Delta} 2\ t\text{-BuO}\bullet \tag{3.13}$$

$$t\text{-BuO}\bullet + R_3SiH \longrightarrow t\text{-BuOH} + R_3Si\bullet \tag{3.14}$$

Rate constants for Reaction (3.14) were measured directly by LFP techniques and are collected in Table 3.4 [23–27]. These values reflect the overall (or molecular) reactivity of the substrates regardless of the site or mechanism of the reaction. Mechanistic studies have shown that the attack of the *t*-BuO• radical on Et_3SiH occurs in about 80 % of the cases at the SiH moiety and in 20 % at the ethyl groups at 27 °C [23], whereas the attack on $(Me_3Si)_3SiH$ occurs in about 95 % of the cases at the SiH moiety and in 5 % at the trimethylsilyl groups at 27 °C [26].

The rate constants for reaction of *t*-BuO• radical with silanes increase along the series $Et_3SiH < Ph_3SiH < (MeS)_3SiH < (Me_3Si)_3SiH$ with the expected intermediate values for silanes having mixed substituents. In particular, the rate constants decrease along the two series $PhSiH_3 < Ph_2SiH_2 < Ph_3SiH$ and $Me_3SiH < Me_3SiSi(H)Me_2 < (Me_3Si)_2Si(H)Me < (Me_3Si)_3SiH$.

The Arrhenius parameters obtained for the reaction of *t*-BuO• radical with Et_3SiH are $\log A/M^{-1}\,s^{-1} = 8.5$ and $E_a = 8.8$ kJ/mol [23]. The Arrhenius parameters for Me_3SiH in the gas phase are also available and were obtained by competition with the *tert*-butoxyl radical decomposition, i.e., $\log A/M^{-1}\,s^{-1} = 8.7$ and $E_a = 10.9$ kJ/mol [28]. These preexponential factors lie in the expected range and, therefore, the activation energy is expected to be the major factor

Table 3.4 Rate constants for reactions of *tert*-butoxyl [23–27] and cumylperoxyl [29] radicals with silicon hydrides

Silane	*t*-BuO• $k/M^{-1}\,s^{-1}$ at 27 °C[a]	$PhMe_2COO\bullet$ $k/M^{-1}\,s^{-1}$ at 73 °C[b]
$n\text{-}C_5H_{11}SiH_3$	1.1×10^7	
$PhSiH_3$	7.5×10^6	0.90
Ph_2SiH_2	1.3×10^7	
Et_3SiH	5.7×10^6	0.10[c]
$PhMe_2SiH$	6.6×10^6	0.21
Ph_2MeSiH		0.43
Ph_3SiH	1.1×10^7	0.67
$(Me_3Si)Me_2SiH$	1.7×10^7	
$(Me_3Si)_2MeSiH$	6.2×10^7[d]	
$(Me_3Si)_3SiH$	1.1×10^8	66.3
$H(PhSiH)_nH$		24.7[e]
$(MeS)_3SiH$	4.4×10^7	
$(i\text{-PrS})_3SiH$	4.5×10^7	

[a] Method: laser flash photolysis, if not otherwise mentioned. Solvent: 2:1 (v/v) di-*tert*-butyl peroxide/benzene or 1:4 (v/v) di-*tert*-butyl peroxide/isooctane.

[b] Method: inhibited hydrocarbon oxidation. Solvent: cumene.

[c] Referring to $t\text{-BuMe}_2SiH$.

[d] Average value of competitive studies $(Me_3Si)Me_2SiH/(Me_3Si)_2MeSiH$ and $(Me_3Si)_2MeSiH/(Me_3Si)_3SiH$ in *tert*-butylbenzene [6].

[e] Value for each SiH moiety.

in determining the *t*-BuO• radical–silane reactivity. The trends outlined above can be entirely attributed to more favorable thermodynamic factors along the series.

3.3.2 PEROXYL RADICALS

The kinetics for the reaction of cumylperoxyl radical with a variety of silanes (Reaction 3.15) have been measured by using inhibited hydrocarbon oxidation methodology (Table 3.4) [29]. The trends in reactivity for the cumylperoxyl radical are similar to those observed for *t*-BuO• radical, although the reactions are about seven orders of magnitude slower. The rate constants increase along the two series $PhSi(H)Me_2 < Ph_2Si(H)Me < Ph_3SiH$ and $PhSiH_3 < Ph_2SiH_2 < Ph_3SiH$, when the statistical number of abstracted hydrogens is taken into account. Furthermore, the rate constants increase about 7 and 660 times on going from *t*-$BuMe_2SiH$ to Ph_3SiH and $(Me_3Si)_3SiH$, respectively.

$$PhMe_2COO\bullet + R_3SiH \longrightarrow PhMe_2COOH + R_3Si\bullet \qquad (3.15)$$

The similar reactivity of peroxyl with $(CF_3)_2NO\bullet$ radicals towards a large variety of substrates has been observed previously and attributed to rather similar thermochemistries and spin density distributions for these two radicals [30]. The Arrhenius parameters for the reaction of the persistent $(CF_3)_2NO\bullet$ radical with *n*-Bu_3SiH determined by kinetic EPR spectroscopy are $\log A/M^{-1}\,s^{-1} = 5.5$ and $E_a = 32.2$ kJ/mol, which corresponds to $k = 4.3\,M^{-1}\,s^{-1}$ at 73 °C [31].

3.3.3 ARYLOXYL AND AROYLOXYL RADICALS

The Arrhenius parameters for the reaction of persistent aryloxyl radical **25** with $(Me_3Si)_3SiH$ (Reaction 3.16) were measured spectrophotometrically and found to be $\log A/M^{-1}\,s^{-1} = 4.3$ and $E_a = 43.8$ kJ/mol [22]. A rate constant of $4.8 \times 10^{-4}\,M^{-1}\,s^{-1}$ can be calculated at 25 °C. It has been suggested that the low preexponential factor is due to geometric constraints on the transition state.

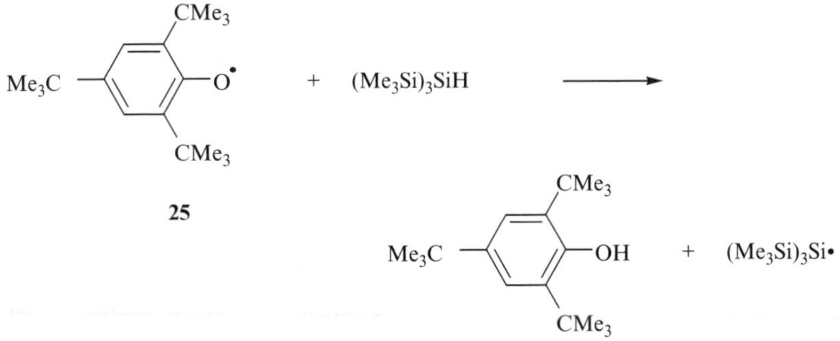

25

(3.16)

Absolute rate constants for the reaction of Et_3SiH with aroyloxyl radicals **26** (Reaction 3.17) were measured by laser flash photolysis and were found to be in the range of $(3.8\text{–}7.4) \times 10^6\ M^{-1}\ s^{-1}$ at 24 °C [32]. The rate constants of *t*-BuO• and PhC(O)O• radicals with Et_3SiH are identical within experimental error. Furthermore, the rate constants were correlated by the Hammett equation using σ^+ substituent constants, and a ρ value of +0.68 was obtained.

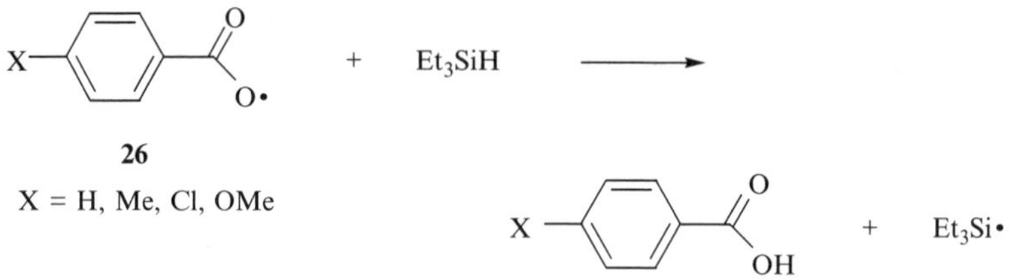

26

X = H, Me, Cl, OMe

(3.17)

3.4 SULFUR-CENTRED RADICALS

The reaction of thiyl radicals with silicon hydrides (Reaction 3.18) is the key step of the so called *polarity-reversal catalysis* in the radical-chain reduction of alkyl halides as well as in the hydrosilylation of olefins using silane–thiol couple (see Sections 4.5 and 5.1) [33]. The reaction is strongly endothermic and reversible (Reaction −3.18).

$$RS\bullet + R_3SiH \underset{k_{SiH}}{\overset{k_{SH}}{\rightleftharpoons}} RSH + R_3Si\bullet \qquad (3.18/-3.18)$$

The rate constants k_{SH} and k_{SiH} were determined in cyclohexane at 60 °C relative to $2k_t$ for the self-termination of the thiyl radicals, using the kinetic analysis of the thiol-catalysed reduction of 1-bromooctane and 1-chlorooctane by silane, respectively [34]. The k_{SH} values are collected in Table 3.5 (third column) the lowest value being $3.2 \times 10^4\ M^{-1}\ s^{-1}$ for the reaction of 1-adamantanethiyl radical (1-AdS•) with Et_3SiH. The reverse rate constants (k_{SiH}) were obtained for the reaction of $Et_3Si\bullet$ with 1-AdSH and $(t\text{-}BuO)_3SiSH$ and found to be $5 \times 10^7\ M^{-1}\ s^{-1}$ for both of them. The equilibrium constants $K_{eq} = k_{SH}/k_{SiH}$ are 6.2×10^{-4} and 2.6×10^{-3} and hence ΔG_r are +20.4 and +16.5 kJ/mol, respectively. Unfortunately, ΔS_r is unknown, which would have allowed one to determine ΔH_r, i.e., the difference in bond dissociation enthalpies of the corresponding silane and thiol in solution. The difference of the simpler silane and thiol, $DH(Me_3Si\text{—}H)$ and $DH(MeS\text{—}H)$, in the gas phase corresponds to $\Delta H_r = +31.9$ kJ/mol (cf. Chapter 2).

Rate constants for the reaction of thiyl radicals with the *t*-BuMePhSiH were also extracted from the kinetic analysis of the thiol-catalysed radical-chain racemization of enantiomerically pure (*S*)-isomer [34]. Scheme 3.2 shows the reaction mechanism that involves the rapid inversion of silyl radicals together with reactions of interest. The values in cyclohexane solvent at 60 °C are collected in the last column of Table 3.5.

Table 3.5 Rate constants for the reactions of thiyl radicals with silicon hydrides in cyclohexane at 60 °C [34]

Silane	RS•	$k_{SH}/M^{-1}\,s^{-1a}$	$k_{SH}/M^{-1}\,s^{-1b}$
Et_3SiH	1-AdS•	3.2×10^4	
	t-$C_{12}H_{25}S$•	3.5×10^4	
	t-BuMePhSiS•	4.1×10^4	
	Ph_3SiS•	4.3×10^4	
	$(t\text{-BuO})_3SiS$•	1.3×10^5	
	i-Pr_3SiS•	1.6×10^5	
Ph_3SiH	1-AdS•	7.5×10^4	
	Ph_3SiS•	8.8×10^4	
t-BuMePhSiH	1-AdS•	3.4×10^4	3.5×10^4
	n-$C_{12}H_{25}S$•		1.6×10^4
	t-BuMePhSiS•		5.5×10^4
	Ph_3SiS•		5.6×10^4
	$(t\text{-BuO})_3SiS$•		2.6×10^5
	i-Pr_3SiS•		2.5×10^5

[a] Method: reduction of 1-bromooctane by silane catalysed by thiols.
[b] Method: thiol-catalysed racemization of (S)-t-BuMePhSiH.

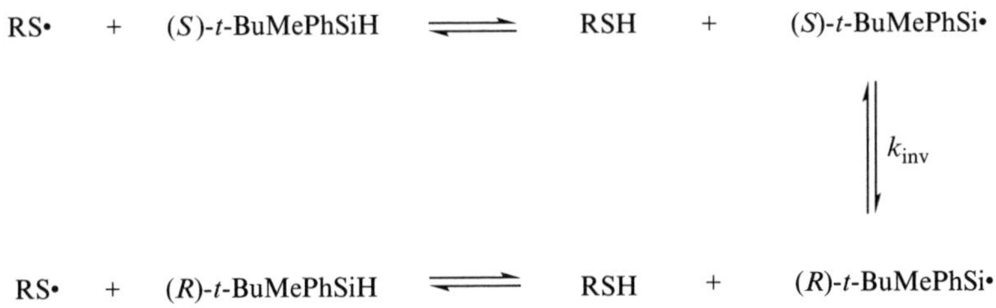

Scheme 3.2 Thiol-catalysed radical chain racemization

The rate constants of 1-AdS• with Et_3SiH and t-BuMePhSiH are relatively fast taking into consideration the endothermicity of these reactions. The silanethiyl radicals Ph_3SiS• and t-BuMePhSiS• abstract hydrogen with similar rate constants, whereas the more hindered $(t\text{-BuO})_3SiS$• and i-Pr_3SiS• radicals show an enhancement in their reactivity, which suggests that entropic effects may play an important role.

3.5 KETONE TRIPLETS

The excited states of carbonyl compounds are often considered to be similar to alkoxyl radicals because of the unpaired electron on the oxygen atom. In particular, the benzophenone n$-\pi^*$ triplet mostly reacts in the same manner as and at a similar rate to t-BuO• radicals.

Absolute rate constants for the reaction of the benzophenone triplet with Et_3SiH, n-$C_5H_{11}SiH_3$, $PhSiH_3$, and Cl_3SiH have been measured by laser flash

Table 3.6 Rate constants for reactions of benzophenone n–π* triplet with silicon hydrides at room temperature [23, 25]

Silane	$k/M^{-1}\,s^{-1a}$	$k/M^{-1}\,s^{-1b}$
n-$C_5H_{11}SiH_3$	8.8×10^6	
$PhSiH_3$	5.0×10^6	
Et_3SiH	9.6×10^6	2.4×10^7
n-Bu_2MeSiH		3.0×10^7
$(Me_3SiCH_2)_2MeSiH$		2.7×10^7
$PhMe_2SiH$		8.8×10^6
$(Me_3SiO)_2MeSiH$		3.4×10^{6c}
$(EtO)_3SiH$		1.7×10^6
$(Me_3Si)_3SiH$		3.4×10^8

[a] Laser flash photolysis; in benzene [23].
[b] Phosphorescence; 9:1 (v/v) acetonitrile/acetone [25].
[c] A value of $2.5 \times 10^6 M^{-1}\,s^{-1}$ was obtained in benzene.

photolysis in benzene (Table 3.6) [23]. Rate constants of phosphorescence quenching of benzophenone with a variety of silicon hydrides have also been measured in 9:1 (v/v) mixture of acetonitrile–acetone as the solvent and are also collected in Table 3.6 [35]. Similarly, the rate constants of benzophenone triplet with Et_4Si and $(EtO)_4Si$ are found to be $< 1 \times 10^5$ and $1 \times 10^6\ M^{-1}\,s^{-1}$, respectively [35], which suggests that the reaction with Et_3SiH involves only the SiH moiety whereas with $(EtO)_3SiH$ involves H atom abstraction from both the SiH moiety and EtO groups, with about 55:45 regioselectivity favoring the SiH abstraction. The Arrhenius parameters obtained from laser flash photolysis experiments for the reaction of benzophenone triplet with Et_3SiH are $\log A/M^{-1}\,s^{-1} = 8.9$ and $E_a = 10.9\,kJ/mol$ [23]. Comparing the data of n-$C_5H_{11}SiH_3$, $PhSiH_3$, Et_3SiH and $(Me_3Si)_3SiH$ from Table 3.6 with the corresponding kinetic data of t-BuO• radicals (Table 3.4), it can be seen that these two transient species have a rather similar reactivity towards silanes. In particular, the rate constants for benzophenone triplet with n-$C_5H_{11}SiH_3$ and $PhSiH_3$ are slightly slower, whereas with Et_3SiH and $(Me_3Si)_3SiH$ are 5 and 3 times faster, respectively.

Excited triplets of other ketones have been studied to a lesser extent [23,35]. The xanthate and the p-methoxyacetophenone triplets were found to be more ($6.0 \times 10^7\ M^{-1}\,s^{-1}$) and less ($\sim 8 \times 10^5\ M^{-1}\,s^{-1}$) reactive, respectively, than the benzophenone triplet with Et_3SiH. The p-methylbenzophenone triplet behaves similarly to benzophenone triplet toward $(Me_3SiO)_2MeSiH$, whereas the p-phenylbenzophenone triplet shows no reaction. Also no reaction was observed for the benzil triplet with Et_3SiH and $(Me_3SiO)_2MeSiH$.

3.6 HYDROGEN ATOM: AN EXAMPLE OF GAS-PHASE KINETICS

The substituent effect on the reactivity of the Si—H bond in $Me_{4-n}SiH_n$, where $n = 0$–3, towards different atoms has been the subject of several experimental studies in the gas phase and a number of theoretical investigations. For attacking species such as H, Cl and Br atoms, the room temperature rate constant corrected for reaction path degeneracy, k/n, is higher for Me_3SiH than H_4Si by a factor of around 4 [36]. However, the interplay of activation energy and preexponential factor is complex and is not fully understood. The reaction of H• atoms is reported in Table 3.7 in detail, for a better understanding of the reactivity trend in this case [36]. Indeed, the methyl substitution leads to an increase in the Si—H bond reactivity when the statistical number of abstracted hydrogens is taken into account. Theory reproduced very well the experimental rate constants at 25 °C (values in parenthesis) [37]. However, the experimental activation energies and A factors are difficult to rationalize because the changes are small and may be also masked by experimental errors. Theoretical calculations suggest that the activation energies for the three reactions are nearly the same (i.e., unaffected by methylation), whereas the increase of reactivity along the series reflects a parallel change in the preexponential factors. The increase in A factor with methylation has been explained in terms of an increase in the entropy of the transition state associated either with the introduction of soft skeletal vibrations, or the freeing of the hindered internal rotation of the methyl groups [38].

3.7 THEORETICAL APPROACHES

A number of theoretical methodologies, spanning from high-level *ab initio* to empirical calculations have been applied to obtain information on the factors that influence the reactivity of Si—H bond towards radicals and atoms.

High-level computational methods are limited, for obvious reasons, to very simple systems. In the previous section we showed the contribution of the theory for a better understanding of the entropic and enthalpic factors that influence the reactions of hydrogen atom with the simplest series of silanes $Me_{4-n}SiH_n$, where $n = 1–3$. Calculated energy barriers for the forward and reverse hydrogen atom abstraction reactions of Me•, Et•, *i*-Pr• and *t*-Bu• radicals with $Me_{4-n}SiH_n$, where $n = 0–3$, and $(H_3Si)_3SiH$ have been obtained at

Table 3.7 The substituent effect on the reactivity of silanes towards H• atoms[a] (in parenthesis theoretical data)[b]

Silane	n(Si—H bonds)	$kn^{-1}/10^7\,M^{-1}\,s^{-1}$[c]	$An^{-1}/10^{10}\,M^{-1}\,s^{-1}$	E_a/kJ mol^{-1}
$MeSiH_3$	3	7.8 (7.7)	1.7 (1.5)	13.2 (13.4)
Me_2SiH_2	2	11.9 (11.4)	2.4 (2.1)	13.2 (13.1)
Me_3SiH	1	16.7 (15.4)	2.0 (3.0)	11.7 (13.2)

[a] From [36].
[b] Using saddle-point structure at the BHLYP/6-311+G** level and single-point *ab initio* at the PMP4/6-311+G(3df,2p)//BHLYP level [37].
[c] At 25 °C.

level of theory as high as QCISD/DZP//MP2/DZP [39]. Transition states in these reactions were found to prefer a collinear arrangement of groups at the hydrogen atom undergoing transfer. These barriers are calculated to decrease, on moving from methyl to primary, secondary and tertiary alkyl radicals, in agreement with the available experimental observations, but their absolute values are overestimated.

A nonparametric method that is related to the original London equation has been used to calculate energies of activation for hydrogen abstraction, which agrees with experiment generally to within 4 kJ/mol [40,41]. Indeed, the activation energies calculated for the hydrogen abstraction from Me_3SiH and $(Me_3Si)_3SiH$ by CH_3CH_2• and from Me_3SiH by CH_3O• are 36.8, 22.6 and 10.5 kJ/mol, respectively, and compare very nicely with the experimental values of 33.5, 18.8 and 8.9 kJ/mol reported in this chapter [40]. In addition to the reaction enthalpy, this method demonstrates the importance of triplet repulsion between the terminal atoms of the reacting three-electron system.

An empirical algorithm that relates the activation energy to four molecules involved in the reaction (an extended form of Evans–Polanyi equation) predicts activation energies with a good accuracy [42]. The E_a for the reactions of *t*-BuO• and Cl_3C• radicals with Et_3SiH are calculated to be 12.6 and 33.9 kJ/mol, respectively, that match well with the values of 8.8 and 33.9 kJ/mol reported in this chapter. Activation energies for the reaction of a large variety of silanes with different type of radicals have been obtained by applying a semi-empirical method using intersecting parabolas [22,43].

3.8 REFERENCES

1. Chatgilialoglu, C., and Newcomb, M., *Adv. Organomet. Chem.*, 1999, **44**, 67.
2. Griller, D., and Ingold, K.U., *Acc. Chem. Res.*, 1980, **13**, 317.
3. Newcomb, M., *Tetrahedron*, 1993, **49**, 1151.
4. Newcomb, M. In *Radicals in Organic Synthesis*, P. Renaud and M.P. Sibi (Eds), Vol. 1, Wiley-VCH, Weinheim, 2001, pp. 317–336.
5. Chatgilialoglu, C., Ferreri, C., and Lucarini, M., *J. Org. Chem.*, 1993, **58**, 249.
6. Ballestri, M., Chatgilialoglu, C., Guerra, M., Guerrini, A., Lucarini, M., and Seconi, G., *J. Chem. Soc., Perkin Trans. 2,* 1993, 421.
7. Oba, M., Kawahara, Y., Yamada, R., Mizuta, H., and Nishiyama, K., *J. Chem. Soc., Perkin Trans. 2,* 1996, 1843.
8. Chatgilialoglu, C., Timokhin, V.I., and Ballestri, M., *J. Org. Chem.*, 1998, **63**, 1327.
9. Chatgilialoglu, C., Guerrini, A., and Lucarini, M., *J. Org. Chem.*, 1992, **57**, 3405.
10. Chatgilialoglu, C., Dickhaut, J., and Giese, B., *J. Org. Chem.*, 1991, **56**, 6399.
11. Curran, D.P., Xu, J., and Lazzarini, E., *J. Chem. Soc., Perkin Trans. 1,* 1995, 3046.
12. Chatgilialoglu, C., Altieri, A., and Fischer, H., *J. Am. Chem. Soc.*, 2002, **124**, 12816.
13. Chatgilialoglu, C., Ferreri, C., Guerra, M., Timokhin, V., Froudakis, G., and Gimisis, T., *J. Am. Chem. Soc.*, 2002, **124**, 10765.
14. Newcomb, M., and Park, S. U., *J. Am. Chem. Soc.*, 1986, **108**, 4132.
15. Chatgilialoglu, C., Ferreri, C., Lucarini, M., Pedrielli, P., and Pedulli, G.F., *Organometallics*, 1995, **14**, 2672.

16. Rong, X.X., Pan, H.-Q., Dolbier, Jr., W.R., and Smart, B.E., *J. Am. Chem. Soc.*, 1994, **116**, 4521.
17. Shtarev, A.B., Dolbier, Jr., W.R., and Smart, B.E., *J. Am. Chem. Soc.*, 1999, **121**, 2110.
18. Dolbier, Jr., W.R., and Li, A.-R., *J. Chem. Soc., Perkin Trans. 2,* 1998, 79.
19. Baruch, G., and Horowitz, A., *J. Phys. Chem.*, 1980, **84**, 2535.
20. Chatgilialoglu, C., Ferreri, C., Lucarini, M., Venturini, A., and Zavitsas A.A., *Chem. Eur. J.*, 1997, **3**, 376.
21. Lucarini, M., Marchesi, E., Pedulli, G.F., and Chatgilialoglu, C., *J. Org. Chem.*, 1998, **63**, 1687.
22. Varlamov, V.T., Denisov, E.T., and Chatgilialoglu, C., *J. Org. Chem.*, 2001, **66**, 6317.
23. Chatgilialoglu, C., Scaiano, J.C ., and Ingold, K.U., *Organometallics*, 1982, **1**, 466.
24. Chatgilialoglu, C., Ingold, K.U., Lusztyk, J., Nazran, A.S., and Scaiano, J.C., *Organometallics*, 1983, **2**, 1332.
25. Lusztyk, J., Maillard, B., and Ingold, K.U., *J. Org. Chem.*, 1986, **51**, 2457.
26. Chatgilialoglu, C., and Rossini, S., *Bull. Soc. Chim. Fr.*, 1988, 298.
27. Chatgilialoglu, C., Guerra, M., Guerrini, A., Seconi, G., Clark, K.B., Griller, D., Kanabus-Kaminska, J., and Martinho-Simões, J.A., *J. Org. Chem.*, 1992, **57**, 2427.
28. Lee, Y.E., and Choo, K.Y., *Int. J. Chem. Kinet.*, 1986, **18**, 267.
29. Chatgilialoglu, C., Timokhin, V.I., Zaborovskiy, A.B., Lutsyk, D.S., and Prystansky, R.E., *J. Chem. Soc., Perkin Trans. 2,* 2000, 577.
30. Chatgilialoglu, C., Malatesta, V., and Ingold, K.U., *J. Phys. Chem.*, 1980, **84**, 3597.
31. Doba, T., and Ingold, K.U., *J. Am. Chem. Soc.*, 1984, **106**, 3958.
32. Chateauneuf, J., Lusztyk, J., and Ingold, K.U., *J. Am. Chem. Soc.*, 1988, **110**, 2886.
33. Roberts, B.P., *Chem. Soc. Rev.*, 1999, **28**, 25.
34. Cai, Y., and Roberts, B.P., *J. Chem. Soc., Perkin Trans. 2,* 2002, 1858.
35. Müller, U., Helmstreit, W., and Timpe, H.-J., *J. Photochem Photobiol. A: chem.*, 1989, **50**, 1.
36. Arthur, N.L., and Miles, L.A., *J. Chem. Soc., Faraday Trans.*, 1998, **94**, 1077.
37. Yu, X., Li, S.-M., Xu, Z.-F., Li, Z.-S., and Sun, C.-C., *J. Phys. Chem. A,* 2001, **105**, 7072.
38. Arthur, N.L., Potzinger, P., Reimann, B., and Steenbergen, H-P., *J. Chem. Soc., Faraday Trans.*, 1989, **85**, 1447.
39. Dakternieks, D., Henry, D.J., and Schiesser, C.H. *J. Chem. Soc., Perkin Trans. 2,* 1998, 591.
40. Zavitsas, A.A., and Chatgilialoglu, C., *J. Am. Chem. Soc.*, 1995, **117**, 10645.
41. Zavitsas, A.A., *J. Am. Chem. Soc.*, 1998, **120**, 6578.
42. Roberts, B.P., and Steel, A.J., *J. Chem. Soc., Perkin Trans. 2,* 1994, 2155.
43. Denisov, E.T., *Russ. Chem. Bull.*, 1998, **47**, 1274.

4 Reducing Agents

4.1 GENERAL ASPECTS OF RADICAL CHAIN REACTIONS

The use of free-radical reactions in organic synthesis started with the reduction of functional groups. The purpose of this chapter is to give an overview of the relevance of silanes as efficient and effective sources for facile hydrogen atom transfer by radical chain processes. A number of reviews [1–7] have described some specific areas in detail. Reaction (4.1) represents the reduction of a functional group by silicon hydride which, in order to be a radical chain process, has to be associated with initiation, propagation and termination steps of the radical species. Scheme 4.1 illustrates the insertion of Reaction (4.1) in a radical chain process.

$$RZ + R'_3SiH \longrightarrow RH + R'_3SiZ \qquad (4.1)$$

A large number of compounds are known to decompose thermally or photolytically to generate free radicals (see Section 4.2). These radicals (In•) participate in the initiation step by abstracting a hydrogen atom from the reducing agent R'_3SiH. In the propagation steps, the removal of the functional group Z in the organic substrate takes place by action of $R'_3Si\bullet$ radical via a reactive intermediate or a transition state represented by $[RZ(\bullet)SiR'_3]$. A site-specific radical R• is generated, which then reacts with the silicon hydride and gives the reduced product, together with 'fresh' $R'_3Si\bullet$ radicals to continue the chain. Chain reactions terminate by radical–radical combination or disproportionation reactions, with the general concept that the more remote the termination reactions the more efficient is the chain reaction. Kinetically speaking, to be efficient the rate of chain transfer steps between radicals and starting materials must be higher than that of chain termination steps between radicals. The

Organosilanes in Radical Chemistry C. Chatgilialoglu
© 2004 John Wiley & Sons, Ltd ISBN: 0-471-49870-X

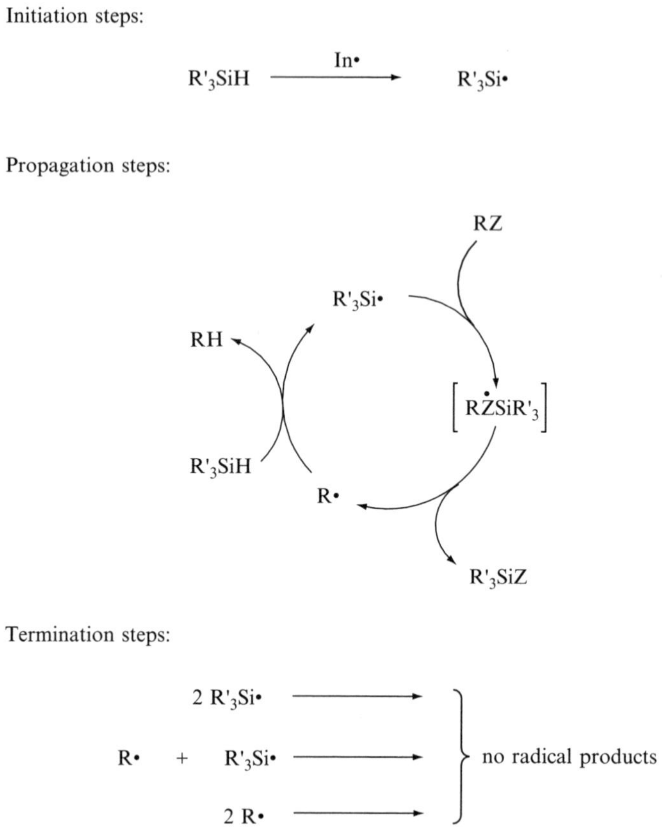

Scheme 4.1 The reaction mechanism for the radical chain removal of a functional group by organosilicon hydrides

following observations have been made [8]: (i) the termination rate constants in the liquid phase are controlled by diffusion, i.e., close to $10^{10}\ M^{-1}\ s^{-1}$ (see below), (ii) in chain reactions radical concentrations are about 10^{-8} M (depending on reaction conditions), and (iii) the concentration of substrates is generally between 0.05 and 0.5 M, indicating that the rate constants for the chain transfer steps must be higher than $10^3\ M^{-1}s^{-1}$. Using the extensive kinetic data available on the hydrogen atom abstraction step from silicon hydrides (see Chapter 3), and on the Z group removal by the silyl radical (see later in this Chapter), one can choose the conditions for efficient radical reductions. Since radical chain reactions require a steady supply of radicals, faster propagation steps (i.e., higher kinetic chain length) require small amounts of radical initiator. To foresee the reaction efficiency and outcome is a powerful feature of radical chemistry, provided that synthetic organic chemists acquire a familiarity with the kinetic data. In my personal experience, the integration of organic synthesis with physical organic chemistry has been a winning horse for introducing some novelty in radical-based organic transformations.

Generally, radical chain reactions are carried out in nonpolar solvents although strong solvent effects on propagation steps are rare. Apart from the polarity, a much more important criterion for the solvent choice is the solvent's inertness towards the chain propagating radicals involved.

4.1.1 RADICAL–RADICAL REACTIONS

The bimolecular rate constant ($2k_t$) for the decay of $Me_3Si^\bullet$ radical has been measured by kinetic EPR studies in the liquid phase and found to be $3.0 \times 10^9\ M^{-1}\ s^{-1}$ at 20 °C [9], whereas a value of $1.5 \times 10^{10}\ M^{-1}\ s^{-1}$ at 22 °C is obtained from the time profile of the transient absorption in the gas phase [10]. The Arrhenius parameters were also measured in the liquid phase and found to be $\log A/M^{-1}\ s^{-1} = 9.9$ and $E = 4.1$ kJ/mol [9]. Trimethylsilyl radicals, generated in the liquid phase by reaction of t-BuO• radicals with Me_3SiH, are shown to react with each other by both combination (Reaction 4.2) and disproportionation (Reaction 4.3), the former being about 10 times faster [11,12].

$$2\ Me_3Si^\bullet \longrightarrow Me_3SiSiMe_3 \tag{4.2}$$

$$2\ Me_3Si^\bullet \longrightarrow Me_3SiH + CH_2{=}SiMe_2 \tag{4.3}$$

The bimolecular rate constants for the self-reaction of the $Et_3Si^\bullet$, $Me_3SiSi(^\bullet)Me_2$ and $(Me_3Si)_3Si^\bullet$ radicals have also been obtained in the liquid phase. A rate constant of $1.4 \times 10^9\ M^{-1}\ s^{-1}$ at −65 °C was measured for $Me_3SiSi(^\bullet)Me_2$ by kinetic EPR spectroscopy [13]. By means of laser flash photolysis experiments, the $Et_3Si^\bullet$ and $(Me_3Si)_3Si^\bullet$ radicals were shown to decay with second-order kinetics with $2k_t/\varepsilon_{308nm} = 1.1 \times 10^7\ cm\ s^{-1}$ and $2k_t/\varepsilon_{300nm} = 5.0 \times 10^6\ cm\ s^{-1}$, respectively, at ambient temperature [14,15]. Estimated values of appropriate molar absorption coefficients allowed for the calculation of the values $2k_t = 1 \times 10^{10}$ and $5 \times 10^9\ M^{-1}\ s^{-1}$ for $Et_3Si^\bullet$ and $(Me_3Si)_3Si^\bullet$ radicals, respectively [14,16]. While the fate of the reaction between two $Et_3Si^\bullet$ radicals is still not known, the termination products of other silyl radicals have been determined. Pentamethyldisilyl radicals, produced by the reaction of $Me_3SiSi(H)Me_2$ with photogenerated t-BuO• radicals at room temperature, behave similarly to the $Me_3Si^\bullet$ radical [17]. That is, products due to the combination (Reaction 4.4) and disproportionation (Reaction 4.5) of these radicals were detected in a ratio of ≥ 2.1.

$$2\ Me_3Si\dot{S}iMe_2 \longrightarrow Me_3SiSiMe_2SiMe_2SiMe_3 \tag{4.4}$$

$$2\ Me_3Si\dot{S}iMe_2 \longrightarrow Me_3SiSi(H)Me_2 + CH_2{=}SiMeSiMe_3 \tag{4.5}$$

On the other hand, $(Me_3Si)_3Si\bullet$ radicals produced by thermal Reaction (4.6) gave only combination products [18]. Under the same conditions, $Me_3SiSi(H)R^1R^2$, where R^1 and R^2 are methyl and/or phenyl groups resulted in less than 1 % yields of the coupling products.

$$2\ (Me_3Si)_3SiH \xrightarrow[130^\circ C]{t\text{-BuOOBu-}t} \underset{92\%}{(Me_3Si)_3Si-Si(SiMe_3)_3} \qquad (4.6)$$

The kinetics data on the reactions of silyl radicals with carbon-centred radicals are also available. The rate constant for the cross-combination of $CH_3\bullet$ with $Me_3Si\bullet$ was measured to be $6.6 \times 10^{10}\ M^{-1}\ s^{-1}$ in the gas phase [19]. Studies on the steady-state and the pulse radiolysis of Et_3SiH in methanol showed that the cross-combination of $Et_3Si\bullet$ with $CH_3O\bullet$ and $HOCH_2\bullet$ occurs with rate constants of 1.1×10^8 and $0.7 \times 10^8\ M^{-1}\ s^{-1}$, respectively [20].

4.2 RADICAL INITIATORS

The most popular thermal initiator is 2′, 2′-azobisisobutyronitrile (AIBN), with a half-life ($t_{1/2}$) of 1 h at 81 °C and 10 h at 65 °C in toluene [8,21]. Generally, 5–10 mol% of initiator is added either all in one portion or by slow addition over a period of time. There are other azo compounds which can be chosen, depending on the reaction conditions. Indeed, the nature of the substituent play an important role as can be seen for 2′, 2′-azobis-(4-methoxy)-3,4-dimethyl-valeronitrile (AMVN), with a $t_{1/2}$ of 1 h at 56 °C and 10 h at 33 °C in toluene. There are also hydrophilic azo compounds, such as 2′, 2′-azobis-(2-methylpropionamidine) dihydrochloride (APPH), with a $t_{1/2}$ of 10 h at 56 °C in water.

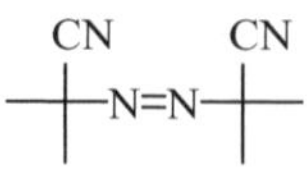

AIBN

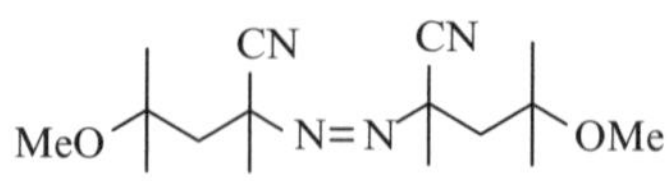

AMVN

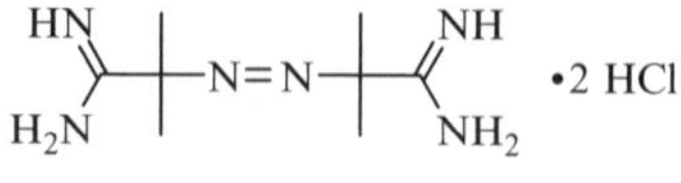

APPH

Peroxides are used when the reaction requires a more reactive initiating species. Thermolysis of dibenzoyl peroxide [PhC(O)O—OC(O)Ph], with a $t_{1/2}$ of 1 h at 95 °C and 7 h at 70 °C, is the most familiar to synthetic chemists. It initially produces acyloxyl radicals that often decarboxylate prior to undergoing bimolecular reactions and affording the equally reactive phenyl radicals.

Thermolysis of *tert*-butyl perbenzoate [PhC(O)OOBu-*t*] and di-*tert*-butyl peroxide (*t*-BuOOBu-*t*), whose $t_{1/2}$ are 1 h at 125 °C and 147 °C, respectively, has been used from time to time.

Photochemically generated radicals in chain reactions are less familiar to synthetic chemists [8,21]. The above mentioned peroxides have been used in the presence of light to initiate radical chain reactions at room or lower temperatures. Azo compounds are also known to decompose photolytically to afford alkyl radicals. AIBN has rarely been used under such conditions.

Triethylborane in the presence of very small amounts of oxygen is an excellent initiator for radical chain reactions. For a long time it has been known that trialkylboranes R_3B react spontaneously with molecular oxygen to give alkyl radicals (Reaction 4.7), but only recently has this approach successfully been applied as the initiation [22]. The reactions can be run at temperatures as low as −78 °C, which allow for a better control of stereoselectivity (see below).

$$R_3B + O_2 \longrightarrow R_2BOO\bullet + R\bullet \tag{4.7}$$

A correct choice of the initiator is generally decided according to the experimental conditions. For example, α-cyanoalkyl radicals derived from AIBN are capable of abstracting a hydrogen atom from $(TMS)_3SiH$, but not from Ph_2SiH_2. Both alkoxyl and aryloxyl radicals are able to abstract a hydrogen atom efficiently even from Et_3SiH, that has one of the strongest Si—H bond energies. It is advisable to add the initiators slowly during the course of the reaction, either in solution or portion-wise, with due regard for the half-life of their decomposition at the operating temperature of the reaction. The direct addition at once of the thermal initiator together with temperatures much higher than those of 1 h half-life are generally not the optimal conditions. At low temperatures, when the thermal initiation is obviously impractical, the choice of Et_3B/O_2 initiation is particular useful in combination with $(TMS)_3SiH$ as reducing agent. Alternatively, photoinitiation using AIBN or dibenzoyl peroxide could be useful.

4.3 TRIS(TRIMETHYLSILYL)SILANE

The synthetic application of free radical reactions has increased dramatically since the mid-1980s. Tributyltin hydride is the most commonly used reagent for the reduction of functional groups. However, it was later shown that organotin compounds are toxic and environmentally not tolerated, and they are difficult to remove from non-polar reaction products [1–7]. It becomes now more evident that it is inappropriate to propose these reagents for application in medicinal chemistry. Therefore, a great deal of research has been devoted to finding new reagents with favourable characteristics from toxicological and environmental points of view, and with chemistry analogous to organotin hydrides, so that the knowledge gained in the field of free radicals could be further applied.

During the 1980s, I started to envisage a suitable approach based on organosilanes. Trialkylsilanes were shown to be poor reducing agents in free-radical chain processes. That is, although trialkylsilyl radicals are among the most reactive species towards various organic functional groups (see below), the corresponding silanes were rather poor H atom donors towards alkyl radicals (see Section 3.1), therefore they do not support chain reactions under normal conditions. I became persuaded from our own and others' work that the hydrogen donor ability of organosilanes could be adjusted by a wise choice of thermodynamic driving forces plugged in by the substituents to the silicon atom. This turned out to be the case.

The discovery of tris(trimethylsilyl)silane as a good radical-based reducing agent is an amusing anecdote. The rationalization of chemical physical data of the silane/silyl radical systems allowed me to write on the paper $(TMS)_3SiH$, as the simplest formula for a good reducing agent [23,24]. By checking the literature, this material was indeed already reported by Gilman and coworkers in 1965 and then was almost totally ignored for the next 20 years [25]. Another case of Sleeping Beauty in chemistry! Sandra Rossini who worked with me as a technician at the Italian National Research Council (CNR) synthesized this material and went to run the proton NMR in CCl_4 to check the purity of the product. She came back to me, saying that she could not see any resonance related to the SiH moiety. After a while, we understood that the reason was the reduction of CCl_4 to $CHCl_3$ by $(TMS)_3SiH$ directly in the NMR tube! At the Gordon Research Conference on Free Radical Reactions in 1987, I gave a lecture on *Novel Silanes as Radical Reducing Agents* followed by two relevant communications [26,27], and in 1990 this discovery was awarded with the Fluka prize as 'Reagent of the Year'.

$(TMS)_3SiH$ has indeed quickly proven to be a valid alternative to tin hydride for the majority of its radical chain reactions [1]. It should have been expected that the efficiency of the silicon reagent, coupled with favourable characteristics of ease of purification and lack of toxicity of $(TMS)_3SiH$ and its byproducts, were overwhelming reasons for it to become more and more the reagent of choice in radical reactions. Nonetheless, tributyltin hydride still remains the reagent primarily used by synthetic chemists who decide to incorporate one or more radical steps in their synthetic schemes. Personally, I believe that Bu_3SnH has long been established as a reliable radical reducing agent, supported by a vast literature which highlights its advantages and also clarifies its limitations, so that it is a 'traditional' reagent and $(TMS)_3SiH$ remains in the background as a radical-based reducing agent. It is also worth underlining that there is a growth of cases where the two reagents behave differently, mainly due to the different behaviour of the corresponding radicals.

4.3.1 DEHALOGENATIONS

The reductive removal of bromine and iodine atoms by $(TMS)_3SiH$ is straightforward. Generally, equimolar or a slight excess of reducing agent is employed. A variety of solvents can be used although aromatic solvents such as benzene or toluene are the most common. The reactions are complete after a short time. A few examples are given in Reactions (4.8) – (4.11) [28–31]. Multiple dehalogenations are possible in a one-pot procedure by using the corresponding equivalent of reducing agent (Reaction 4.11). The efficiency depends to a limited extent on the substituent; in alkyl bromides the rate constants decrease along the series benzyl > tertiary alkyl > secondary alkyl > primary alkyl > phenyl (see Table 4.1) [32]. Relative rate constants for the reaction of $(TMS)_3Si\bullet$ with 13 substituted benzyl bromides have been measured at 80 °C, and analysis of the data in terms of a Hammett-type plot shows this silyl radical to be strongly nucleophilic [33].

Table 4.1 Rate constants at 20 °C for the reaction of $(TMS)_3Si\bullet$ radicals with some bromides [32]

Bromide	$k/M^{-1}\,s^{-1}$
$PhCH_2Br$	9.6×10^8
$(CH_3)_3CBr$	1.2×10^8
$CH_3CH_2CH(CH_3)Br$	4.6×10^7
$CH_3(CH_2)_4Br$	2.0×10^7
C_6H_5Br	4.6×10^6

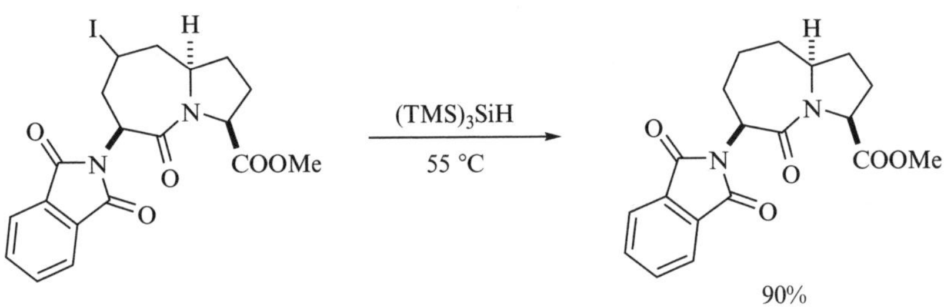

(4.8)

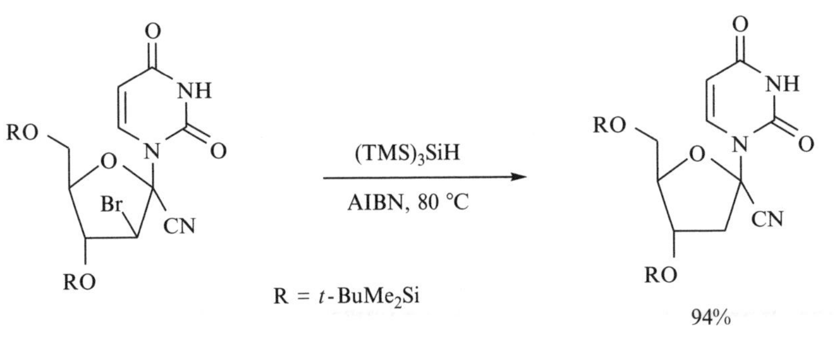

(4.9)

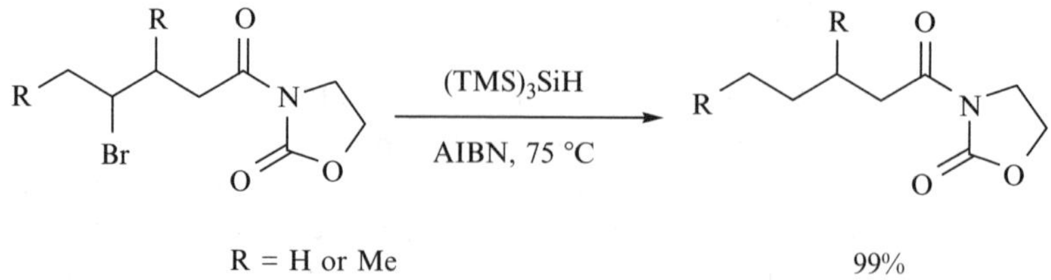

(4.10)

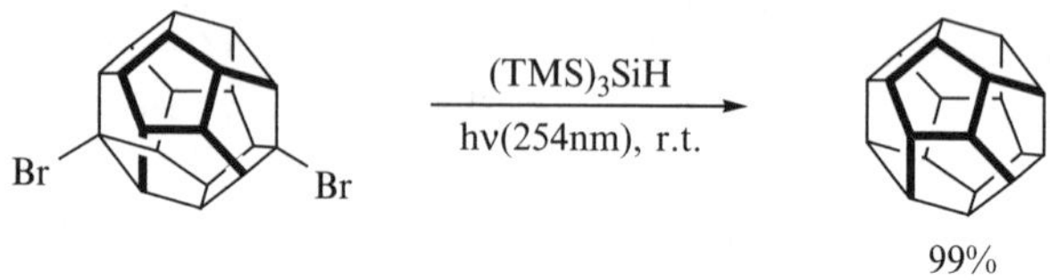

(4.11)

A highly diastereoselective and efficient method for the synthesis of (2′*S*)- and (2′*R*)-2′-deoxy[2′-^{2}H]ribonucleoside derivatives starting from the corresponding bromide at the 2′ position has been developed [34]. Reactions (4.12) and (4.13) show the flexibility of this methodology to obtain a single diastereoisomer, (2′*S*) and (2′*R*), respectively. The high stereoselectivity in these reactions is due to the transfer of hydrogen or deuterium atoms from the less hindered side of the ring. It is worth mentioning that the solvent employed in Reaction (4.13) was either THF-d_8 or 2,2,5,5,-tetramethyl-THF, in order to eliminate a small percentage of hydrogen donation from the reaction medium.

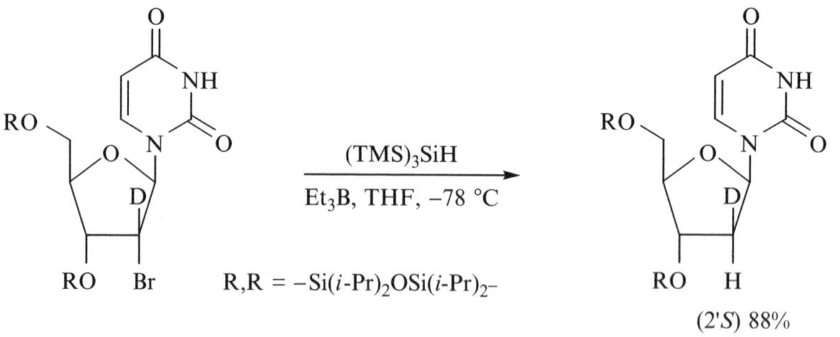

(4.12)

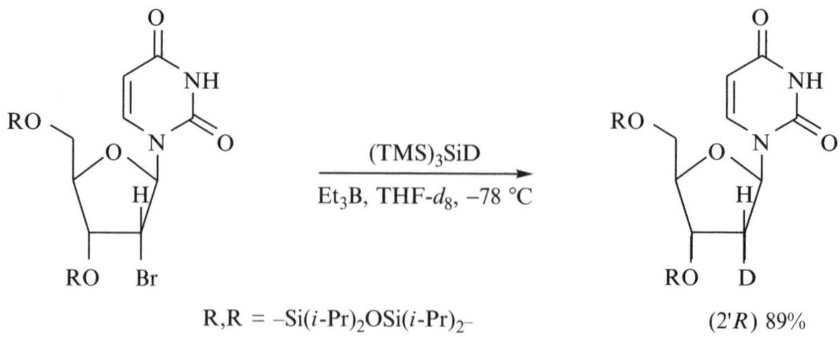

(4.13)

A good diastereocontrol is obtained for the debromination of Reaction (4.14) and it is attributed to the bulky reducing agent, which approaches the radical intermediate from the less hindered face *anti* to the two vicinal substituents [35].

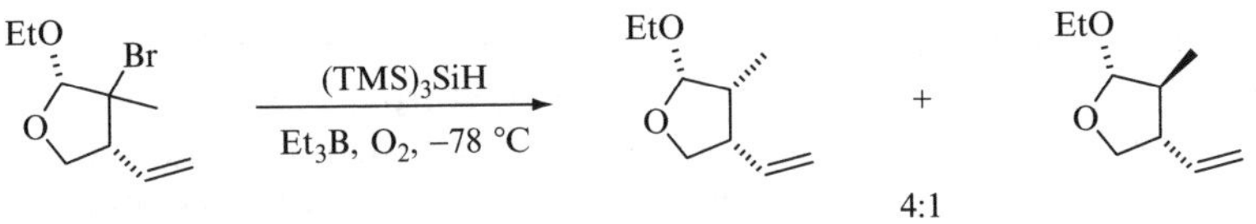

(4.14)

The reduction of bromides bearing a β-sulfinyl group by $(TMS)_3SiH$ is an emerging method for the preparation of substituted allenes. Reaction (4.15) was tested with different α-bromo vinyl sulfoxides and yields ranged from satisfactory to good ones [36]. The reaction involves a bromine atom removal followed by β-elimination of a sulfinyl radical. The use of large excess of initiator suggests that these transformations do not involve a properly maintained radical chain.

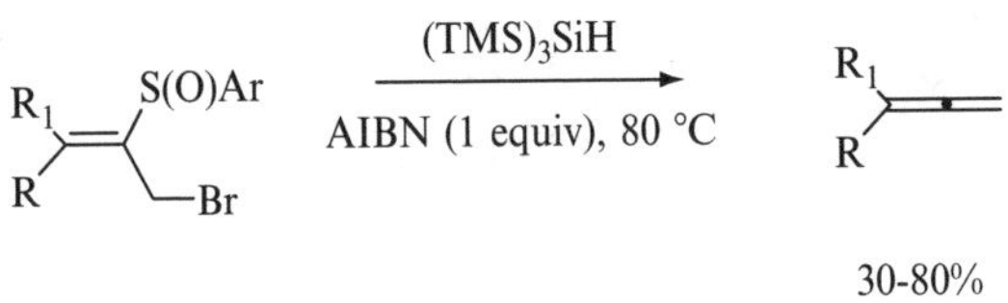

(4.15)

$(TMS)_3SiH$ can be also used as a reagent for driving the reduction of iodides and bromides through a radical mechanism together with sodium borohydride, the reductant that is consumed [37]. For example, 1-bromonaphthalene is treated with an excess of $NaBH_4$ (50 equiv) and a small amount of $(TMS)_3SiH$ (0.1 equiv), under photochemical initiation conditions, to give the reduced product in 88 % yield.

For tertiary, secondary, and primary chlorides the reduction becomes increasingly difficult due to shorter chain lengths. On the other hand, the replacement of a chlorine atom by hydrogen in polychlorinated substrates is much easier. Table 4.2 shows the rate constants for the reaction of $(TMS)_3Si\bullet$ radical with some chlorides [32]. The comparison with the analogous data of Table 4.1 shows that for benzyl and tertiary alkyl substituents the chlorine atom abstraction is 2–3 orders of magnitude slower than for the analogous bromides.

Table 4.2 Rate constants at 20 °C for the reaction of $(TMS)_3Si\bullet$ radicals with some chlorides [32]

Chloride	$k/M^{-1}\,s^{-1}$
CCl_4	1.7×10^8
$CHCl_3$	6.8×10^6
$PhCH_2Cl$	4.6×10^6
$CH_3(CH_2)_5C(CH_3)_2Cl$	4.0×10^5
$RCH_2C(O)Cl$	7×10^{5a}

[a] At 80 °C.

To illustrate the efficiency of dechlorination reactions, a few examples have been chosen from different areas. Reaction (4.16) shows chlorine atom removal from the 2′-position of sugar moiety whereas Reaction (4.17) shows the formation of a 3-unsubstituted azetidinone [38,39]. The reduction of the dichloride depicted in Reaction (4.18) represented the key step in the total synthesis of dactomelynes, a group of natural products isolated from a marine organism. The use of $(TMS)_3SiH$ at room temperature allowed the monochloride to be obtained in high yield and stereoselectively ($\beta:\alpha = 13:1$), whereas other radical-based reducing systems failed [40]. In this respect, the stereoselective reduction of *gem*-dichlorides by $(TMS)_3SiH$ and Bu_3SnH was previously reported and evidenced that the silane has a stronger preference than tin to transfer a hydrogen atom from the less hindered side of the molecule, due to the different spatial shapes of the two reagents [41].

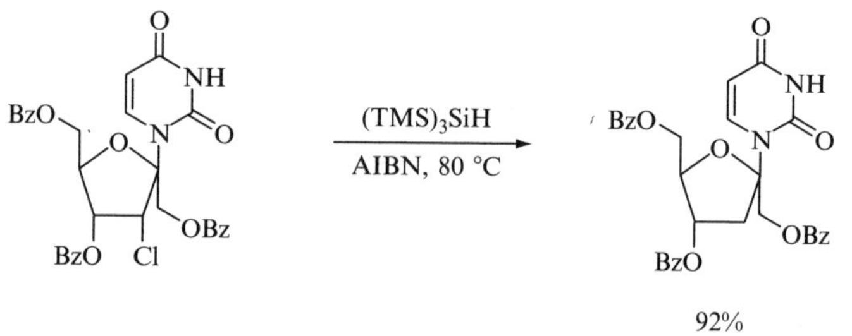

(4.16)

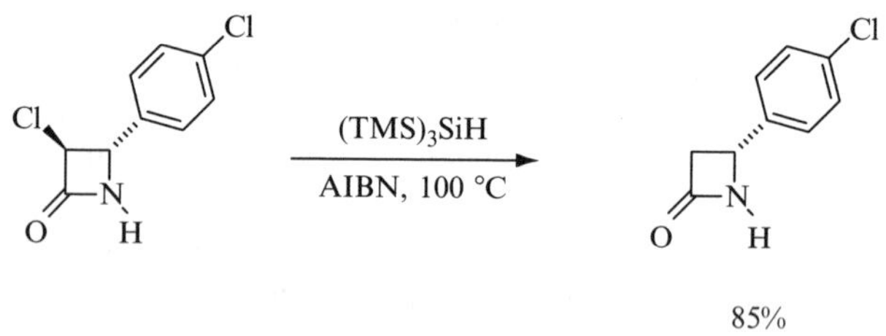

(4.17)

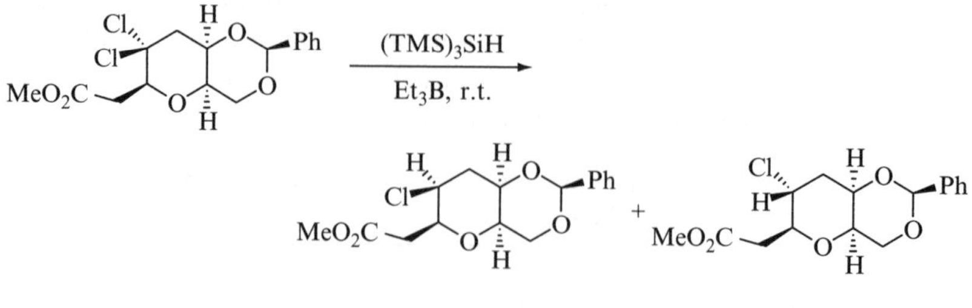

(4.18)

Under free-radical conditions, the reaction of $(TMS)_3SiH$ with acid chlorides, RC(O)Cl, gives the corresponding aldehydes and/or the decarbonylation products depending on the nature of substituent R [42]. The reduction of 1-adamantanecarbonyl chloride is given in Reaction (4.19).

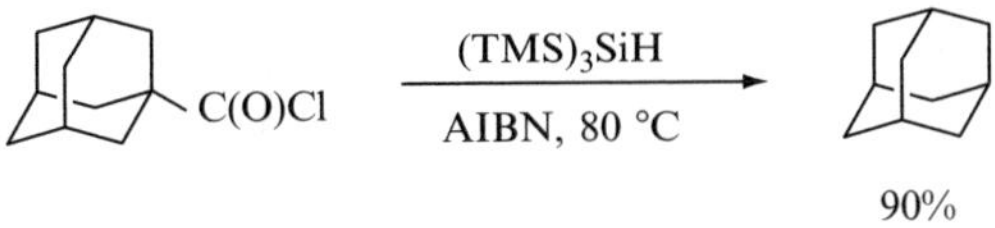

(4.19)

4.3.2 REDUCTIVE REMOVAL OF CHALCOGEN GROUPS (RS AND RSe)

The removal of PhS from n-$C_{10}H_{21}$—SPh by $(TMS)_3SiH$ is a sluggish reaction, the reason probably being that the displacement of a primary alkyl radical by $(TMS)_3Si\bullet$ radical is not an efficient step, in accordance with the available upper limit rate constant (Table 4.3) [43]. However, the reaction changes its efficiency when the resulting carbon-centred radicals are further stabilized by an α heteroatom. Indeed, $(TMS)_3SiH$ can induce the efficient radical chain monoreduction of 1,3-dithiolanes, 1,3-dithianes, 1,3-oxathiolanes, 1,3-oxathiolanones and 1,3-thiazolidines derivatives [44–46]. Three examples are outlined in Reaction (4.20). The $(TMS)_3Si$ group incorporated in the adducts could be conveniently deprotected with fluoride ions generating a thiolate anion which could be used for further synthetic transformations [47]. The reduction of chiral 1,3-thiazolidine derivatives (as 1:1 mixture of diastereoisomers) was used as a model for studying the stereoselectivity in the hydrogen abstraction of α-aminoalkyl radicals [48]. These intermediates abstract a hydrogen atom giving different *anti*/*syn* ratios, depending on the substituents. An example is given in Reaction (4.21).

Table 4.3 Rate constants at 20 °C for the reaction of $(TMS)_3Si\bullet$ radicals with a variety of substrates [43]

Substrate	$k/M^{-1}\,s^{-1}$
RCH_2SPh	$< 5 \times 10^6$
RCH_2SePh	9.6×10^7
$RCH_2C(O)SePh$	2×10^{8a}
c-$C_6H_{11}OC(S)SPh$	1.1×10^9
c-$C_6H_{11}NC$	4.7×10^7
$(CH_3)_3CNO_2$	1.2×10^7

[a] At 80 °C.

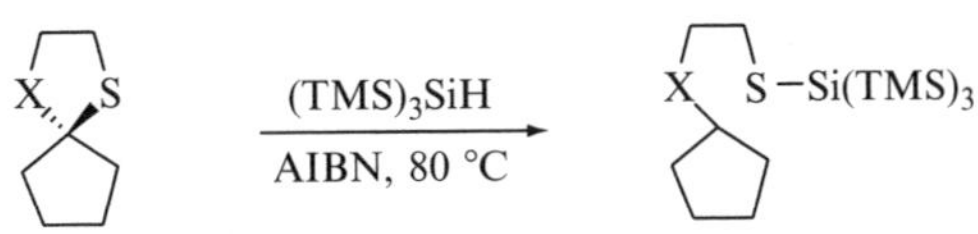

(4.20)

X = S 85%
X = O 84%
X = NC(O)OEt 79%

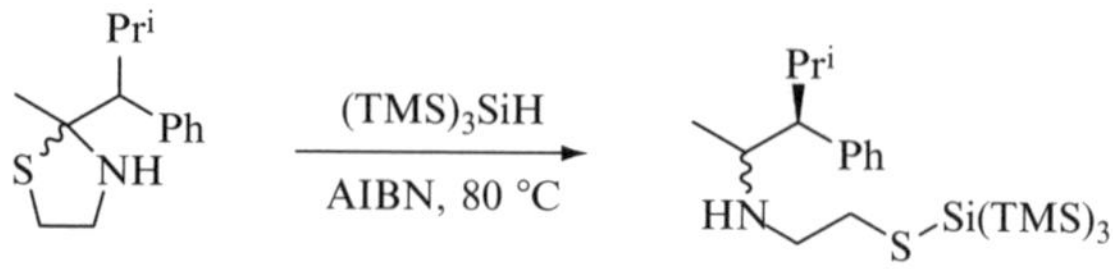

(4.21)

92% (*anti*:*syn* = 84:16)

The removal of the tritylthio (Ph_3CS) protecting group used in guanine derivatives has been accomplished in a good yield by the action of $(TMS)_3SiH$ under free-radical conditions (Reaction 4.22) [49].

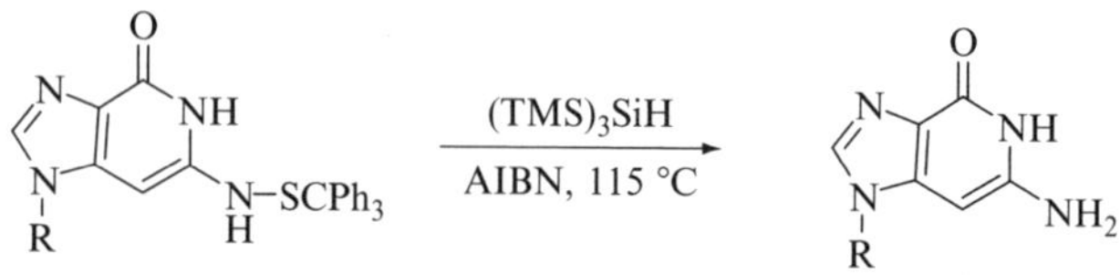

(4.22)

81%

Secondary alkyl selenides are reduced by $(TMS)_3SiH$, as expected in view of the affinity of silyl radicals for selenium-containing substrates (Table 4.3) [40]. Reaction (4.23) shows the phenylseleno group removal from the 2′ position of nucleoside [50]. Similarly to 1,3-dithiolanes and 1,3-dithianes, five- and six-membered cyclic selenoacetals can be monoreduced to the corresponding selenides in the presence of $(TMS)_3SiH$ [51]. The silicon hydride preferentially approached from the less hindered equatorial position to give *trans*/*cis* ratios of 30/70 and 25/75 for the five-membered (Reaction 4.24) and six-membered cyclic selenoacetals, respectively.

(4.23)

87%

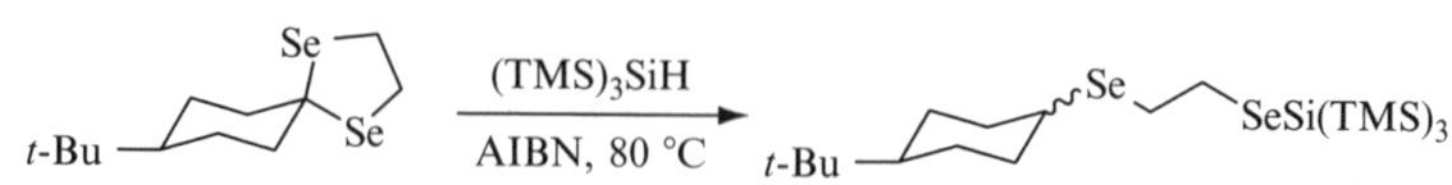

(4.24)

trans/*cis* = 30/70

Phenyl selenoesters have been reported to undergo reduction to the corresponding aldehydes and/or alkanes in the presence of $(TMS)_3SiH$ under free-radical conditions [52]. The decrease of aldehyde formation observed along the primary, secondary and tertiary substituted series, under the same conditions, indicated that a decarbonylation of acyl radicals takes place. An example is shown in Reaction (4.25), in which the phenylselenoester affords the decarbonylated β-lactam in a good yield [53]. Similar chemistry is expected by the replacement of PhSe with the MeSe group. Indeed, Reaction (4.26) shows that the methylselenoester, when treated by similar conditions, gave the reduction product in a 78% yield [54].

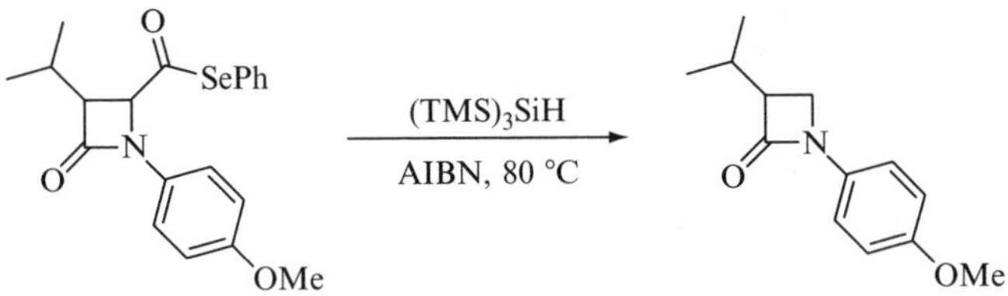

(4.25)

85%

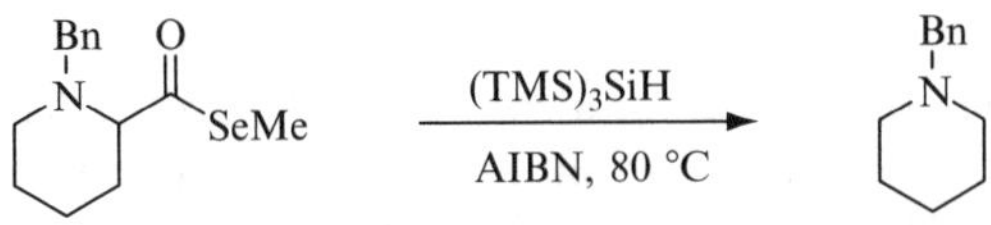

(4.26)

78%

The removal of the hydroxy group has been achieved from an appropriate selenocarbonate by heating in the presence of $(TMS)_3SiH$ and AIBN in benzene [55]. Reaction (4.27) was inserted in the multistep synthesis of an alkaloid. The deoxygenation was achieved with 87% efficiency of the two steps.

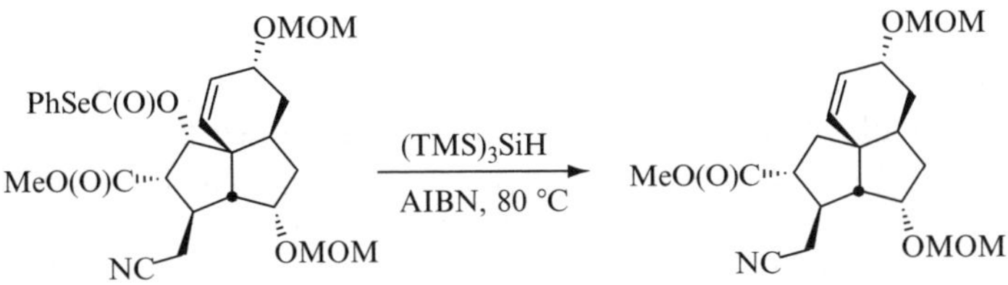

(4.27)

>87%

In the above-described displacement reactions (S_H2) of sulfur- and selenium-containing compounds, the mechanism could either be a synchronous or a stepwise process (Scheme 4.2). Thus, in the S_H2 stepwise mechanism an intermediate sulfuranyl or selenyl radical is formed, followed by α-cleavage [56]. Based on competitive studies, a stepwise process was suggested to occur in the reaction of $(TMS)_3Si$• radical with *n*-decylphenylselenide [43]. On the other hand, *ab initio* calculations supported the synchronous process and the same observations were explained in terms of an overall reversible reaction [57].

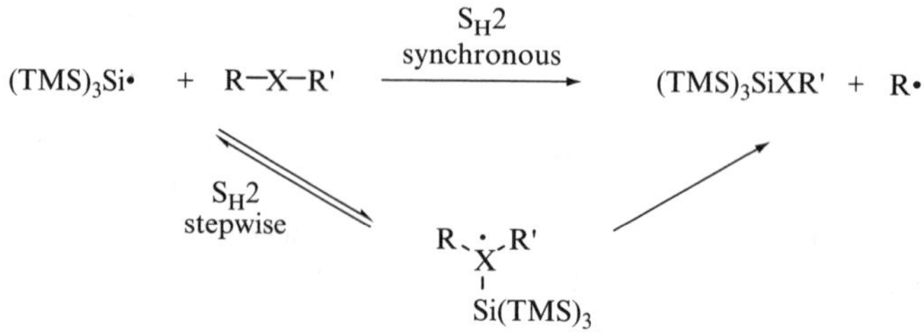

Scheme 4.2 Possible mechanistic paths for the displacement of a chalcogen group (R′X, where X = S or Se) by $(TMS)_3Si\bullet$ radicals

4.3.3 DEOXYGENATION OF ALCOHOLS (BARTON–MCCOMBIE REACTION)

A relevant reductive process, which has found wide application in organic synthesis, is the deoxygenation of alcohols introduced in 1975 by Barton and McCombie [58]. Reaction (4.28) shows that the thiocarbonyl derivatives, easily obtained from the corresponding alcohol, can be reduced in the presence of Bu_3SnH under free radical conditions. The reactivity of xanthates and thiocarbonyl imidazolides [58] was successfully extended to *O*-arylthiocarbonates [59] and *O*-thioxocarbamates [60]. Several reviews have appeared on this subject, thus providing an exhaustive view of this methodology and its application in natural product synthesis [61–64].

$$\mathrm{ROH} \longrightarrow \mathrm{ROC(S)X} \xrightarrow{\mathrm{Bu_3SnH}} \mathrm{RH} \qquad (4.28)$$

X = SR, OAr, NHPh, 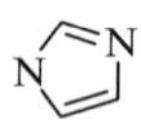

$(TMS)_3SiH$ replaced successfully tin hydrides and often proved to be a superior reagent [65]. In the initial work, it has been shown that $(TMS)_3SiH$ and its silylated by-products are not toxic [66] and that the reaction is efficient also at room temperature [67]. The reduction of thiono esters of cholesterol to cholestene (Reaction 4.29) has been used for evaluation of by-products in biological assays [66]. The effect of substituents on the phenyl ring of thiono ester has also been evaluated, see for example Reaction (4.30) [67]. An improved procedure for the homolytic deoxygenation of aminoacid derivatives has also been reported (Reaction 4.31). It consists of the treatment with $(TMS)_3SiH$ (without purification), followed by an acid treatment by TFA and an aqueous extraction, which was sufficient to separate the product from the silyl and sulfur-containing by-products [68]. The dideoxygenation of 1,6-anhydro-D-glucose with $(TMS)_3SiH$ has been described to afford the desired product in a

86% yield (Reaction 4.32), whereas other radical-based reducing systems give much poorer yields [69].

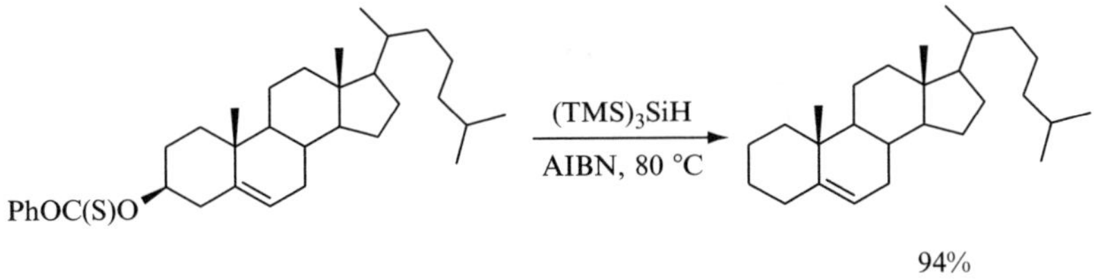

(4.29)

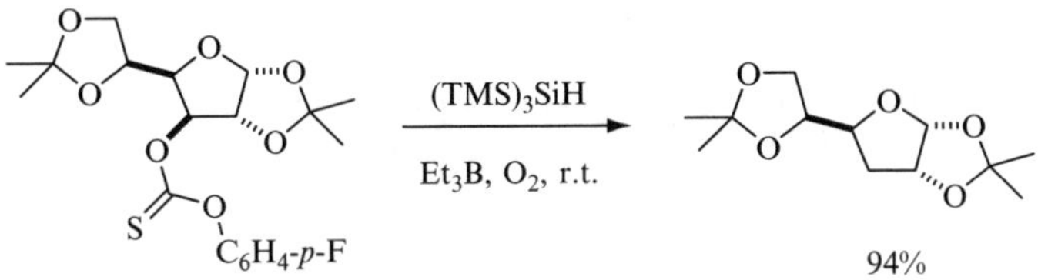

(4.30)

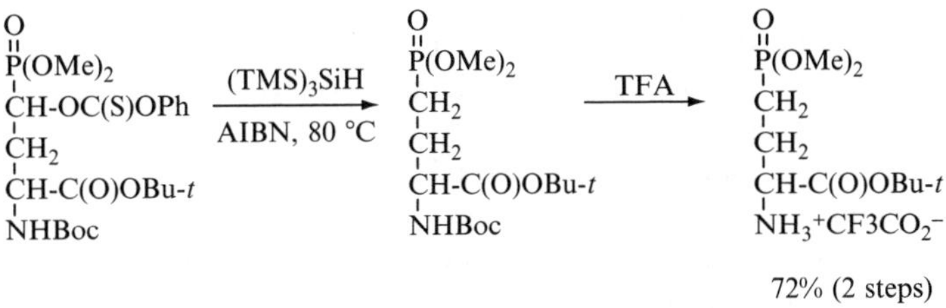

(4.31)

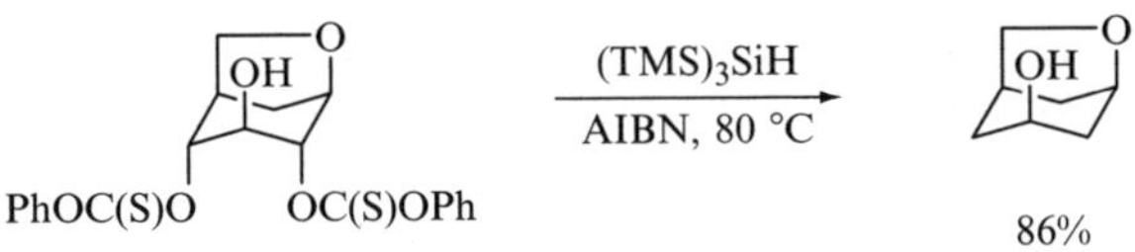

(4.32)

Radical deoxygenation via thioimidazolyloxy derivatives is also found to be efficient process. Two examples are reported in Reactions (4.33) and (4.34) [70,71]. In particular, dideoxygenation was useful to prove the structure of the stemodane ring system, achieved by other routes.

(4.33)

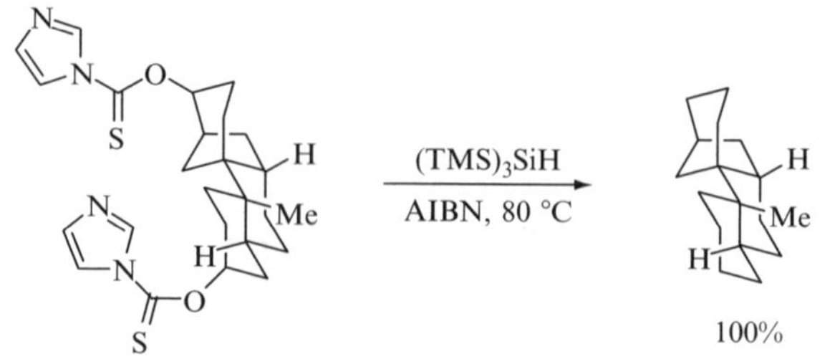

(4.34)

In the case of nucleosides, deoxygenation represents an important procedure, when occurring on the hydroxyl group in the position 2′. In fact, this is the simplest method to convert ribonucleosides to deoxyribonucleosides (Reaction 4.35) which provide the units to be incorporated in DNA oligomers [59]. Also many new potent anti-HIV and antiviral drugs, recently introduced in therapy, have the 2′- or 3′-deoxy as well as 2′,3′-dideoxy nucleoside skeleton. The formation of thiocarbonyl derivatives of the alcoholic function on the sugar moiety and its subsequent radical reduction give the easiest access to these substrates, avoiding any other side reactions. 2′-Deoxyapio-β-D-furanosyl nucleosides were prepared from the corresponding thiocarbonate derivatives (Reaction 4.36), and tested against HIV, Herpes simplex and other viruses [72]. Analogously, in Reaction (4.37) the removal of 2′-hydroxyl group has been accomplished via the phenoxythiocarbonyl derivative under normal conditions to produce the protected spironucleoside in good yield [73]. Pharmaceutically important 2′- and/or 3′-deoxynucleosides were described in a patent and the key step for the synthesis was achieved by treating the new and economically more convenient (cyanoethylthio)-thiocarbonyl derivatives of nucleosides, instead of classical xanthates, with (TMS)$_3$SiH in very mild conditions (Reaction 4.38) [74]. In the field of nucleoside mimics, β-2′-deoxypseudouridine [75] and β-2′-deoxyzebularine [76] were prepared by radical-based deoxygenation from the corresponding β-pseudouridine and β-zebularine using (TMS)$_3$SiH as reducing agent.

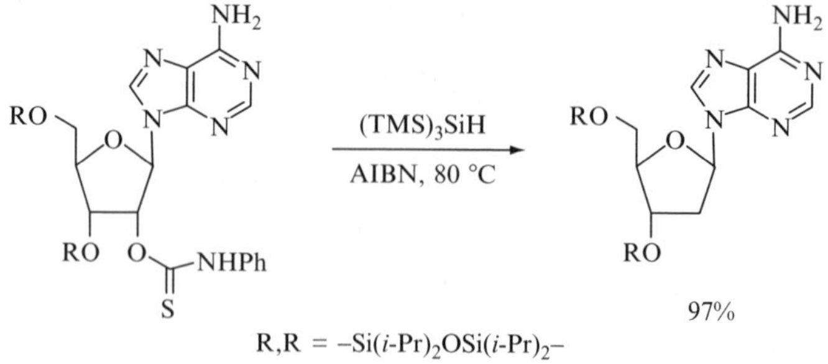

(4.35)

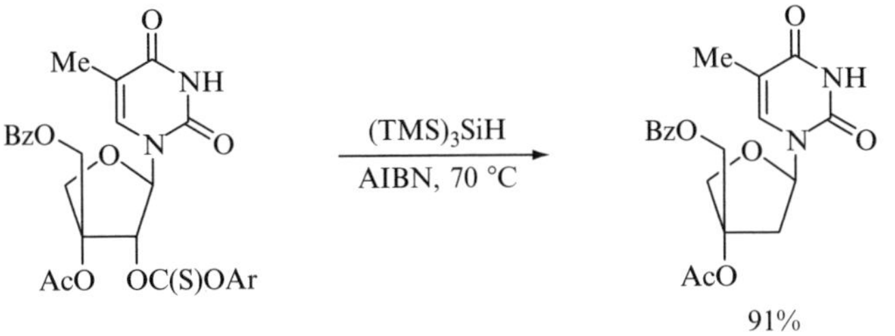

(4.36)

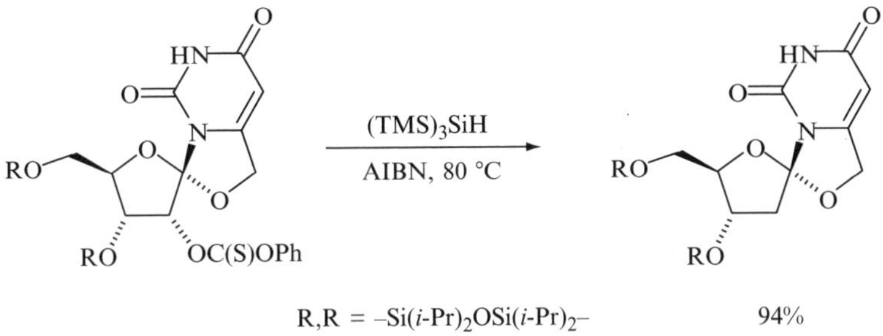

(4.37)

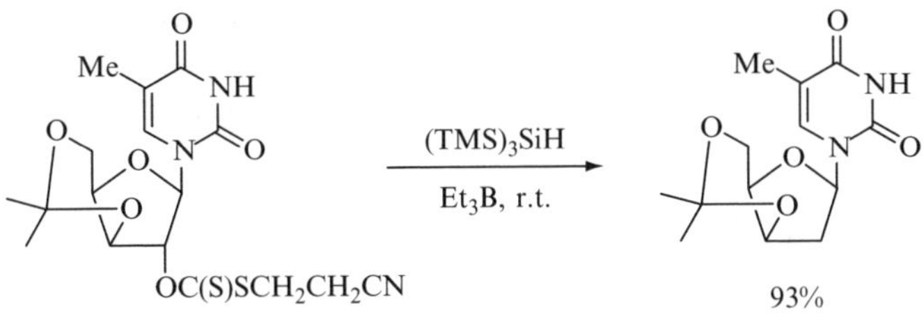

(4.38)

Cyclohexyl xanthate has been used as a model compound for mechanistic studies [43]. From laser flash photolysis experiments the absolute rate constant of the reaction with $(TMS)_3Si\bullet$ has been measured (see Table 4.3). From a competition experiment between cyclohexyl xanthate and *n*-octyl bromide, xanthate was ca 2 times more reactive than the primary alkyl bromide instead of ca 50 as expected from the rate constants reported in Tables 4.1 and 4.3. This result suggests that the addition of silyl radical to thiocarbonyl moiety is reversible. The mechanism of xanthate reduction is depicted in Scheme 4.3: $(TMS)_3Si\bullet$ radicals, initially generated by small amounts of AIBN, attack the thiocarbonyl moiety to form in a reversible manner a radical intermediate that undergoes β-scission to form alkyl radicals. Hydrogen abstraction from the silane gives the alkane and $(TMS)_3Si\bullet$ radical, thus completing the cycle of this chain reaction.

The application of $(TMS)_3SiH$ has also been extended to the dideoxygenation of *vic*-diols that readily affords the corresponding olefins. Reaction (4.39) shows an example of radical-based dideoxygenation of a bis-*O*-thioxocarbamate derivative by this silane under standard conditions [77]. In this way, a

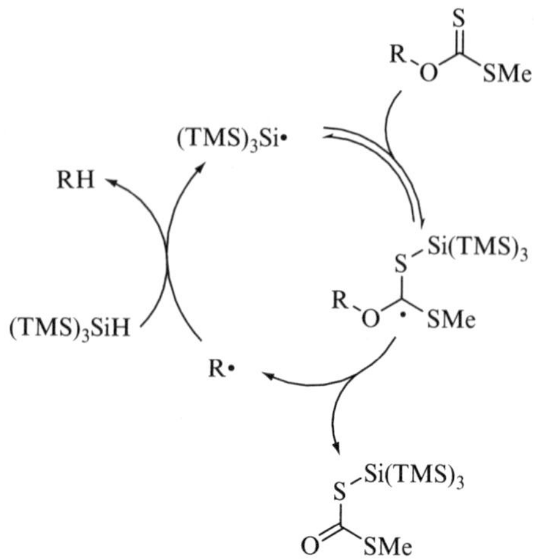

Scheme 4.3 Propagation steps for the reaction of cyclohexyl xanthate with $(TMS)_3$ SiH

variety 2′, 3′-didehydro-2′, 3′-dideoxy derivatives of ribonucleosides were prepared. From a mechanistic point of view, the initial propagation steps are similar to Scheme 4.3 until the formation of R•, followed by a β-elimination to give the olefin and a radical fragment that continues the chain by hydrogen abstraction from the silane.

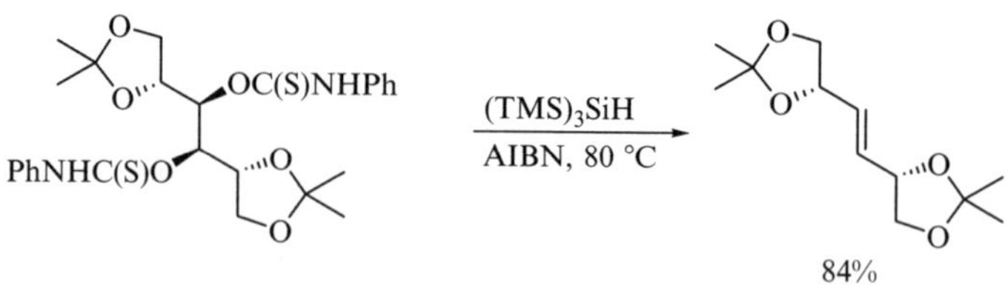

(4.39)

4.3.4 MISCELLANEOUS REACTIONS

Isocyanides can be reduced to the corresponding hydrocarbon by $(TMS)_3SiH$ [43]. The reaction can be considered a smooth route for the deamination of primary amines, through the preparation of isocyanides via formylation and dehydration. The efficiency of the reduction is independent from the nature of the alkyl substituent. That is, primary, secondary, and tertiary isocyanides at 80 °C gave the corresponding hydrocarbons in good yields. Two examples are given in Reactions (4.40) and (4.41) [1,43]. The key step for these chain reactions is expected to be the fragmentation of the intermediate radical derived from the fast addition of $(TMS)_3Si$• radical to the specific substrate (Table 4.3).

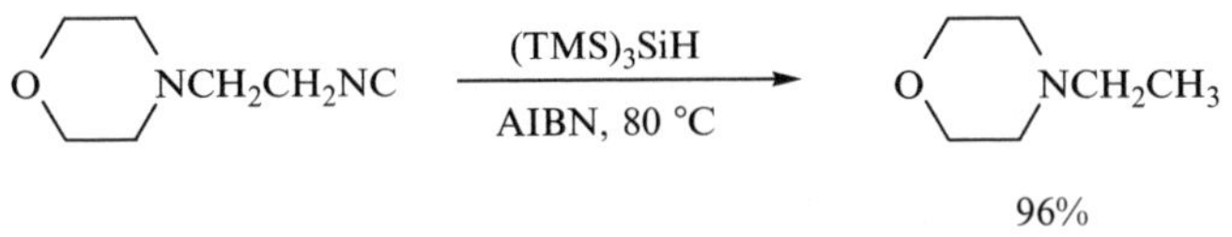

(4.40)

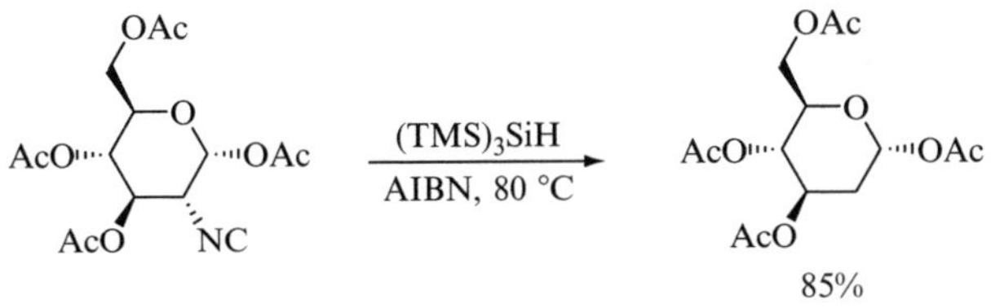

(4.41)

Some unusual transformations will follow. The term 'unusual' refers to unexpectedly smooth processes of H atom replacements obtained with $(TMS)_3SiH$ under radical conditions. Reaction (4.42) reports the replacement of a pyridinium moiety by hydrogen, with $(TMS)_3SiH$ under standard experimental conditions using *t*-BuOH as the solvent. In fact the two substrates (R = Me, Et) afforded 3-fluoro-2-aminopyridine derivatives in good yields [78], leaving the fluorine substituent untouched.

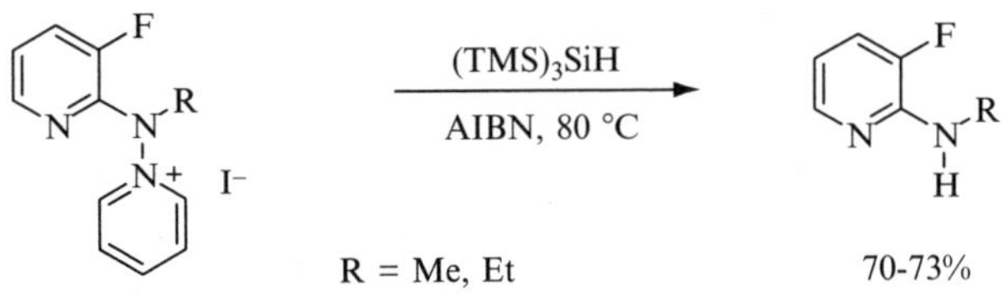

(4.42)

The deoxygenation of nitroxides by $(TMS)_3SiH$ is shown in Reaction (4.43) [79]. Indeed, the reaction of this silane with TEMPO, in the presence of thermal or photochemical radical initiators, afforded the corresponding amine in quantitative yield, together with the siloxane $(TMS)_2Si(H)OSiMe_3$. The apparently unexpected detection of the siloxane can be accounted for by the reaction mechanism shown in Scheme 4.4.

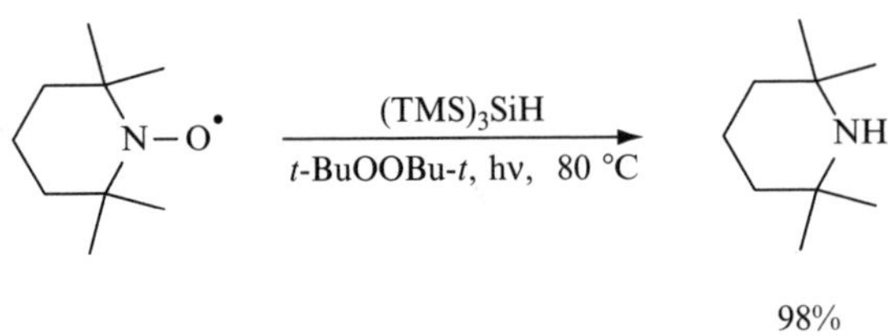

(4.43)

$(TMS)_3SiH$ is not able to reduce tertiary nitroalkanes to the corresponding hydrocarbons whereas tin hydrides are efficient agents. This 'anomalous' behaviour is due to the fact that the nitroxide adducts, formed by the addition of $(TMS)_3Si\bullet$ radical to the nitro compounds, fragment preferentially at the nitrogen—oxygen bond rather than at the carbon—nitrogen bond, as in the

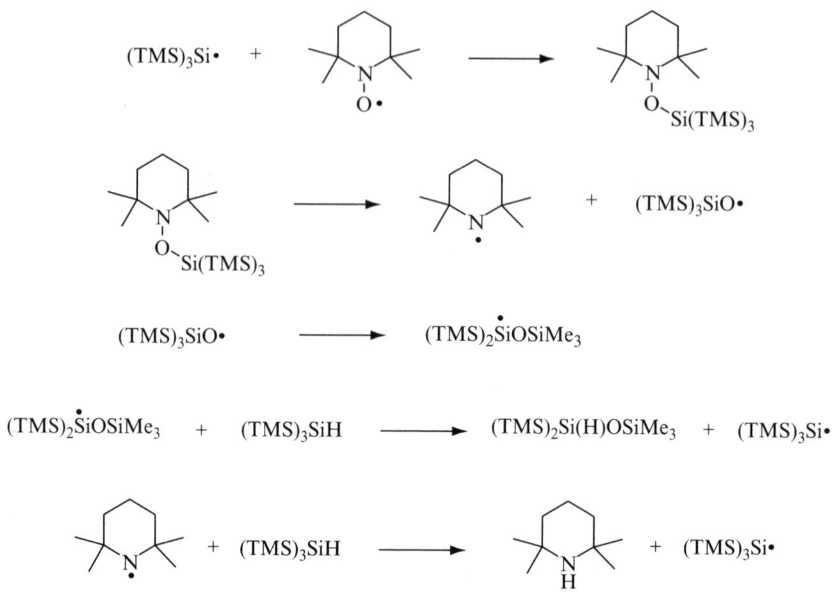

Scheme 4.4 Reaction mechanism for the reduction of nitroxide to corresponding amines

analogous tin adduct (Scheme 4.5) [80]. Indeed, the reactions of *N*-nitroamines with $(TMS)_3SiH$ under normal radical conditions afford the corresponding *N*-nitrosoamines in good yield (Reaction 4.44) [81]. For the fate of $(TMS)_3SiO\bullet$ radical refer to Scheme 4.4.

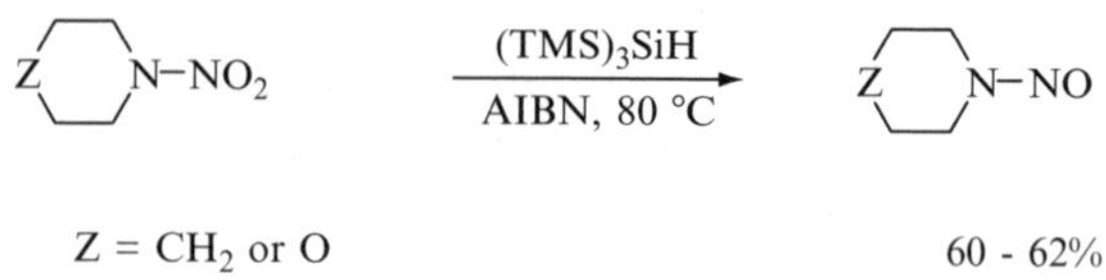 (4.44)

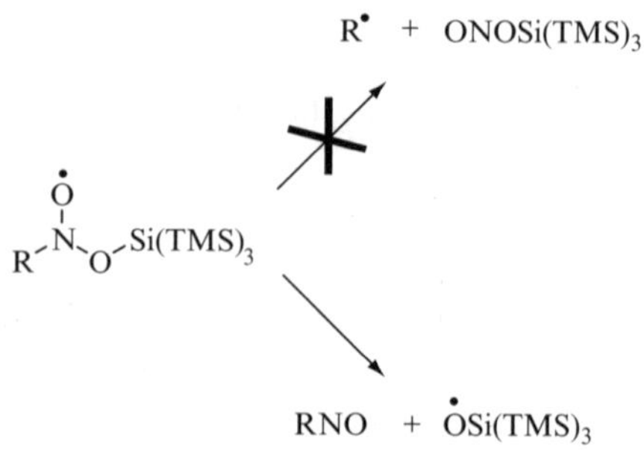

Scheme 4.5 Possible mechanistic paths for the fragmentation of nitroxide

Tris(trimethylsilyl)silane reacts with phosphine sulfides and phosphine selenides under free radical conditions to give the corresponding phosphines or, after treatment with BH_3–THF, the corresponding phosphine–borane complex in good to excellent yields (Reaction 4.45) [82]. Stereochemical studies on P-chiral phosphine sulphides showed that these reductions proceed with retention of configuration. An example is given in Reaction (4.46).

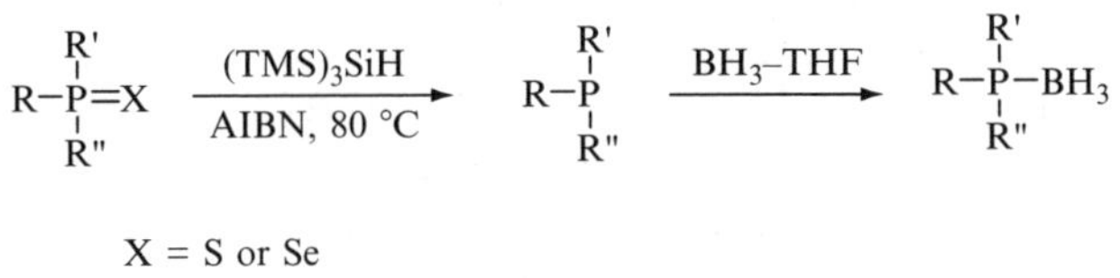

(4.45)

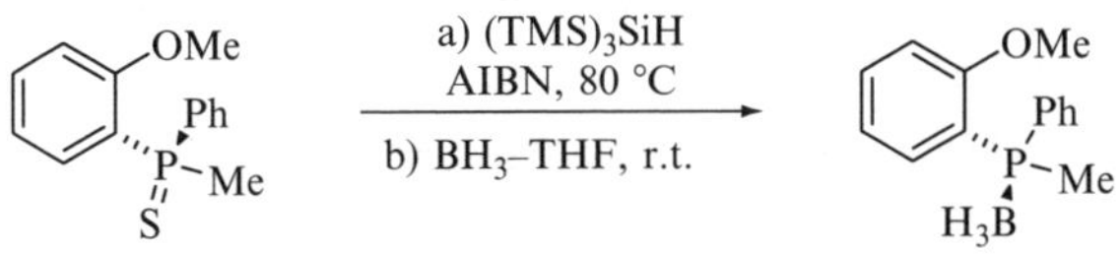

(4.46)

The photochemical reduction of 1-methylquinolinium ions by $(TMS)_3SiH$ proceeds regioselectively to afford the corresponding 1,4-dihydroquinones in a water–acetonitrile solvent system (Reaction 4.47) [83]. Mechanistic studies demonstrated that the reactions are initiated by photoinduced electron transfer from the silane to the singlet excited states of 1-methylquinolinium ions to give the silane radical cation–quinolinyl radical pairs, followed by hydrogen transfer in the cage to yield 1,4-dihydroquinones and silicenium ion. Silyl cations are quenched by water.

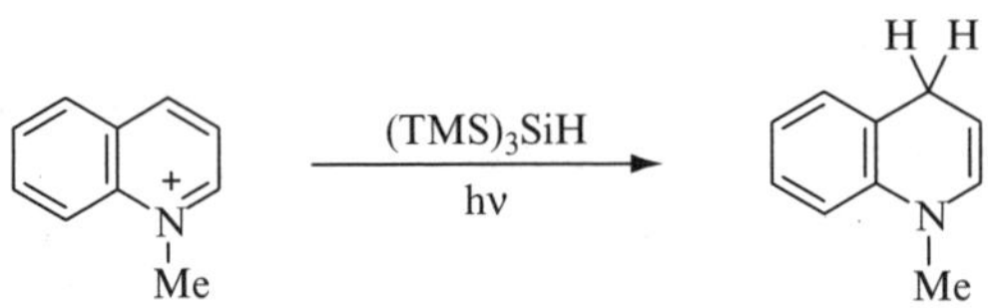

(4.47)

4.3.5 APPENDIX

It is worth pointing out that the replacements of a variety of functional groups by a hydrogen described so far are not only efficient and straightforward processes but the work-up is rather simple: in most cases, the reaction mixtures are concentrated by evaporation of the solvent and then flash-chromatographed to isolate the products. Furthermore, it has been shown that $(TMS)_3SiH$ and its silylated by-products are not toxic [66]; this is very important for the pharmaceutical application of the silane reagent, since the biological assays on the final compounds are not affected by any remaining silylated materials. An increase in the choice of $(TMS)_3SiH$ as a reducing agent it is expected in the future, especially in the synthesis of natural products and pharmacologically active compounds.

$(TMS)_3SiH$ reacts spontaneously and slowly at ambient temperature with molecular oxygen to form siloxane as the sole product (Reaction 4.48) [84]. This reaction, which occurs via a radical cascade reaction, will be discussed in detail in Chapter 8. Two aspects of the autoxidation of $(TMS)_3SiH$ are of interest to synthetic chemists: (i) for radical reactions having long chain lengths, traces of molecular oxygen can serve to initiate reactions, therefore no additional radical initiator is needed, and (ii) the oxidized product does not interfere with radical reactions, therefore the reagent can be used even if partially oxidized, taking into account the purity of the material. In our experience, commercially available materials are ca 98 % pure by GC analysis. This analytical method can also be used for a titration of the reagent, thus giving the exact concentration of silane for kinetic purposes.

$$(Me_3Si)_3SiH + O_2 \longrightarrow (Me_3SiO)_2Si(H)SiMe_3 \quad (4.48)$$

4.4 OTHER SILICON HYDRIDES

4.4.1 TRIALKYLSILANES

Trialkylsilanes are not capable of donating the hydrogen atom at a sufficient rate to propagate the chain. As reported in Section 3.1.1, the attack of primary alkyl radicals on Et_3SiH occurs in about 60 % of cases at the SiH moiety and 40 % at the ethyl groups at 130 °C. Therefore, chain reactions are not supported under normal conditions, although trialkylsilyl radicals are among the most reactive species toward various organic functional groups (see below). Early work based on the reduction of polychloroalkanes by trialkylsilanes, using dibenzoyl peroxide as the initiator, gathered the relative rates of hydrogen abstraction from Si—H moiety and structural information about silyl radicals (see Chapter 1) [85]. Triethylsilane reacts with polyfluorinated halocarbons under free radical conditions initiated by thermal decomposition of peroxides. For example, 1,2-dichlorohexafluorocyclobutane reacts with an excess of Et_3SiH, affording the dihydro derivative in excellent yield (Reaction 4.49) [86]. Primary or secondary substituted acyl chlorides, RC(O)Cl, in the presence of di-*tert*-butyl peroxide at 140–170 °C gave the corresponding RH in 50–70 % yields [87]. Similarly, chloroformates, ROC(O)Cl afforded the corresponding RH in good yields [88].

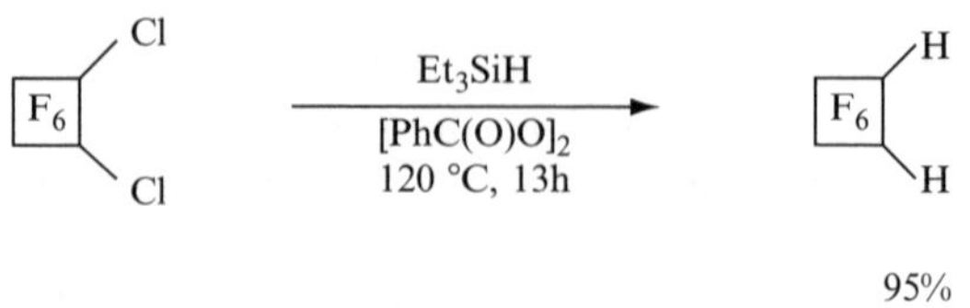

(4.49)

The reduction of thiocarbonyl derivatives by Et_3SiH can be described as a chain process under 'forced' conditions (Reaction 4.50) [89,90]. Indeed, in Reaction (4.51) for example, the reduction of phenyl thiocarbonate in Et_3SiD as the solvent needed 1 equiv of dibenzoyl peroxide as initiator at 110 °C, and afforded the desired product in 91 % yield, where the deuterium incorporation was only 48 % [90]. Nevertheless, there are some interesting applications for these less reactive silanes in radical chain reactions. For example, this method was used as an efficient deoxygenation step (Reaction 4.52) in the synthesis of 4,4-difluoroglutamine [91]. 1,2-Diols can also be transformed into olefins using the Barton–McCombie methodology. Reaction (4.53) shows the olefination procedure of a bis-xanthate using Et_3SiH [89].

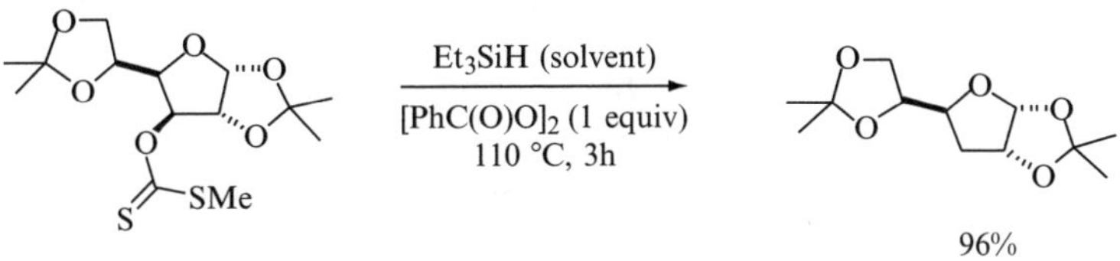

(4.50)

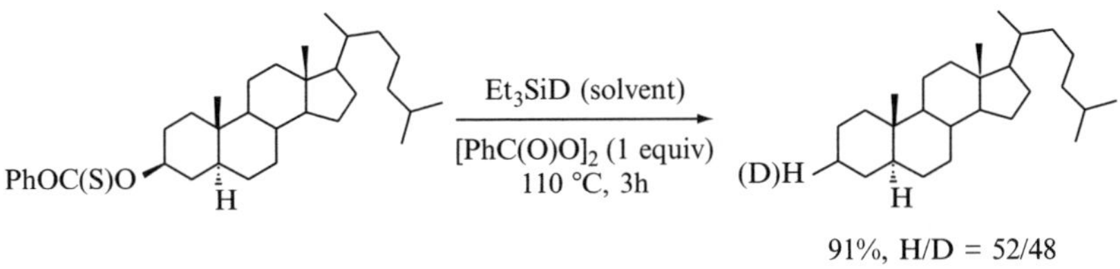

(4.51)

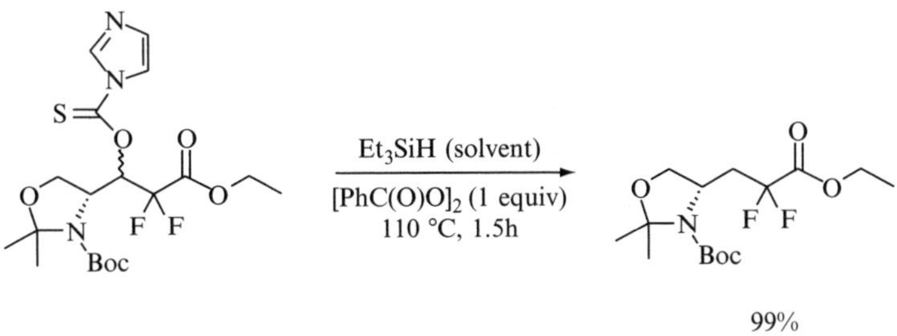

(4.52)

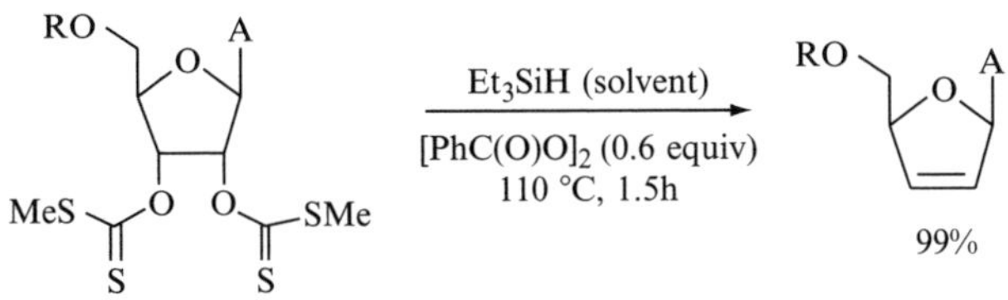

(4.53)

A large body of absolute kinetic data, obtained by laser flash photolysis techniques, for the reactions of $Et_3Si\bullet$ radicals with organic halides is available

(Reaction 4.54). The rate constants at room temperature and the available Arrhenius parameters are given in Table 4.4 [92]. The reactivity trends are the following: (i) for a particular R group the rate constant decreases along the series X = I > Br > Cl > F, and (ii) for a particular X atom the rate constant decreases along the series R = benzyl > *tert*-alkyl > *sec*-alkyl > primary alkyl > phenyl. Although these trends in reactivity are attributable in large part to thermodynamic factors, it was found that the higher reactivity of polychlorinated alkanes relative to monochlorinated alkanes is also due to higher preexponential factors [92]. It has been suggested that a charge-transfer interaction in the transition state (Equation 4.55) can have a greater influence on the preexponential factor than on the activation energy, and the following explanation has been proposed: the greater the polar contribution, the smaller the restriction on the orientation of the $Et_3Si\bullet$ radical with respect to the carbon–halogen bond being broken. In the limiting case of complete electron transfer, the resulting ion pair would not be subject to any restriction in its relative rotational motion; this gain of two rotational degrees of freedom in the transition state would enhance the preexponential factor by ca 10^2. Therefore, electron transfer may be extensive in the $Et_3Si\bullet/CCl_4$ transition state.

$$R_2-\underset{R_3}{\overset{R_1}{|}}\!\!\!Si\cdot \quad + \quad RX \longrightarrow R_2-\underset{R_3}{\overset{R_1}{|}}\!\!\!Si-X \quad + \quad R\cdot \tag{4.54}$$

$$Et_3\dot{Si}\ \dot{X}\ \dot{R} \leftrightarrow Et_3\overset{+}{\dot{Si}}\ X^{-}\ \dot{R} \leftrightarrow Et_3\overset{+}{\dot{Si}}\ \dot{X}\ R^{-} \tag{4.55}$$

Table 4.4 Kinetic data for the reaction of $Et_3Si\bullet$ radicals with a variety of halides [92]

Halide	$k/M^{-1}\,s^{-1}$ at 27 °C	$\log A/M^{-1}\,s^{-1}$	$E_a/kJ\,mol^{-1}$
$CH_3(CH_2)_4Cl$	3.1×10^5		
$(CH_3)_3CCl$	2.5×10^6	8.7	13.4
$PhCH_2Cl$	2.0×10^7	8.9	8.8
CH_2Cl_2	7.1×10^7	9.4	8.8
$CHCl_3$	2.5×10^8		
CCl_4	4.6×10^9	10.2	3.3
Cl_3CCCl_3	4.1×10^9	10.4	4.6
C_6H_5Br	1.1×10^8		
$CH_3(CH_2)_4Br$	5.4×10^9	9.3	2.9
$(CH_3)_3CBr$	1.1×10^9	9.7	3.6
$PhCH_2Br$	2.4×10^9		
C_6H_5I	1.5×10^9		
CH_3CH_2I	4.3×10^9	10.4	4.2
$(CH_3)_2CHI$	1.4×10^{10}		

The decrease in reactivity for monohalogenated compounds along the series I > Br > Cl is also due in part to a decrease of the preexponential factor, which has been attributed to the diminished importance of the charge-transfer interactions along this series. Rate constants for the reaction of the $Et_3Si\bullet$ radical with a series of ring-substituted benzyl chlorides were also obtained. A Hammett plot gives a $\rho = +0.64$, one of the largest positive ρ values reported for a free-radical reaction that is indicative of the strong nucleophilic character of silyl radicals [93].

The reactions of trichloro- and tetrachloroethylene with $Et_3Si\bullet$ have been studied in detail [94]. With the former, $Et_3Si\bullet$ radicals undergo addition as well as Cl abstraction, and at 65 °C these reactions proceed at almost equal rates. With tetrachloroethylene, $Et_3Si\bullet$ radicals proceed more that 95 % via Reaction (4.56) with a rate constant of $1.0 \times 10^7 M^{-1} s^{-1}$ at 27 °C [95].

(4.56)

Absolute rate constants for the reaction of $Et_3Si\bullet$ with a variety of sulfonyl halides (Reaction 4.57) have been measured by laser flash photolysis [96,97]. For sulfonyl chlorides, the rate constants are very high, approaching the diffusion-controlled limit and are almost independent of the substituent, i.e., the rate constants at ca 25 °C are $3.2 \times 10^9, 4.6 \times 10^9, 5.4 \times 10^9$ and $7.5 \times 10^9 M^{-1} s^{-1}$ for $MeSO_2Cl$, $PhSO_2Cl$, Me_2NSO_2Cl and $EtOSO_2Cl$, respectively. Polar contributions, similar to those in Equation (4.55) have also been proposed for such fast processes. Sulfonyl fluorides also react with $Et_3Si\bullet$ radicals in a similar fashion. Rate constants for $MeSO_2F$ and p-Me-$C_6H_4SO_2F$ are found to be 1.3×10^7 and $0.9 \times 10^7 M^{-1} s^{-1}$, respectively, at room temperature.

$$Et_3Si\bullet + RSO_2X \longrightarrow Et_3SiX + RSO_2\bullet \qquad (4.57)$$

4.4.2 PHENYL SUBSTITUTED SILICON HYDRIDES

Phenyl or mixed alkyl/phenyl substituted silicon hydrides show similar reactivities to trialkylsilanes. Indeed, by replacing one alkyl by a phenyl group the effect on the hydrogen donating ability of SiH moiety increases, only slightly (see Section 3.1). As an example, the rate constant for hydrogen abstraction by primary alkyl radical from Ph_3SiH is about one order of magnitude higher than from Et_3SiH. The rate constants for halogen abstraction by the phenyl substituted silyl radicals are collected in Table 4.5. The reactivity trends of t-$BuPh_2Si\bullet$ and $Ph_3Si\bullet$ are similar and for a particular RX the rate constants range between $Et_3Si\bullet$ (smaller) and $(TMS)_3Si\bullet$ (higher).

Although under canonical radical chain conditions these silanes are poor reducing agents, there are some interesting applications, in particular, using Ph_2SiH_2 [100]. For example, the reaction of a variety of substituted adamantanes

Table 4.5 Rate constants ($M^{-1}\,s^{-1}$) at 27 °C for the reaction of phenyl substituted silyl radicals with a variety of halides [98,99]

Halide	t-$BuPh_2Si\bullet$	$Ph_3Si\bullet$
$CH_3(CH_2)_4Cl$		6.3×10^4
$(CH_3)_3CCl$	7.9×10^5	8.0×10^5
CH_2Cl_2	5.4×10^6	3.4×10^6
$CHCl_3$	3.0×10^8	1.1×10^8
CCl_4	3.6×10^9	
$CH_3(CH_2)_4Br$	2.9×10^8	
$(CH_3)_2CHBr$	3.9×10^8	
$(CH_3)_3CBr$	6.0×10^8	

with Ph_2SiH_2 have been studied in some detail (Reaction 4.58). For X = Cl, I, SePh, the reactions in refluxing toluene with at least 1 equiv of dibenzoyl peroxide after a few hours gave mainly the recovery of starting material. For X = Br, the reduction proceeded with a 91% yield under similar conditions, whereas for X = NC the yield is quantitative in dioxane, which can be envisaged to be the real hydrogen donor.

The deoxygenation reaction in the majority of the cases works equally well with Ph_2SiH_2. Reactions (4.59) and (4.60) show that Ph_2SiH_2 gave the reduction of the thiocarbonyl derivative of a secondary alcohol in a radical chain process at room temperature, as well as the deoxygenation of a primary neopentyl alcohol moiety of the hederagenin in refluxing chlorobenzene [100]. Today, it is known that the Barton–McCombie deoxygenation reaction gives excellent results even without a reducing agent, by using only a stoichiometric amount of lauroyl or benzoyl peroxide and a solvent able to act as the hydrogen donor, such as 2-propanol [64]. We can conclude that some deoxygenation reactions work well in the presence of silanes with a low ability of hydrogen donation, despite of their mechanistically complex radical process (see also Sections 4.5 and 5.5). Deoxygenation of alcohols has also been reported via trifluoroacetate derivatives with Ph_2SiH_2 (Reaction 4.61) [101]. The application of Ph_2SiH_2 to radical method for the formation of olefins from 1,2-diols [101] and from β-hydroxy sulfones [102] has been performed. As an example, the deoxygenation of the bis-xanthate occurred in refluxing toluene or dioxane in excellent yield (Reaction 4.62).

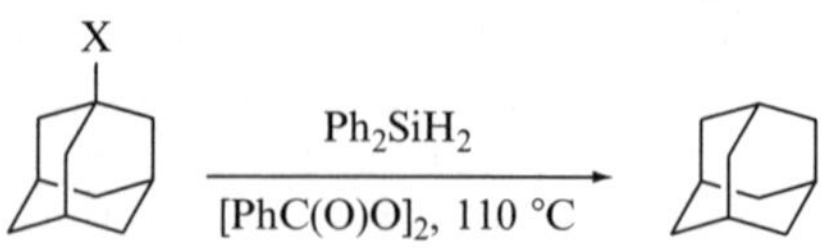

(4.58)

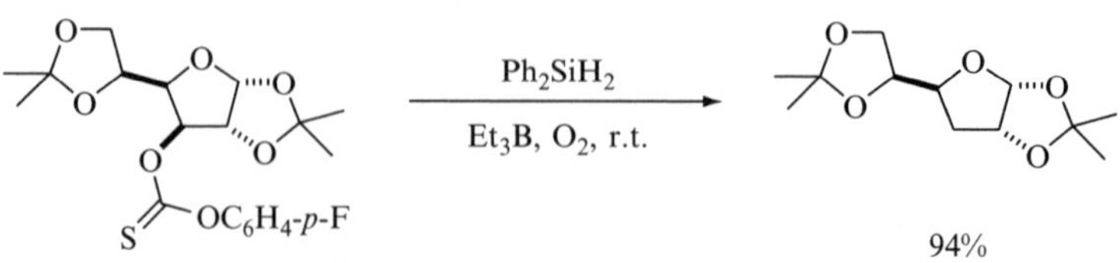

(4.59)

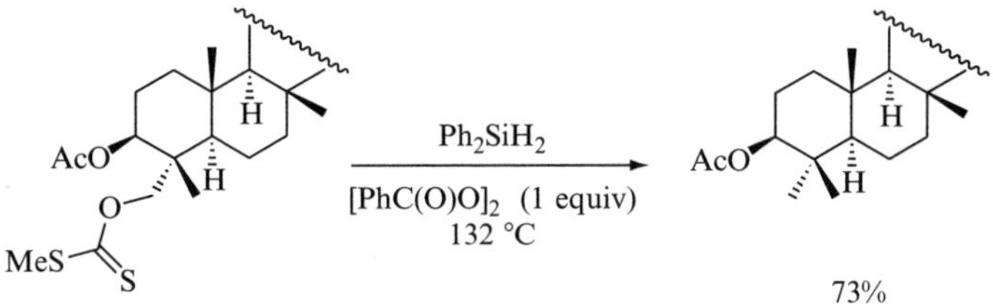

(4.60)

73%

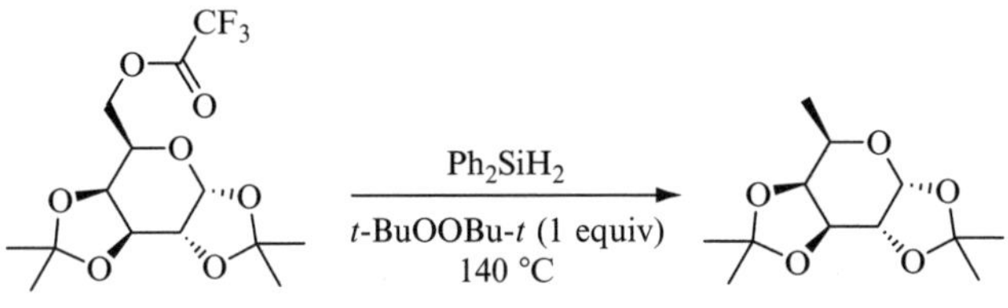

(4.61)

73%

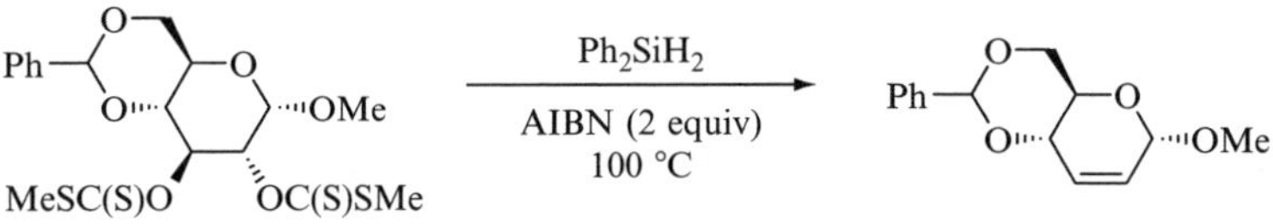

(4.62)

97%

A variety of arylsilanes with hydrophilic groups have been prepared for testing radical dehalogenations in aqueous media [103]. Yields varied from poor to good ones depending of the substrate and the silicon hydride. Reaction (4.63) gives an example that afforded the sugar product in good yields, using Et_3B as the radical initiator.

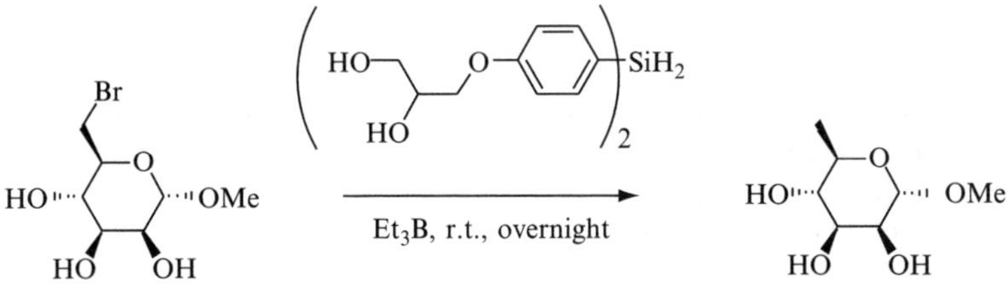

(4.63)

88%

9,10-Disilaanthracenes as the alternative silanes for the reduction of halides and Barton–McCombie type deoxygenation have been proposed [104,105]. Two examples are given in Reactions (4.64) and (4.65). As discussed in Section 3.1.1, these silanes having conformationally locked phenyl substituents at the silicon atoms show enhanced hydrogen donor abilities towards alkyl radicals. Indeed, the rate constants for hydrogen abstraction from these substrates depend from the number of available hydrogens and can reach values as high as $(TMS)_3SiH$.

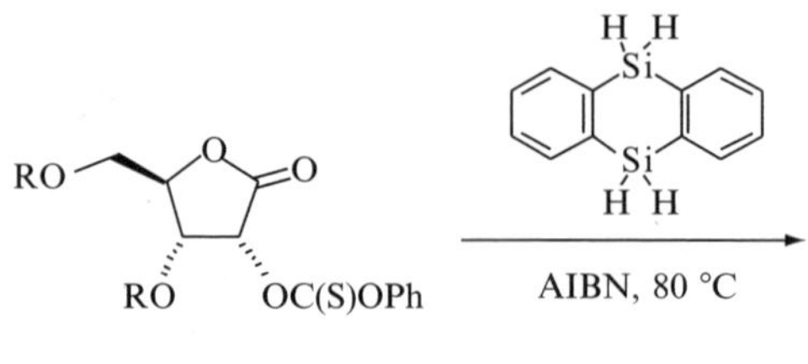

(4.64)

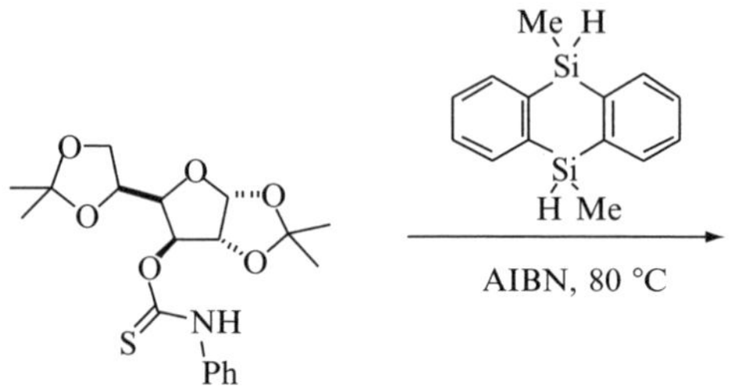

(4.65)

4.4.3 SILYL SUBSTITUTED SILICON HYDRIDES

Following the success of $(TMS)_3SiH$, a systematic investigation of organosilanes having different silyl substituents at the SiH moiety has been carried out with the aim of discovering other reagents and testing how to tune the reactivity by the choice of substituents. The idea that they might be capable of sustaining analogous radical chain reactions turned out to be correct. Indeed, the reductions of organic chlorides, bromides, iodides, phenyl selenides, isonitriles and thionocarbonates by several other silyl substituted silicon hydrides have emerged.

By replacing one alkyl group by a R_3Si group in trialkylsilanes the effect on the hydrogen donating ability of SiH moiety increases by about one order of magnitude, the effect being cumulative (see Chapter 3). Reductions of a variety of organic derivatives were carried out by using $RSi(H)Me_2$, where R = $SiMe_3$, $SiMe_2SiMe_3$, or $Si(SiMe_3)_3$, and the initiation given by benzoyl peroxide or *tert*-butyl perbenzoate (AIBN was found not to be efficient in these cases) [106]. An example for the simplest silane is reported in Reaction (4.66). A few rate constants for the halogen abstraction by $Me_3SiSi(\bullet)Me_2$ radicals are also known and are smaller than that of $Et_3Si\bullet$ or higher than that of $(TMS)_3Si\bullet$ (Table 4.6).

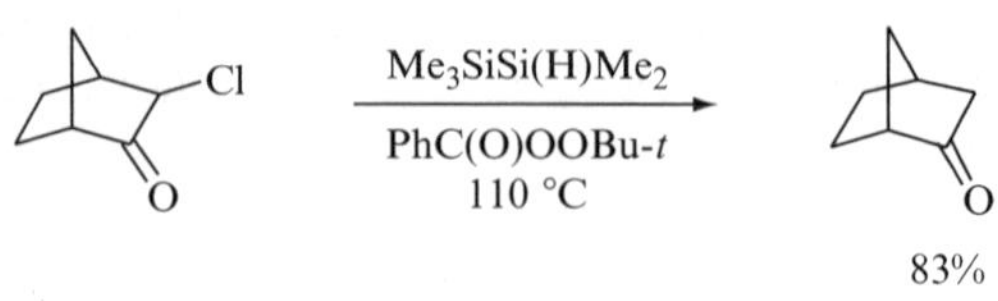

(4.66)

Table 4.6 Rate constants at 27 °C for the reaction of $Me_3SiSi(\bullet)Me_2$ radicals with a few halides [107]

Halide	$k/M^{-1}\,s^{-1}$
$(CH_3)_3CCl$	4.2×10^5
$CH_3(CH_2)_4Br$	1.6×10^8
$(CH_3)_3CBr$	2.6×10^8

Reduction of a variety of organic derivatives was carried out by using $(Me_3Si)_2Si(H)Me$ under normal conditions, i.e., AIBN at 80 °C [108]. The mono-reduction of a *gem*-dichloride is shown as an example in Reaction (4.67). With two silyl substituents, $(Me_3Si)_2Si(H)Me$ is an effective reducing agent which allows the formation of the desired product to be favoured due to a slower hydrogen transfer.

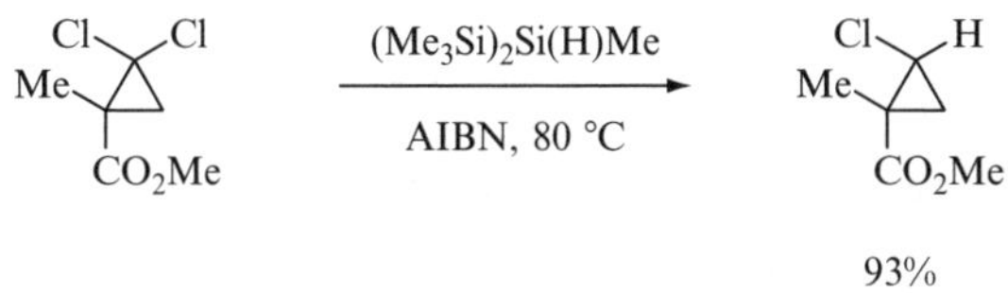

(4.67)

Poly(phenylsilane)s of the type $H(RSiH)_nH$, where R = *n*-hexyl or phenyl, have been used as radical-based reducing agents for organic halides [109]. They rival the effectiveness of $(TMS)_3SiH$ in dehalogenation reactions. A few examples are given in Reactions (4.68) and (4.69). The repetitive hydrogen transfer from the same molecule of $H(RSiH)_nH$ allows these compounds to be used in small quantities. Work up can be done by adding *n*-pentane and filtering off the polymeric material, which precipitates from the crude mixture, and in a few cases the isolation of the product was readily obtained. This approach has also been used as a procedure for the partial or total functionalization of the SiH bond (see Section 8.3).

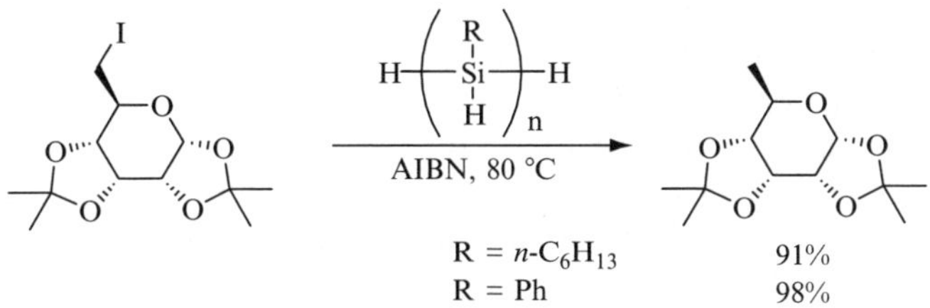

(4.68)

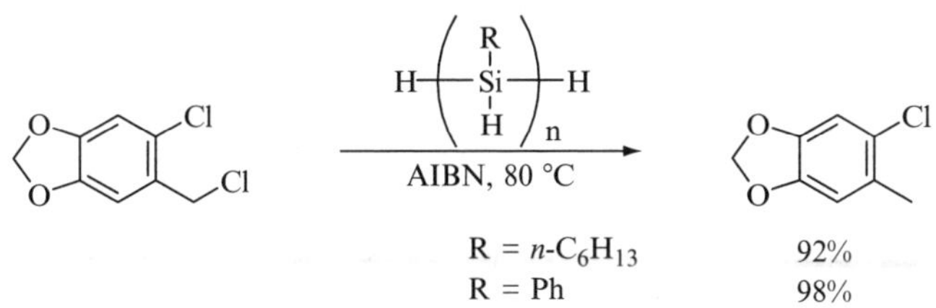

(4.69)

An interesting variant of the above mentioned reaction is the use of 1,1,2,2-tetraphenyldisilane. The reductions of alkyl bromides [110], phenyl chalcogenides [111] and xanthates [112] have been performed using either AIBN in refluxing ethanol or Et_3B/O_2 at room temperature. Yields varied from moderate to excellent, depending on the experimental conditions. Two examples with 3-cholestanyl bromide and 2-phenylseleno sugar are given in Reaction (4.70) and (4.71). Disilanes having *para*-fluoro or *para*-methoxy phenyl substituents showed similar reactivities [110–112].

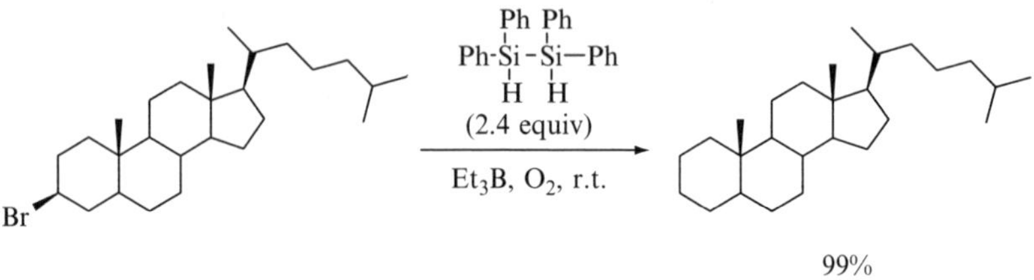

(4.70)

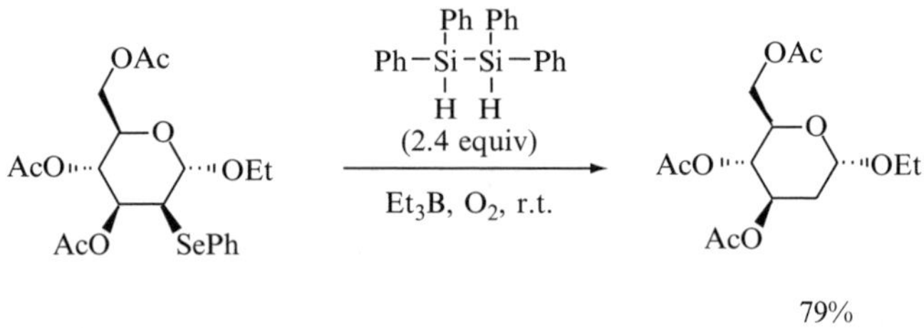

(4.71)

4.4.4 ALKYLTHIO SUBSTITUTED SILICON HYDRIDES

Substitution at the SiH moiety has been carried out with alkylthio groups, such as MeS and *i*-PrS. Tris(alkylthio)silanes, $(RS)_3SiH$, are radical-based reducing agents which can effect the reduction of bromides, iodides, xanthates, phenylselenides and isocyanides in toluene, using AIBN at 85 °C as the initiator (Reaction 4.71) [113]. The reduction of chlorides is efficient only when α-stabilizing groups or *gem*-dichlorides are present. It is interesting to note that under the reduction conditions and in the absence of substrates the reagent is consumed, forming $(RS)_4Si$, probably by a S_H2 reaction at the sulfur atom [113]. By replacing a Me_3Si group in $(TMS)_3SiH$ with the thiol group Me_3SiS, the resulting silicon hydride behaves similarly (Reaction 4.72) [114]. Thiol-substituted silicon hydrides have not found an application in synthetic procedures so far, probably due to their unpleasant smell.

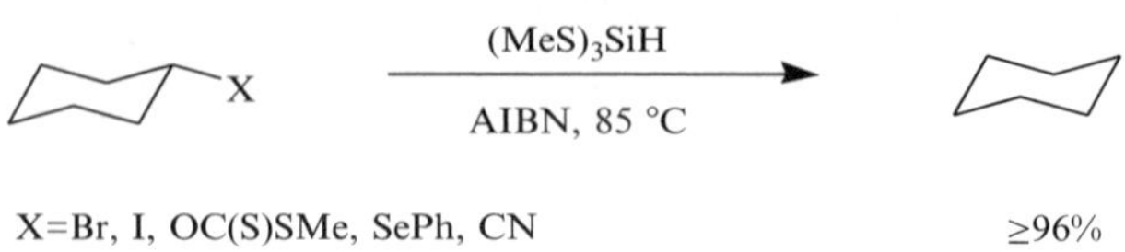

(4.71)

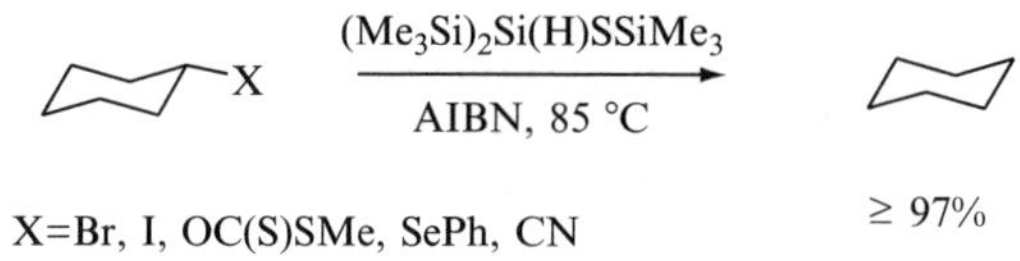

(4.72)

4.5 SILICON HYDRIDE / THIOL MIXTURE

The low reactivity of alkyl and/or phenyl substituted organosilanes in reduction processes can be ameliorated in the presence of a catalytic amount of alkanethiols. The reaction mechanism is reported in Scheme 4.6 and shows that alkyl radicals abstract hydrogen from thiols and the resulting thiyl radical abstracts hydrogen from the silane. This procedure, which was termed polarity-reversal catalysis, has been applied to dehalogenation, deoxygenation and desulfurization reactions [115].

$$R\bullet + XSH \longrightarrow RH + XS\bullet$$

$$XS\bullet + Et_3SiH \longrightarrow XSH + Et_3Si\bullet$$

$$Et_3Si\bullet + RZ \longrightarrow Et_3SiZ + R\bullet$$

Scheme 4.6 Propagation steps for polarity-reversal catalysis

For example, 1-bromoadamantane is quantitatively reduced with 2 equiv of triethylsilane and in the presence of a catalytic amount of *tert*-dodecanethiol (Reaction 4.73). Similarly desulfurization Reaction (4.74) occurs readily. Although generally these reactions do not work in the absence of thiol, the reduction of xanthate derivatives give similar yields in the presence or absence of thiol. To account for this result, it was suggested that Et_3SiSH is formed *in situ* by side reactions, thus obviating the need to add a separate thiol catalyst [116]. Indeed, a variety of alkyl and/or aryl substituted silanes (R_3SiH) react under normal free radical conditions with carbonyl sulfide (O═C═S), a by-product of Barton–McCombie deoxygenation, to give the corresponding silanethiol (R_3SiSH) (see also Section 5.5) [117].

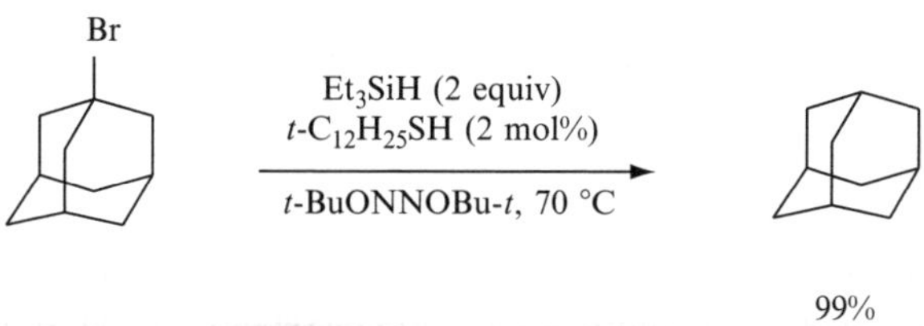

(4.73)

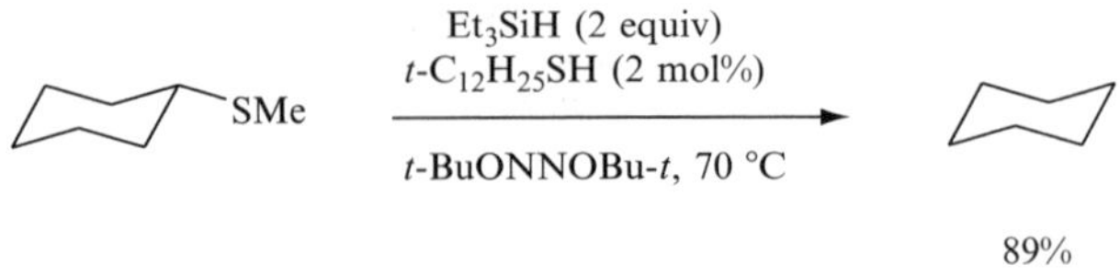

(4.74)

The rate constants for the reaction of $Et_3Si\bullet$ radicals with dialkyl sulfides R_2S were measured by the laser flash photolysis technique and decrease in the order R = primary > secondary > tertiary, viz., 1.1×10^7, 8.8×10^6, and 3.3×10^6 M^{-1} s^{-1} for n-Bu_2S, s-Bu_2S and t-Bu_2S, respectively [118].

4.6 SILANETHIOLS

The fact that thiols are good H atom donors toward alkyl radicals and that silyl radicals are among the most reactive known species for abstraction and addition reactions, suggests that any class of compounds, which allows for the transformation of a thiyl to a silyl radical via a fast intramolecular rearrangement, will potentially be a good radical-based reducing agent. Applying this strategy, the compounds $(Me_3Si)_3SiSH$ and $(Me_3Si)_2Si(Me)SH$ were designed [114, 119]. The reductions of bromides, iodides and isocyanides by $(Me_3Si)_3SiSH$ are extremely efficient processes, as the reactions are completed in 5 min using AIBN at 85 °C. The reaction mechanism is outlined in Scheme 4.7.

4.7 SILYLATED CYCLOHEXADIENES

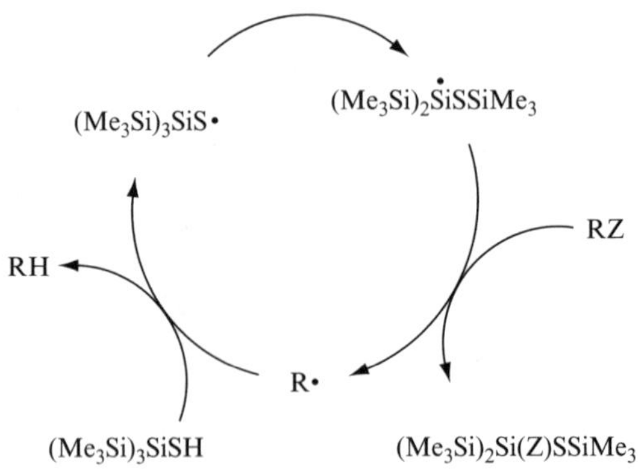

Scheme 4.7 Propagation steps for the radical chain removal of a functional group by $(Me_3Si)_3SiSH$

Silylated 1,4-cyclohexadienes were introduced as reducing agents in radical chain reactions such as dehalogenation, deoxygenation via thionocarbonate ester and deselenization [120]. Two examples are given in Reactions (4.75) and (4.76), which show that these reactions work well also with different substituted silyl moieties.

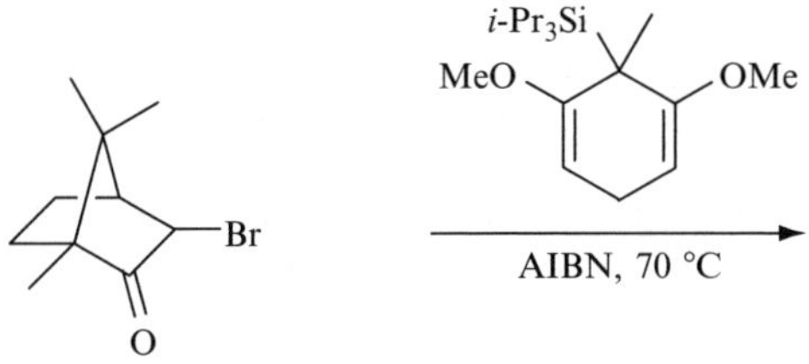

(4.75)

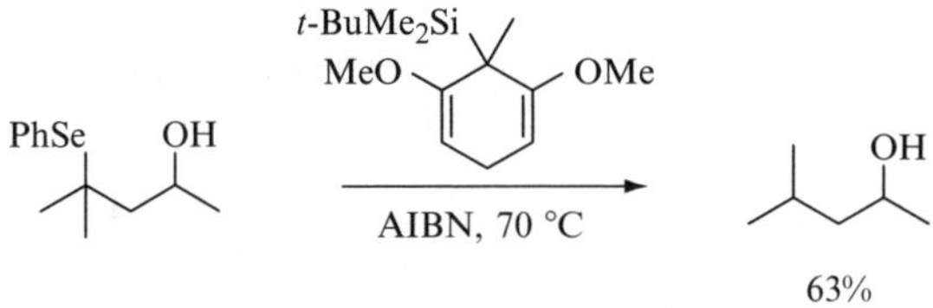

(4.76)

Scheme 4.8 shows that the CH_2 of cyclohexadiene moiety acts as the H donor with formation of cyclohexadienyl radical as the intermediate, which rapidly ejects the silyl radical upon re-aromatization. The silyl radical is able to propagate the chain by reaction with a starting halide. The hydrogen donation of silylated cyclohexadienes toward primary alkyl radicals is reported to be $1 \times 10^5\ M^{-1}s^{-1}$ at 70 °C [120], which is in accord with the reported range of 10^3–$10^4\ M^{-1}\ s^{-1}$ at room temperature for the reaction of primary and secondary alkyl radicals with differently substituted cyclohexadienes [121].

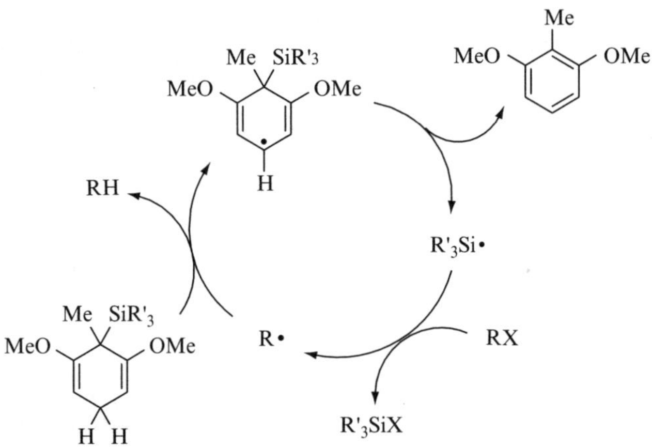

Scheme 4.8 Silylated cyclohexadienes as radical-based reducing agents

4.8 REFERENCES

1. Chatgilialoglu, C., *Acc. Chem. Res.*, 1992, **25**, 188.
2. Chatgilialoglu, C., *Chem. Rev.*, 1995, **95**, 1229.
3. Chatgilialoglu, C., Ferreri, C., and Gimisis, T. In *The Chemistry of Organic Silicon Compounds*, Z. Rappoport, and Y. Apeloig (Eds), Wiley, Chichester, 1998, pp. 1539–1579.
4. Baguley, P.A., and Walton, J.C., *Angew. Chem. Int. Ed.*, 1998, **37**, 3072.
5. Chatgilialoglu, C. In *Radical in Organic Synthesis*, Volume 1, P. Renaud, and M. P. Sibi (Eds), Wiley-VCH, Weinheim, 2001, pp. 28–49.
6. Gilbert, B.C., and Parson, A.F., *J. Chem. Soc., Perkin Trans.,* 2 2002, 367.
7. Studer, A., and Amrein, S., *Synthesis*, 2002, 835.
8. Walling, C., *Tetrahedron*, 1985, **19**, 3887.
9. Gaspar, P.P., Haizlip, A.D., and Choo, K.Y., *J. Am. Chem. Soc.*, 1972, **94**, 9032.
10. Shimo, N., Nakashima, N., and Yoshihara, K., *Chem. Phys. Lett.*, 1986, **125**, 303.
11. Cornett, B.J., Choo, K.Y., and Gaspar, P.P., *J. Am. Chem. Soc.*, 1980, **102**, 377.
12. Safarik, I., Jodhan, A., Strausz, O.P., and Bell, T.N., *Chem. Phys. Lett.*, 1987, **142**, 115.
13. Watts, G.B., and Ingold, K.U., *J. Am. Chem. Soc.*, 1972, **94**, 491.
14. Chatgilialoglu, C., Scaiano, J.C., and Ingold, K.U., *Organometallics*, 1982, **1**, 466.
15. Chatgilialoglu, C., and Rossini, S., *Bull. Soc. Chim. Fr.*, 1988, 298.
16. Alberti, A., and Chatgilialoglu, C., *Tetrahedron* 1990, **46**, 3963.
17. Hawari, J.A., Griller, D., Weber, W.P., and Gaspar, P.P., *J. Organomet. Chem.*, 1987, **326**, 335.
18. Ishikawa, M., Nakamura, A., and Kumada, M., *J. Organometal. Chem.*, 1973, **59**, C11.
19. Niiranen, J. T., and Gutman, D., *J. Phys. Chem.*, 1993, **97**, 9392.
20. Getoff, N., Ritter, A., Schwörer, F., and Bayer, P., *Radiat. Phys. Chem.*, 1993, **41**, 797.
21. Kita, Y., and Matsugi, M. In *Radical in Organic Synthesis*, Volume 1, P. Renaud, and M.P. Sibi, (Eds), Wiley-VCH, Weinheim, 2001, pp. 1–10.
22. Yorimitsu, H., and Oshima, K. In *Radical in Organic Synthesis*, Volume 1, P. Renaud, and M.P., Sibi (Eds), Wiley-VCH, Weinheim, 2001, pp. 11–27.
23. Chatgilialoglu, C. In *Paramagnetic Organometallic Species in Activation / Selectivity, Catalysis*, M. Chanon, M. Julliard, and J. C. Poite (Eds), Kluwer, Dordrecht, 1989, pp. 119–129. (Proceedings of a NATO ARW held in St. Maximin, France, 4–9 October 1987).
24. Chatgilialoglu, C. In *Free Radicals in Synthesis and Biology*, F. Minisci (Ed), Kluwer, Dordrecht, 1989, pp. 115–123. (Proceedings of a NATO ARW held in Bardolino, Italy, 8–13 May 1988).
25. Gilman, H., Atwell, W.H., Sen, P.K., and Smith, C.L., *J. Organomet. Chem.*, 1965, **4**, 163.
26. Kanabus-Kaminska, J.M., Hawari, J.A., Griller, D., and Chatgilialoglu, C., *J. Am. Chem. Soc.*, 1987, **109**, 5267.
27. Chatgilialoglu, C., Griller, D., and Lesage, M., *J. Org. Chem.*, 1988, **53**, 3641.
28. Robl, J.A., *Tetrahedron Lett.*, 1994, **35**, 393.
29. Chatgilialoglu, C., and Gimisis, T., *Chem. Commun.*, 1998, 1249.
30. Mero, C.L., and Porter, N.A., *J. Am. Chem. Soc.*, 1999, **121**, 5155.
31. Wahl, F., Wörth J., and Prinzbach, H., *Angew. Chem. Int. Ed. Engl.*, 1993, **32**, 1722.
32. Chatgilialoglu, C., Griller, D., and Lesage, M., *J. Org. Chem.*, 1989, **54**, 2492.
33. Jiang, X.-K., Ding, W.F.-X., and Zhang, Y.-H., *Tetrahedron*, 1997, **53**, 8479.

34. Kawashima, E., Uchida, S., Miyahara, M., and Ishido, Y., *Tetrahedron Lett.*, 1997, **38**, 7369.
35. Villar, F., Andrey, O., and Renaud, P., *Tetrahedron Lett.*, 1999, **40**, 3375.
36. Mouriès, V., Delouvrié, B., Lacôte, E., Fensterbank, L., and Malacria, M., *Eur. J. Org. Chem.*, 2002, 1776.
37. Lesage, M., Chatgilialoglu, C., and Griller, D., *Tetrahedron Lett.*, 1989, **30**, 2733.
38. Chatgilialoglu, C., Costantino, C., Ferreri, C., Gimisis, T., Romagnoli A., and Romeo, R., *Nucleosides Nucleotides*, 1999, **18**, 637.
39. Bandini, E., Favi, G., Martelli, G., Panunzio, M., and Piersanti, G., *Org. Lett.*, 2000, **2**, 1077.
40. Lee, E., Park, C.M., and Yun, J.S., *J. Am. Chem. Soc.*, 1995, **117**, 8017.
41. Apeloig, Y., and Nakash, M., *J. Am. Chem. Soc.*, 1994, **116**, 10781.
42. Ballestri, M., Chatgilialoglu, C., Cardi, N., and Sommazzi, A., *Tetrahedron Lett.*, 1992, **33**, 1787.
43. Ballestri, M., Chatgilialoglu, C., Clark, K.B., Griller, D., Giese, B., and Kopping, B., *J. Org. Chem.*, 1991, **56**, 678.
44. Arya, P., Samson, C., Lesage, M., and Griller, D., *J. Org. Chem.*, 1990, **55**, 6248.
45. Arya, P., Lesage, M., and Wayner, D.D.M., *Tetrahedron Lett.*, 1991, **32**, 2853.
46. Lesage, M., and Arya, P., *Synlett*, 1996, 237.
47. Arya, P., and Wayner, D.D.M., *Tetrahedron Lett.*, 1991, **32**, 6265.
48. Curran, D.P., and Sun, S., *Tetrahedron Lett.*, 1993, **34**, 6181.
49. Sekine, M., and Seio, K., *J. Chem. Soc. Perkin Trans. 1,* 1993, 3087.
50. Gimisis, T., Ialongo, G., Zamboni, M., and Chatgilialoglu, C., *Tetrahedron Lett.*, 1995, **36**, 6781.
51. Krief, A., Badaoui, E., and Dumont, W., *Tetrahedron Lett.*, 1993, **34**, 8517.
52. Chatgilialoglu, C., Ferreri, C., Lucarini, M., Pedrielli, P., and Pedulli, G.F., *Organometallics*, 1995, **14**, 2672.
53. Alcaide, B., Rodriguez-Vicente, A., and Sierra, M.A., *Tetrahedron Lett.*, 1998, **39**, 163.
54. Quirante, J., Escolano, C., and Bonjock, J., *Synlett*, 1997, 179.
55. Paquette, L.A., Friedrich, D., Pinard, E., Williams, J.P., Laurent, D.St., and Roden, B.A., *J. Am. Chem. Soc.*, 1993, **115**, 4377.
56. Chatgilialoglu, C. In *The Chemistry of Sulphenic Acids and Their Derivatives*, S. Patai (Ed), Wiley, Chichester, 1990, pp. 549–569.
57. Schiesser C.H., and Smart, B.A., *Tetrahedron* 1995, **51**, 6051.
58. Barton, D.H.R., and McCombie, S.W., *J. Chem. Soc., Perkin Trans. 1,* 1975, 1574.
59. Oba, M., and Nishiyama, K., *Tetrahedron*, 1994, **50**, 10193.
60. Robins, M.J., Wilson, J.S., and Hannske, F., *J. Am. Chem. Soc.*, 1983, **105**, 4059.
61. Hartwig, W., *Tetrahedron*, 1983, **39**, 2609.
62. Motherwell, W.B., and Crich, D., *Free Radical Chain Reactions in Organic Synthesis*, Academic Press, London, 1992.
63. Barton, D.H.R., *Half a Century of Free Radical Chemistry*, Cambridge University Press, Cambridge, 1993.
64. Zard, S.Z. In *Radical in Organic Synthesis*, Volume 1, P. Renaud, and M.P. Sibi (Eds), Wiley-VCH, Weinheim, 2001, pp. 90–108.
65. Chatgilialoglu, C., and Ferreri, C., *Res. Chem. Intermed.*, 1993, **19**, 755.
66. Schummer, D., and Höfle, G., *Synlett*, 1990, 705.
67. Barton, D.H.R., Jang, D.O., and Jaszberenyi, J. Cs., *Tetrahedron Lett.*, 1990, **31**, 4681.
68. Perich, J.W., *Synlett*, 1992, 595.
69. Boquet, P., Loustau Cazalet, C., Chapleur, Y., Samreth, S., and Bellamy, F., *Tetrahedron Lett.*, 1992, **33**, 1997.

70. Mathé, C., Imbach, J.-L., and Gosselin, G., *Carbohydr. Res.*, 2000, **323**, 226.
71. Toyota, M., Seishi, T., Yokoyama, M., Fukumoto, K., and Kabuto, C., *Tetrahedron*, 1994, **50**, 1093.
72. Hammerschmidt, F., Öhler, E., Polsterer, J.-P., Zbiral, E., Balzarini, J., and DeClercq, E., *Liebigs Ann.*, 1995, 551.
73. Chatgilialoglu, C., Gimisis, T., and Spada, G.P., *Chem. Eur. J.*, 1999, **5**, 2866.
74. Chu, C.K., and Chen, Y. US Patent 5,384,396 (1995). *Chem. Abstr.*, 1995, **112**, 265937.
75. Grierson, J.R., Shields, A.F., Zheng, M., Kozawa, S.M., and Courter, J.H., *Nucl. Med. Biol.*, 1995, **22**, 671.
76. Barchi, Jr., J.J., Haces, A., Marquez, V.E., and McCormack, J.J., *Nucleosides Nucleotides*, 1992, **11**, 1781.
77. Oba, M., Suyama, M., Shimamura, A., and Nishiyama, K., *Tetrahedron Lett.*, 2003, **44**, 4027.
78. García de Viedma, A., Martínez-Barrasa, V., Burgos, C., Izquiedro, M.L., and Alvarez-Builla, J., *J. Org. Chem.*, 1999, **64**, 1007.
79. Lucarini, M., Marchesi, E., Pedulli, G.F., and Chatgilialoglu, C., *J. Org. Chem.*, 1998, **63**, 1687.
80. Ballestri, M., Chatgilialoglu, C., Lucarini, M., and Pedulli, G.F., *J. Org. Chem.*, 1992, **57**, 948.
81. Imrie, C., *J. Chem Res. (S)*, 1995, 328.
82. Romeo, R., Wozniak, L.A., and Chatgilialoglu, C., *Tetrahedron Lett.*, 2000, **41**, 9899.
83. Fukuzumi, S., and Noura, S., *J. Chem, Soc., Chem. Commun.*, 1994, 287.
84. Chatgilialoglu, C., Guarini, A., Guerrini, A., and Seconi, G., *J. Org. Chem.*, 1992, **57**, 2207.
85. Sakarai, H. In *Free Radicals*, Volume 2, J.K. Kochi (Ed), Wiley, New York, 1973, pp. 741–808.
86. Petrov, V.A., *J. Org. Chem.*, 1998, **63**, 7294.
87. Billingham, N.C., Jackson, R.A., and Malek, F., *J. Chem. Soc., Perkin Trans. 1,* 1979, 1137.
88. Jackson, R.A., and Malek, F., *J. Chem. Soc., Perkin Trans. 1,* 1980, 1207.
89. Barton, D.H.R., Jang, D.O., and Jaszberenyi, J.Cs., *Tetrahedron Lett.*, 1991, **32**, 7187.
90. Chatgilialoglu, C., Ferreri, C., and Lucarini, M., *J. Org. Chem.*, 1993, **58**, 249.
91. Meffre, P., Dave, R.H., Leroy, J., and Badet, B., *Tetrahedron Lett.*, 2001, **42**, 8625.
92. Chatgilialoglu, C., Ingold, K.U., and Scaiano, J.C., *J. Am. Chem. Soc.*, 1982, **104**, 5123.
93. Chatgilialoglu, C., Ingold, K.U., and Scaiano, J.C., *J. Org. Chem.*, 1987, **52**, 938.
94. Horowitz, A., *J. Am. Chem. Soc.*, 1985, **107**, 318.
95. Chatgilialoglu, C., Ingold, K.U., and Scaiano, J.C., *J. Am. Chem. Soc.*, 1983, **105**, 3292.
96. Chatgilialoglu, C., Lunazzi, L., and Ingold, K.U., *J. Org. Chem.*, 1983, **48**, 3558.
97. Chatgilialoglu, C., Griller, D., and Rossini, S., *J. Org. Chem.*, 1985, **54**, 2734.
98. Sluggett, G.W., and Leigh, W.J., *Organometallics,* 1992, **11**, 3731.
99. Ito, O., Hoteiya, K., Watanabe, A., and Matsuda, M. *Bull. Chem. Soc. Jpn.*, 1991, **64**, 962.
100. Barton, D.H.R., Jang, D.O., Jaszberenyi, J.Cs. *Tetrahedron*, 1993, **49**, 7193.
101. Jang, D.O., Kim, J., Cho, D.H., Chung, C.-M., *Tetrahedron Lett.*, 2001, **42**, 1073.
102. Barton, D.H.R., Tachdjian, C., *Tetrahedron,* 1992, **48**, 7109.
103. Yamazaki, O., Togo, H., Nogami, G., and Yokoyama, M., *Bull. Chem. Soc. Jpn.,* 1997, **70**, 2519.

104. Gimisis, T., Ballestri, M., Ferreri, C., Chatgilialoglu, C., Boukherroub, R., and Manuel, G., *Tetrahedron Lett.*, 1995, **36**, 3897.
105. Oba, M., Kawahara, Y., Yamada, R., Mizuta, H., Nishiyama, K., *J. Chem. Soc., Perkin Trans. 2,* 1996, 1843.
106. Ballestri, M., Chatgilialoglu, C., Guerra, M., Guerrini, A., Lucarini, M., and Seconi. G., *J. Chem. Soc., Perkin Trans. 2,* 1993, 421.
107. Lusztyk, J., Maillard, B., and Ingold, K.U., *J. Org. Chem.,* 1986, **51**, 2457.
108. Chatgilialoglu, C., Guerrini, A., Lucarini, M., *J. Org. Chem.*, 1992, **57**, 3405.
109. Chatgilialoglu, C., Ferreri, C., Vecchi, D., Lucarini, M., and Pedulli, G.F., *J. Organomet. Chem.*, 1997, **545/546**, 455.
110. Yamazaki, O., Togo, H., Matsubayashi, S., and Yokoyama, M., *Tetrahedron*, 1999, **55**, 3735.
111. Yamazaki, O., Togo, H., and Yokoyama, M., *J. Chem. Soc., Perkin Trans. 1,* 1999, 2891.
112. Togo, H., Matsubayashi, S., Yamazaki, O., and Yokoyama, M., *J. Org. Chem.*, 2000, **65**, 2816.
113. Chatgilialoglu, C., Guerra, M., Guerrini, A., Seconi, G., Clark, K.B., Griller, D., Kanabus-Kaminska, J., and Martinho-Simões, J.A., *J. Org. Chem.*, 1992, **57**, 2427.
114. Ballestri, M., Chatgilialoglu, C., and Seconi, G., *J. Organomet. Chem.,* 1991, **408**, C1.
115. Cole, S.J., Kirwan, J.N., Roberts, B.P., and Willis, C.R., *J. Chem. Soc., Perkin Trans. 1,* 1991, 103.
116. Roberts, B.P., *Chem. Soc. Rev.*, 1999, **28**, 25.
117. Cai, Y., and Roberts, B.P., *Tetrahedron Lett.*, 2001, **42**, 763.
118. Soundararajan, N., Jackson, J.E., and Platz, M.S., *J. Phys. Org. Chem.*, 1988, **1**, 39.
119. Daroszewski, J., Lusztyk, J., Degueil, M., Navarro, C., and Maillard, B., *J. Chem Soc., Chem. Commun.*, 1991, 587.
120. Studer, A., Amrein, S., Schleth, F., Schulte, T., and Walton, J.C., *J. Am. Chem. Soc.*, 2003, **125**, 5726.
121. Jackson, L., and Walton, J.C., *Tetrahedron Lett.*, 1999, **40**, 7019.

5 Addition to Unsaturated Bonds

Hydrosilylation is a term describing an addition of organic and inorganic silicon hydrides to carbon–carbon or carbon–heteroatom multiple bonds. The hydrosilylation of a carbon–carbon double bond was first reported in 1947 as a free-radical process [1] and became important both in industry and in the laboratory during the fifties and sixties for the production of organosilicon compounds [2]. Good yields were obtained for chlorinated silanes such as Cl_3SiH and $Cl_2Si(H)Me$ in the presence of thermal decomposition of various peroxides [1,2], whereas the failure of alkyl or aryl substituted silanes to give similar products was due to their inefficiency of hydrogen donation to alkyl radicals [3]. The radical-based methodology was later superseded by the introduction of transition metal catalysts [2,4,5]. Catalytic hydrosilylation has been exploited in all directions and today represents a fundamental and elegant methodology for both academic and industrial synthetic chemists [6,7]. Nevertheless, the introduction of a variety of organosilanes in the nineties with different hydrogen donor abilities (see Chapter 3) has made the radical-based hydrosilylation flourish again not as an alternative to catalytic methods, but as a complementary approach. Furthermore, the radical-based hydrosilylation of silicon surfaces has also been studied intensively in the last few years because of its crucial importance in modern technology. The most salient aspects of monolayer formation on silicon using this methodology are discussed in Chapter 8.

The initiation and propagation steps for the radical-based hydrosilylation of carbon–carbon double bonds are reported in Scheme 5.1 [8]. The initially generated radicals (In•) abstract a hydrogen atom from the R_3SiH. The resulting R_3Si• radical adds to the double bond to give a radical adduct, which then reacts with the silicon hydride and gives the addition product, together with 'fresh' R_3Si• radicals to continue the chain. Chain reactions

Organosilanes in Radical Chemistry C. Chatgilialoglu
© 2004 John Wiley & Sons, Ltd ISBN: 0-471-49870-X

terminate by radical–radical combination or disproportionation reactions. For more details on the radical initiators and radical chain reactions see Sections 4.2 and 4.1, respectively. The information in Scheme 5.1 can be extrapolated to any hydrosilylation reaction of carbon–carbon, carbon–heteroatom, or heteroatom–heteroatom multiple bonds.

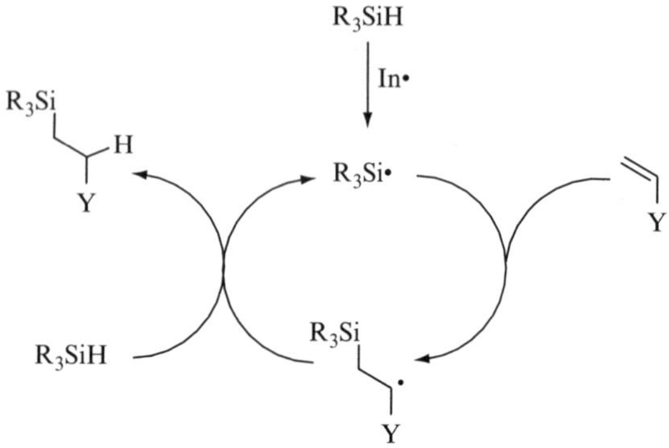

Scheme 5.1 Initiation and propagation steps for radical-based hydrosilylation

5.1 CARBON–CARBON DOUBLE BONDS

5.1.1 FORMATION OF SILYL RADICAL ADDUCTS

The EPR spectroscopic examination of the adduct radicals obtained by the addition of silyl radicals to a large number of alkenes has been carried out, and structural information has been derived [9,10]. β-Trialkylsilyl substituted alkyl radicals adopt an eclipsed conformation with a sizeable rotational barrier. An example is given for the adduct **1**, obtained by the addition of $Me_3Si•$ radical to acrylonitrile, in which the Arrhenius parameters for the hindered rotation about the C—C bond is $\log A/s^{-1} = 12.5$ and $E_a = 15.5$ kJ/mol [11]. Theoretical studies at the UMP2/DZP + BF level of theory on the structural properties of a variety of β substituted ethyl radicals (e.g., $H_3SiCH_2CH_2•$) have also been performed [12]. The high population of the eclipsed conformation is a characteristic of radicals having third-row substituents.

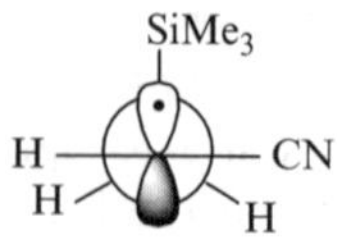

1

In Table 5.1 some representative rate constants for the addition of silyl radicals to alkenes are reported [13–15]. Inspection of these data reveals that the addition of silyl radicals to double bonds is a remarkably facile process. The rate constants for the reaction of $Et_3Si\bullet$ radical with non-activated olefins are ca $10^6\ M^{-1}\ s^{-1}$, whereas olefins having the double bond next to a π-electron system or to an electron-withdrawing substituent are substantially more reactive. Kinetic data for silyl radicals other than triethylsilyl are scant. In Table 5.1 these limited data are reported, and the reactivity trend, i.e., $Et_3Si\bullet > Me_3SiSi(\bullet)Me_2 > (TMS)_3Si\bullet$, observed for a group removal from a particular alkyl halide (cf Chapter 4) seems to hold for the addition to a particular alkene, too.

Arrhenius parameters are available for the reaction of $Et_3Si\bullet$ radicals with $H_2C{=}CH(CH_2)_3CH_3$, $H_2C{=}CHPh$ and $H_2C{=}CCl_2$ [13]. Their preexponential factors are between $10^{9.0}$ and $10^{9.4}\ M^{-1}\ s^{-1}$, which implies that the transition states for these reactions are fairly 'loose'. The high reactivity of silyl radicals towards olefins has been explained in terms of polar effects in some detail [16]. The addition of silyl radicals to alkenes is also a strongly exothermic reaction. It is estimated from thermochemical background that the addition of $Me_3Si\bullet$ radicals to simple $CH_2{=}CH_2$ is exothermic by 105 kJ/mol [16]. The addition of $Me_3Si\bullet$ and $Me_3SiSi(\bullet)Me_2$ radicals to olefinic monomers using semiempirical

Table 5.1 Rate constants ($M^{-1}\ s^{-1}$) for the addition of silyl radicals to alkenes

Substrate	$Et_3Si\bullet$ [a]	$Me_3SiSi(\bullet)Me_2$ [b]	$(TMS)_3Si\bullet$ [c]
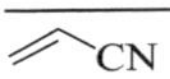	1.1×10^9		6.3×10^7
			9.7×10^7
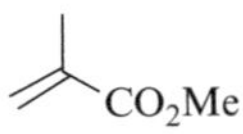	4.6×10^8		
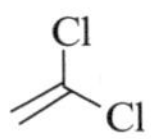	2.7×10^8		
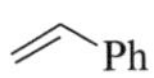	2.2×10^8		5.9×10^7
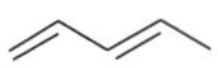	1.4×10^{8}[b]	7.6×10^7	
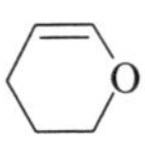	1.2×10^6		
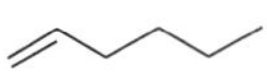	4.8×10^6	3.9×10^6	
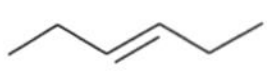	9.6×10^5		
	9.4×10^5		

[a] From Reference [13].
[b] From Reference [14].
[c] From Reference [15].

approaches has been reported [17]. The reaction enthalpy (ΔH_r) at AM1 level of theory for $Me_3Si\bullet$ with acrylonitrile, butyl vinyl ether and styrene was calculated to be −96.2, −100.4 and −115.9 kJ/mol, respectively, whereas the analogous ΔH_f values for $Me_3SiSi(\bullet)Me_2$ were calculated to be −94.6, −70.3 and −106.7 kJ/mol, respectively. Transition states for $Me_3SiSi(\bullet)Me_2$ attack at the terminus of acrylonitrile and butyl vinyl ether were calculated to have attack distances of 3.096 and 3.008 Å, respectively. Due to the high exothermicity of these reactions, it is not surprising to see that the addition of $Me_3Si\bullet$ radicals to 1-hexene is irreversible at temperatures as high as 140 °C [18].

However, $(TMS)_3Si\bullet$ radicals are found to add to a variety of double bonds reversibly and therefore to isomerize alkenes [19]. An example is shown for the interconversion of (*E*)- to (*Z*)-3-hexen-1-ol and vice versa by $(TMS)_3Si\bullet$ radicals (Reaction 5.1). Figure 5.1 shows the time profile of this reaction under standard experimental conditions (AIBN, 80 °C). The equilibration of the two geometrical isomers is reached in ca 10 h, and the percentage of $Z/E = 18/82$ after completion corresponds to an equilibrium constant of $K = 4.5$. The difference in the stability of the two isomers in 2-butenes, i.e., $\Delta G^\circ(E\text{-isomer}) - \Delta G^\circ(Z\text{-isomer}) = -3.1$ kJ/mol, corresponds to $K = 3.5$, since $\Delta G = -RT \ln K$. Therefore, the observed final composition should account for the difference in the stability of the two isomers [20]. It is worth noting that care must be taken in synthetic strategies, since isomerization can occur *in situ*, while accomplishing other reactions [19].

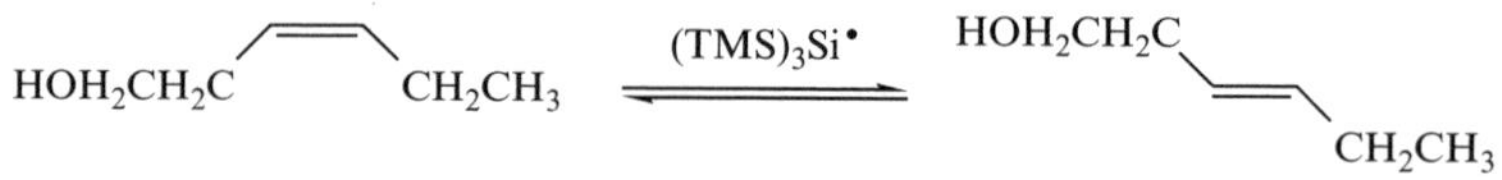

(5.1)

The addition of silyl radicals to double bonds in benzene or substituted benzenes (Reaction 5.2) is the key step in the mechanism of homolytic aromatic substitution with silanes [8,9]. The intermediate cyclohexadienyl radical **2** has been detected by both EPR and optical techniques [21,22]. Similar cyclohexadienyl-type intermediates have also been detected with heteroaromatics like furan and thiophene [23].

2

(5.2)

The addition of a trialkylsilyl radical to benzene is much less exothermic than the addition to a non-activated alkene; the ΔH_r of these reactions has been evaluated to be ca −50 kJ/mol [9,18]. However, the rate constants for the addition of Et_3Si radicals to aromatic and heteroaromatic compounds are similar to those of non-activated alkenes, i.e., $\sim 10^6\ M^{-1}\ s^{-1}$ [24]. Furthermore, both electron-withdrawing groups and electron-donating groups accelerate the addition reaction thus indicating that reactivity is determined mainly by the

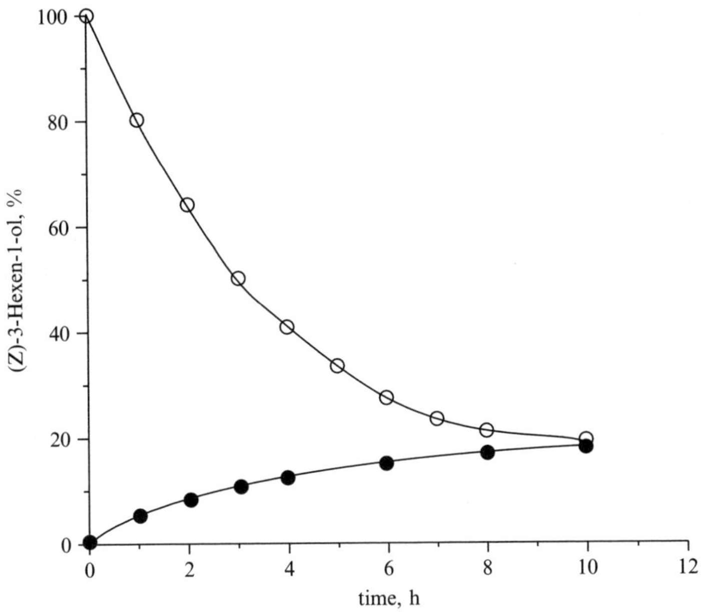

Figure 5.1 Reaction profile for the isomerization of (*Z*)-3-hexen-1-ol to (*E*)-3-hexen-1-ol and vice versa by $(TMS)_3Si\bullet$ radical in refluxing benzene

degree of stabilization of the cyclohexadienyl-type radical rather than by polar effects. Kinetic studies have shown that all phenyl substituted silyl radicals can add to an aromatic ring of their precursor ($\sim 10^6\ M^{-1}\ s^{-1}$), in contrast with trialkylsilyl radicals which decay by combination–disproportionation reactions under similar conditions [24].

The reversibility has been studied in radical **3**, generated from the corresponding cyclohexadiene and either photogenerated (0 °C) or thermally-generated (130 °C) *t*-BuO• radicals (Scheme 5.2). The elimination of $Me_3Si\bullet$ radical is a primary step at high temperatures but unimportant at 0 °C [22]. This mechanistic concept has been used for designing a class of silylating agents, namely, silylated cyclohexadienes that will be further described in this chapter (see also Section 4.7).

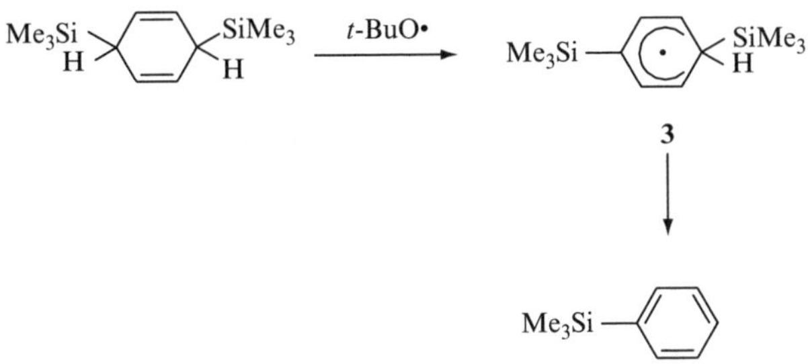

Scheme 5.2 Prove of the reversibility of the $Me_3Si\bullet$ radical addition to substituted benzenes

5.1.2 HYDROSILYLATION OF ALKENES

Hydrosilylation of monosubstituted and *gem*-disubstituted olefins (Reactions 5.3 and 5.4) are efficient processes and have been shown to occur with high regioselectivity (*anti*-Markovnikov) in the case of both electron-rich and electron-poor olefins [25]. For *cis* or *trans* disubstituted double bonds, hydrosilylation is still an efficient process, although it requires slightly longer reaction times and an activating substituent (Reaction 5.5) [25]. Any hydrosilylation product has been observed with 1,2-dialkyl-and 1,2-diaryl-substituted olefins, due to the predominant reversible addition of $(TMS)_3Si\bullet$ radical to the double bond [19].

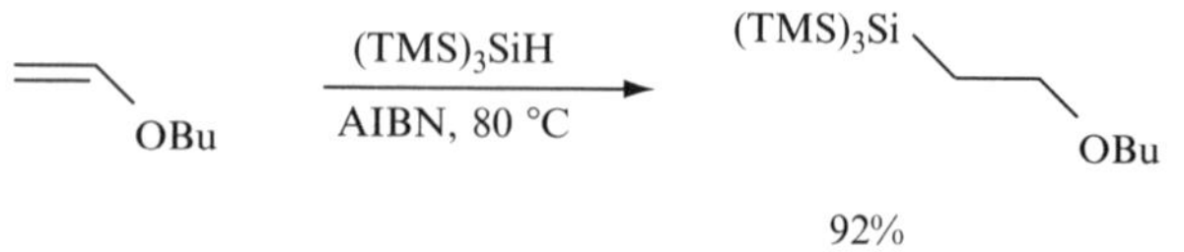

(5.3)

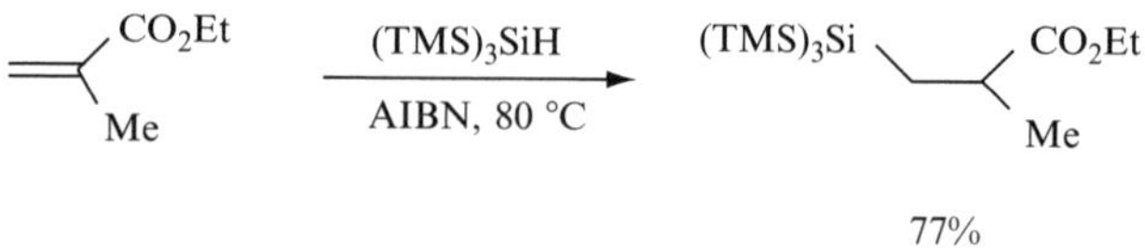

(5.4)

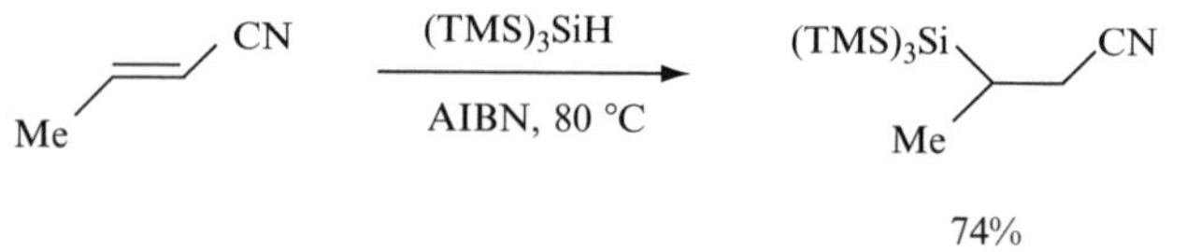

(5.5)

The reversible addition of $(TMS)_3Si\bullet$ radical to the double bonds can be noteworthy also from a synthetic point of view. An example based on the isomerization of the industrially important 1,5,9-cyclododecatriene (**4**) is shown in Scheme 5.3 [19]. The final isomeric composition of 78:20:2 for (*E,E,E*)-**4**:(*Z,E,E*)-**4**:(*Z,Z,E*)-**4**, which is independent of the starting isomer or isomeric mixture, is reached in 5 h, by using $(TMS)_3SiH/t$-BuOOBu-t/143 °C and the yield is ca 80 %.

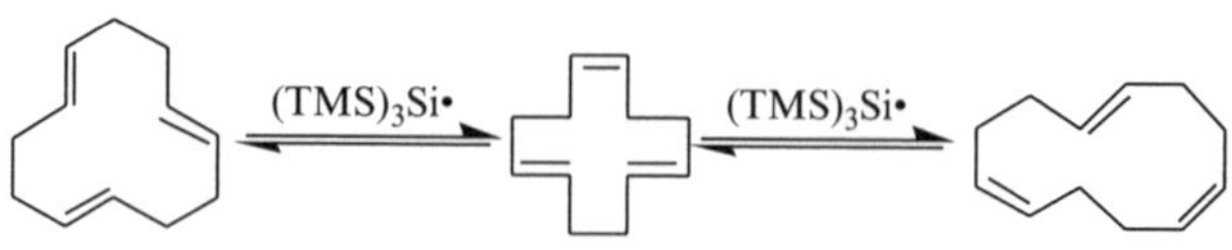

(*E,E,E*)-**4** (*Z,E,E*)-**4** (*Z,Z,E*)-**4**

Scheme 5.3 (*Z*)–(*E*) interconversion catalysed by $(TMS)_3Si\bullet$ radicals

Depending upon the choice of substrates, the hydrosilylation of alkenes with $(TMS)_3SiH$ can also be highly stereoselective. The reaction of $(TMS)_3SiH$ with methylmaleic anhydride, proceeded regiospecifically to the less substituted side, but also diastereoselectively to afford the thermodynamically less stable *cis* isomer (Reaction 5.6) [25]. Stereoselectivity increased by decreasing the reaction temperature, indicating the difference in enthalpy of activation for *syn* vs *anti* attack.

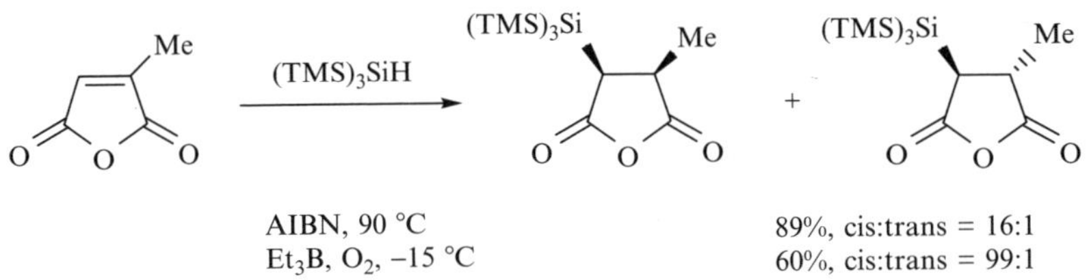

(5.6)

The addition of $(TMS)_3SiH$ to prochiral diethyl methyl fumarate (**5**) gave both diastereoisomers with preferential formation of the threo isomer (Reaction 5.7) [25]. This suggests that the intermediate adduct **6** adopts a preferred conformation due to the allylic strain effect, in which the silyl moiety shields one face of the prochiral radical center, favouring hydrogen transfer to the opposite face, and therefore the threo product is predominantly formed.

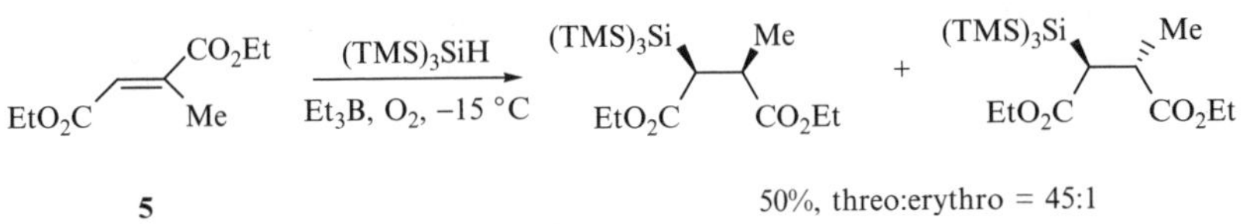

(5.7)

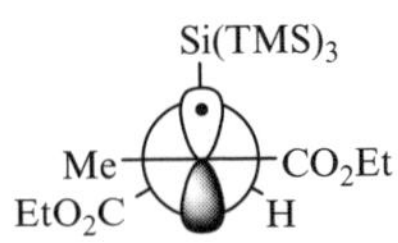

6

Addition of $(TMS)_3SiH$ to α-chiral (*E*)-alkene **7** was found to take place with a complete Michael-type regioselectivity (Reaction 5.8) [26]. A complete *syn* stereoselectivity was observed for R = Me, and it was rationalized in terms of Felkin–Ahn transition state **8**, which favours the *syn* product similar to nucleophilic addition.

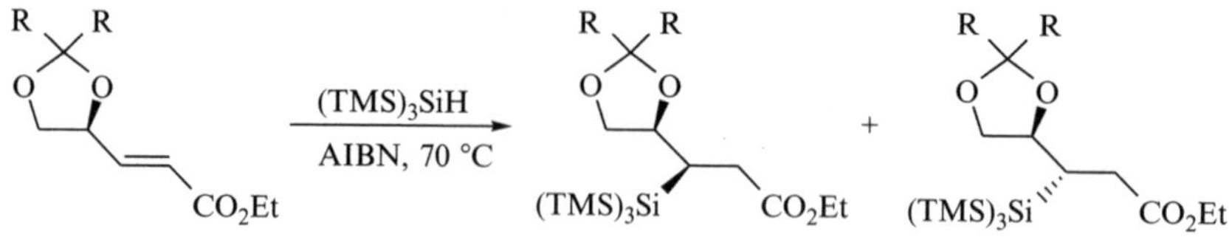

(5.8)

7

R = H 65%, syn:anti = 85:15
R = Me >95%, syn:anti = 100:0

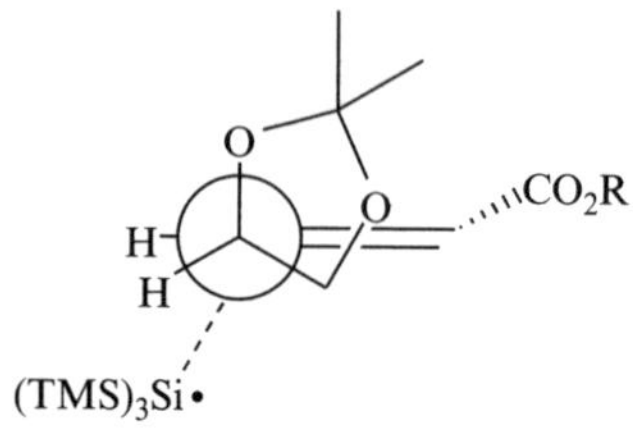

8

Radical chain hydrosilylation of alkenes using R_3SiH, where R = alkyl and/ or phenyl, is not very useful in synthesis because the hydrogen abstraction step is slow under standard experimental conditions (cf. Section 3.1). These reactions proceed under drastic conditions (Reaction 5.9) [3], but can be promoted under milder conditions by the presence of catalytic amounts of a thiol [27]. Thus, the propagation steps in Scheme 5.1 are replaced by those reported in Scheme 5.4, where the thiol acts as the catalyst and the H transfer agent (see also Section 4.5).

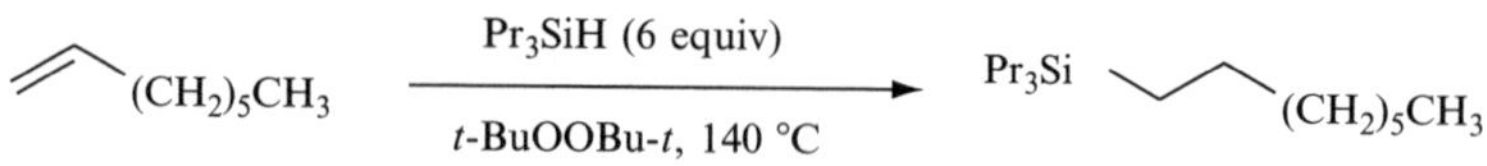

(5.9)

90%

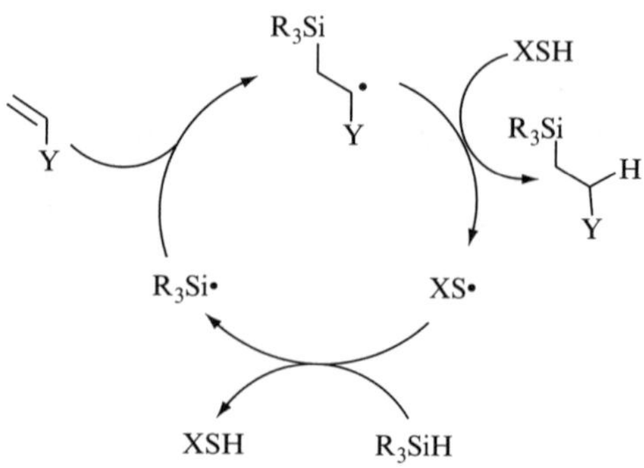

Scheme 5.4 Propagation steps for radical-based hydrosilylation catalysed by thiols

However, poor to moderate yields were obtained for the reaction of Et_3SiH at 60 °C [28]. Thiol catalysis is more effective for the addition of arylsilanes than trialkylsilanes, presumably because the hydrogen abstraction by thiyl radical is more rapid from aromatic silanes (Reactions 5.10 and 5.11) [29,30]. This methodology has also been extended to enantioselective hydrosilylation either using 1,2-asymmetric induction [29] or optically active thiols [30,31]. Reaction (5.10) shows the hydrosilylation of α-chiral (*E*)-and (*Z*)-alkenes **9** with Ph_2SiH_2. The reaction of (*E*)-alkene could occur via a Felkin–Ahn transition state similar to **8** (see above), whereas an allylic-strain transition state **10** might explain the 1,2-stereoinduction with (*Z*)-alkene, since the silyl radical attacks preferentially from the face without significant steric hindrance from the neighbouring dioxolane ring. Reaction (5.11) shows the hydrosilylation of methylenelactone **11** with a few silicon hydrides and thioglucose tetraacetate **12** as the catalyst. Both yields and enantiomeric purities increase with the degree of phenyl substitution at silicon. Thiols have also been shown to catalyse the addition of $(TMS)_3SiH$ to alkenes [31].

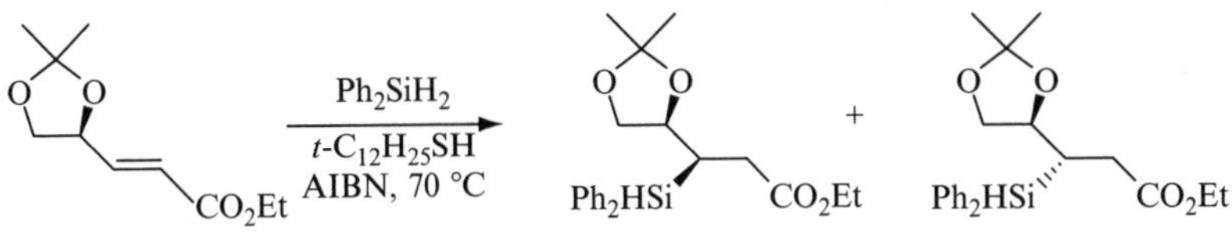

(5.10)

9

(*E*)-**9** 69%, syn:anti = 70:30

(*Z*)-**9** 75%, syn:anti = 91:9

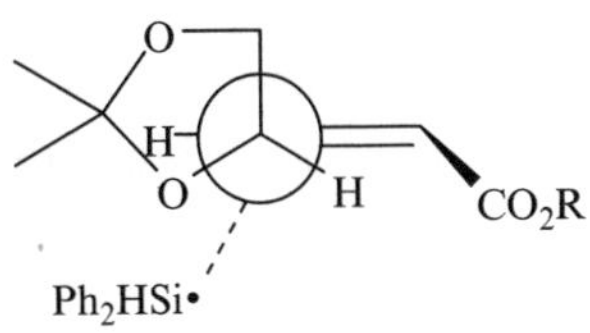

10

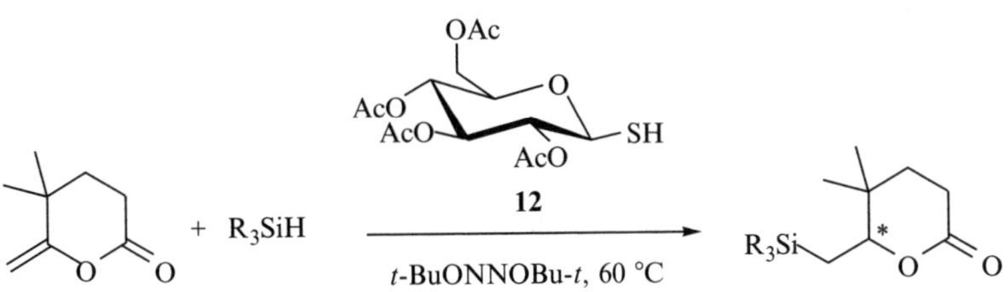

(5.11)

11

$PhMe_2SiH$	52%, ee 23%
Ph_2MeSiH	65%, ee 32%
Ph_3SiH	72%, ee 50%
$(TMS)_3SiH$	92%, ee 47%

Higher enantioselectivities were generally found when β-mannose thiol **13** was used as hydrogen donor. The extra bulkiness provided by the *gem*-β-diphenyl groups in alkene **14** compared to alkene **11** is indicated to be responsible for the high enantiomeric purity observed (Reaction 5.12) [31].

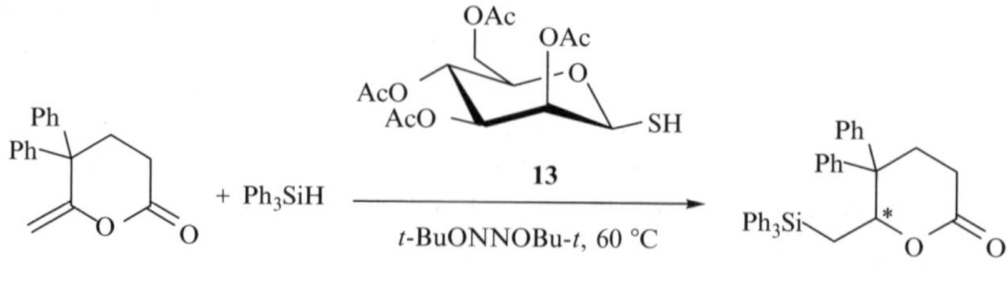

14 ee 95% (5.12)

Thiol-promoted addition of a variety of silyl radicals to the aromatic moiety of camptothecin (**15**) is also reported [32]. The addition occurs predominantly at C7 and C12 positions depending on temperature. At 105 °C, mixture of 7-silyl (favoured) and 12-silyl camptothecins are formed alongside substantial amounts of recovered camptothecin. At 160 °C, 12-silyl isomers are formed preferentially, but the total mass balance is substantially reduced.

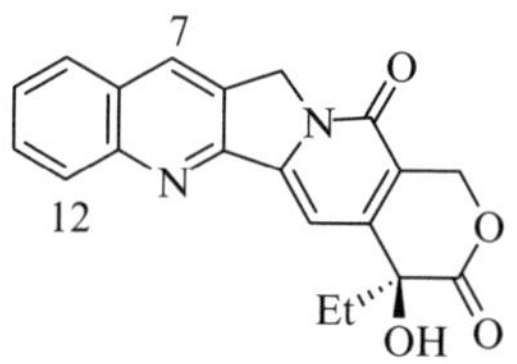

15

The hydrosilylation mechanism and the formation of β-silyl alkyl radical intermediate have been used for accomplishing other synthetically useful radical reactions. The reactions of unsubstituted and 2-substituted allyl phenyl sulfides with $(TMS)_3SiH$ give a facile entry to allyl tris(trimethylsilyl)silanes in high yields (Reaction 5.13, for X = SPh). In this case, we have the addition of $(TMS)_3Si$• radical to the double bond giving rise to a radical intermediate, but the β-scission with the ejection of a thiyl radical is inserted in the mechanism, thus affording the transposed double bond. Hydrogen abstraction from $(TMS)_3SiH$ by PhS• radical completes the cycle of these chain reactions [33]. By an analogous mechanism, allylsilanes can be obtained from allyl phenyl sulfones, although in a lower yields (Reaction 5.13, for X = SO_2Ph).

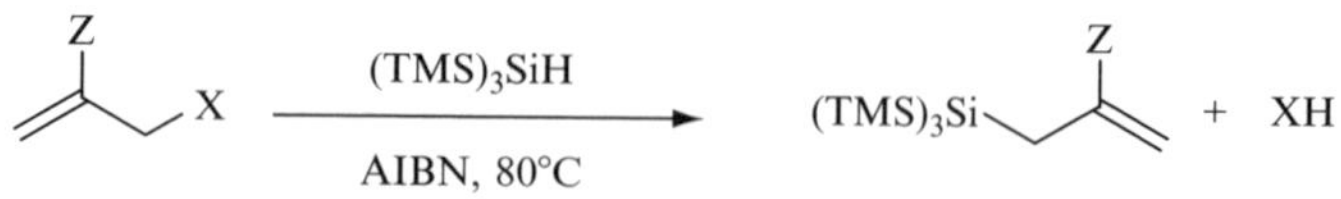

Z = H, Me, Cl, CN, CO_2Et (5.13)

X = SPh 90 - 98%
X = SO_2Ph 45 - 82%

As we anticipated in Section 5.1.1 the class of silylated cyclohexadienes has recently been used as radical transfer hydrosilylating agents for some alkenes

[34]. Reaction (5.14) shows that the silane-addition products of terminal alkenes are obtained in moderate yields. The reaction mechanism starts from the bis-allylic hydrogen of cyclohexadiene **16**, which is abstracted by the initiator, leading to a cyclohexadienyl radical. The aromatization of the intermediate (which can be read as the reversibility of silyl radical addition to aromatic species) then provides the *t*-$BuMe_2Si$• radical. Another source of silyl radicals for hydrosilylation reactions, although not in a high yield, has been envisaged as the organosilylborane $PhMe_2Si$—$B[N(CHMe)_2]_2$ treated by continuous irradiation with a high-pressure mercury lamp (cf. Section 1.1) [35].

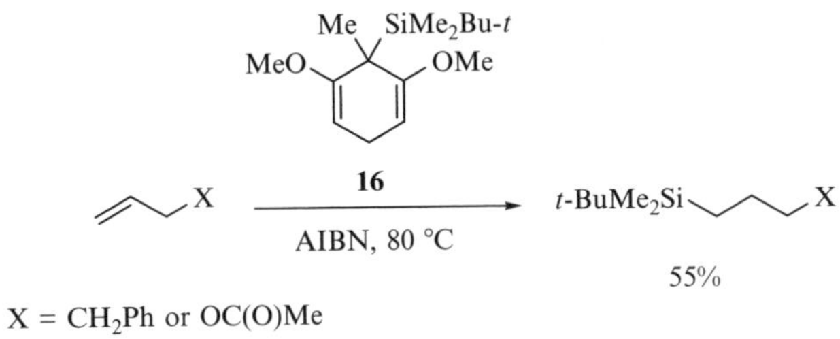

(5.14)

The reaction of Et_3SiH with [1.1.1]propellane under photolytical decomposition of di-*tert*-butyl peroxide afforded products **17** and **18** in 1:3 ratio (Reaction 5.15) [36]. A rate constant of $6.0 \times 10^8\ M^{-1}s^{-1}$ at 19 °C for the addition Et_3Si• radical to [1.1.1]propellane was determined by laser flash photolysis [37]. Thus, it would appear that [1.1.1]propellane is slightly more reactive toward attack by Et_3Si• radicals than is styrene, and significantly more reactive than 1-hexene (cf. Table 5.1).

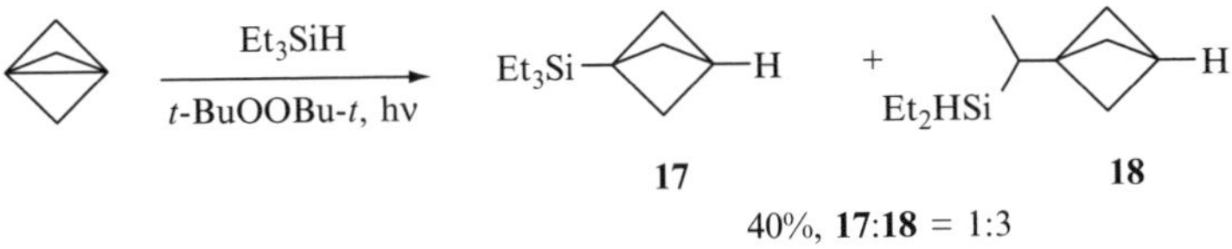

(5.15)

5.2 CARBON–CARBON TRIPLE BONDS

5.2.1 FORMATION OF SILYL RADICAL ADDUCTS

As with alkenes, the addition of silyl radicals to a carbon–carbon triple bond (Reaction 5.16) is also the key step in the hydrosilylation of alkynes [9,10].

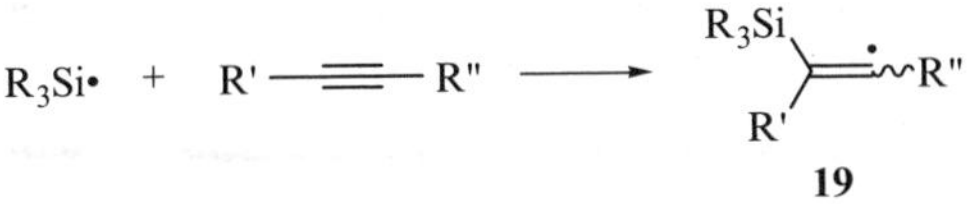

(5.16)

EPR studies of the radical adduct **19** gave an indication for a generally σ-type structure of the radicals (**20**) in which the degree of bending as well as the inversion barrier depend on the α-substituent [38]. However, a π-type structure (**21**) has been identified with certainty by EPR in the case of the vinyl intermediate derived from the addition of $Me_3Si\bullet$ radical to the appropriately silyl substituted acetylene. Analogous linear structures are highly probable for α-phenyl substituted vinyl radicals. Rate constants for the reaction of $Et_3Si\bullet$ radicals with HC≡CBu-*t* and HC≡CPh are reported to be 2.3×10^6 and 2.3×10^8 M^{-1} s^{-1} at room temperature, respectively [13]. Comparison with the olefin analogues in Table 5.1 shows that the reactivity of acetylenes towards hydrosilylation is only slightly less.

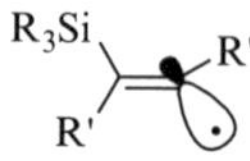

20

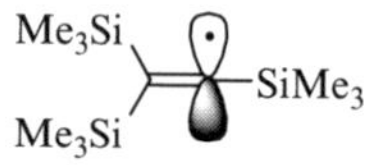

21

5.2.2 HYDROSILYLATION OF ALKYNES

The addition reaction of alkyl and/or phenyl substituted silicon hydrides to acetylenes has limitations mainly due to the hydrogen donation step (cf. Scheme 5.1). Reaction (5.17) shows that the replacement of Ph by Me_3Si group in silanes made the reaction easier, the effect being cumulative. Indeed, the reaction time decreased from 88 h for Ph_3SiH to 3 h for $(TMS)_3SiH$ [39], together with the amelioration of yields, and a slightly better *cis* stereoselectivity.

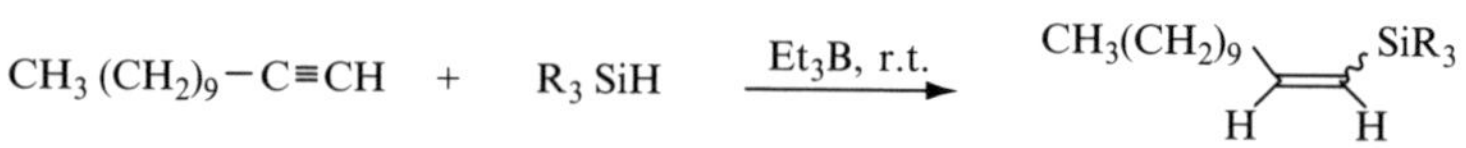

(5.17)

Ph_3SiH	42%, Z:E = 12:1
$Me_3SiSi(H)Ph_2$	78%, Z:E = 16:1
$(Me_3Si)_2Si(H)Ph$	74%, Z:E = 15:1
$(TMS)_3SiH$	98%, Z:E = 17:1

The addition of $(TMS)_3SiH$ to a number of monosubstituted acetylenes has also been studied in some detail (Reaction 5.18) [25,39]. These reactions are highly regioselective (*anti*-Markovnikov) and give terminal $(TMS)_3Si$ substituted alkenes in good yields. High *cis* or *trans* stereoselectivity is also observed, depending on the nature of the substituents at the acetylenic moiety. Normally, (Z)-alkenes were formed. This is because in σ-radicals **22** and **23** at equilibrium the bulky $(TMS)_3Si$ group hinders *syn* attack (Reaction 5.19). Instead *tert*-butylacetylene afforded the (E)-alkene, which suggests that the radical **22** is so strained that only **23** can play a role in the hydrogen abstraction step.

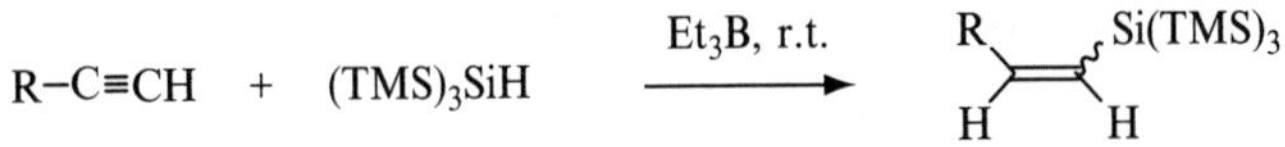

(5.18)

R = $CH_2CH_2CH_2CH_3$ — 85%, Z:E = 19:1
R = C_6H_{11} — 83%, Z:E = 24:1
R = Ph — 85%, Z:E = 99:1
R = CO_2Et — 88%, Z:E = 99:1
R = t-Bu — 88%, Z:E < 1:99

(5.19)

22 **23**

Although $(TMS)_3Si\bullet$ has been proven to isomerize alkenes (see above), the post-isomerization of the hydrosilylation adduct could not be observed due to steric hindrance. Only with $Ph_3Ge\bullet$ radical, the (Z)–(E) interconversion of $(TMS)_3Si$ substituted alkenes was achieved [39].

The addition of $(TMS)_3SiH$ to a number of 1,2-disubstituted acetylenes has also been studied [25]. Yields varied from moderate at room temperature to good at 90 °C (Reaction 5.20). The shielding effect of substituents X is the major contribution for the observed stereoselectivity.

(5.20)

X = CHO — 82%, Z:E = 1:1
X = CO_2Et — 85%, Z:E < 1:99

Also silylated cyclohexadienes succeeded as hydrosilylating agents for some alkynes [34]. Reaction (5.21) shows the silane-addition products provided from two compounds, by using the t-$BuMe_2Si$ substituted reagent. The methodology gave poor yields with other silyl reagents and failed to work at room temperature in the presence of Et_3B.

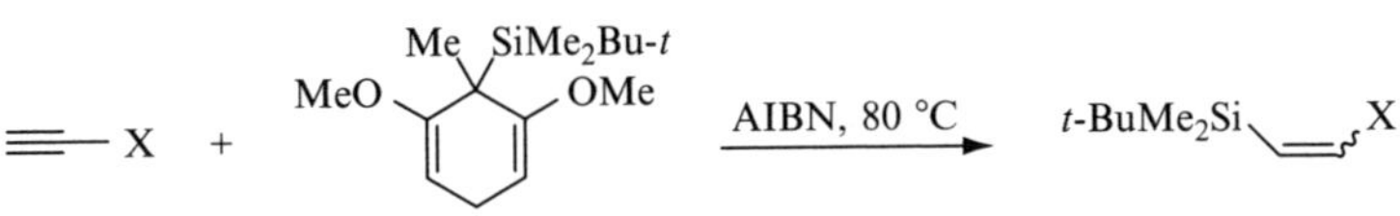

(5.21)

X = Ph — 73%, Z:E = 25:1
X = $CH_2CH_2CH_2CH_3$ — 88%, Z:E = 2:1

5.3 CARBON-OXYGEN DOUBLE BONDS

5.3.1 FORMATION OF SILYL RADICAL ADDUCTS

In the late 1950s it was established that silanes add across the carbonyl moiety of ketones to give the corresponding silyl ethers either by heating at ca 250 °C or by UV irradiation [40]. The occurrence of Reaction (5.22) as the key step in the hydrosilylation of a carbonyl group was substantiated in 1969 when the EPR spectra of the first adduct radicals were observed for acetone and trifluoroacetic acid [41]. At present, a large number of EPR parameters for this class of radicals are available and structural information has been obtained [42,43].

(5.22)

The addition of silyl radicals to α-diones [44,45] or *ortho*-quinones [46,47] presents some special features. That is, an intramolecular migration of the silyl group between the two oxygens occurs in the EPR time scale. Activation parameters for this fluxional motion have been obtained by simulating the line width alternation as the temperature changes. The E_a for the benzil adducts (**24**,**25**) of Et_3Si, Ph_3Si, and $(TMS)_3Si$ groups are 39.3, 24.7 and 36.8 kJ/mol. The preexponential factors are between $10^{11.6}$ and $10^{12.3}\,s^{-1}$, which implies negative activation entropies characterizing these migrations. Therefore, it was suggested that the reaction coordinates proceed through a cyclic transition state **26** [44,45].

24 **25**

(5.23)

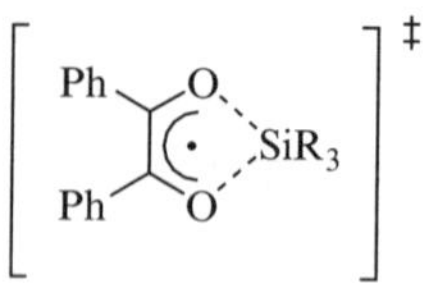

26

The addition of silyl radicals to various 2,6-disubstituted quinones takes place at two different sites, i.e., at the less hindered C=O and at the C=C double bond, the former being ca 4 times slower (Scheme 5.5) [48]. However, kinetic studies showed that radical adducts **27** are prone to rearrange to the thermodynamically more stable isomers **28** via a four-membered transition state. For $X = t$-Bu, the 1,3 carbon to oxygen shift occurs with $\log A/\mathrm{s}^{-1} = 13.8$ and $E_a = 76.1$ kJ/mol.

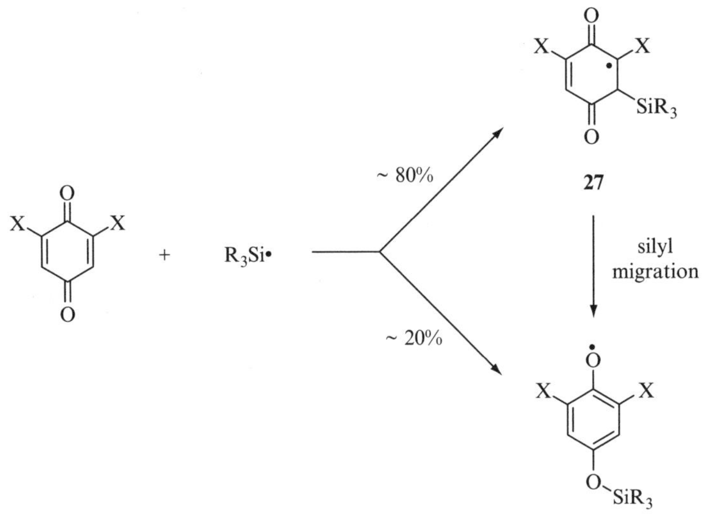

Scheme 5.5 Reaction paths for the addition of silyl radicals to 2,6-disubstituted quinones

Rate constants and Arrhenius parameters for the reaction of Et_3Si• radicals with various carbonyl compounds are available. Some data are collected in Table 5.2 [49]. The ease of addition of Et_3Si• radicals was found to decrease in the order: 1,4-benzoquinone $\geq$ cyclic diaryl ketones, benzaldehyde, benzil, perfluoro propionic anhydride > benzophenone $\gg$ alkyl aryl ketone, alkyl aldehyde > oxalate > benzoate, trifluoroacetate, anhydride > cyclic dialkyl ketone > acyclic dialkyl ketone > formate > acetate [49,50]. This order of reactivity was rationalized in terms of bond energy differences, stabilization of the radical formed, polar effects, and steric factors. Thus, a phenyl or acyl group adjacent to the carbonyl will stabilize the radical adduct whereas a perfluoroalkyl or acyloxy group next to the carbonyl moiety will enhance the contribution given by the canonical structure with a charge separation to the transition state (Equation 5.24).

$$Et_3Si\bullet\ O{=}CRR' \longleftrightarrow Et_3Si{-}O{-}\dot{C}RR' \longleftrightarrow Et_3Si^+O^-{-}\dot{C}RR' \tag{5.24}$$

The data for other silyl radicals are scant. Rate constants for $Me_3SiSi(\bullet)Me_2$ radicals with benzil and propionaldehyde are 4.4×10^7 and $2.0 \times 10^7\ M^{-1}\,s^{-1}$, respectively [14], whereas $(TMS)_3Si$• radicals are at least an order of magnitude slower than Et_3Si• radicals in the analogous additions to duroquinone ($1.0 \times 10^8\ M^{-1}s^{-1}$ at 25 °C) and 3-pentanone ($\sim 8 \times 10^4\ M^{-1}s^{-1}$ at 45 °C) [45].

Table 5.2 Kinetic data for the reaction of $Et_3Si\bullet$ radicals with a variety of carbonyl-containing compounds [49]

Substrate	$k/M^{-1}\,s^{-1}$ at 27 °C	$\log A/M^{-1}\,s^{-1}$	E_a/kJ mol^{-1}
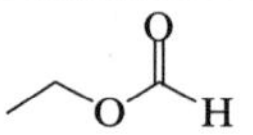	3.5×10^4	8.3	21.3
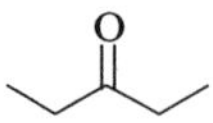	2.8×10^5		
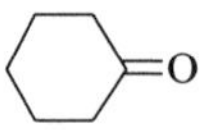	6.6×10^5		
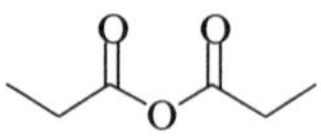	1.6×10^6	8.0	10.5
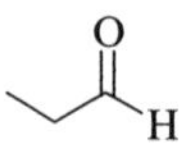	1.2×10^7	7.8	4.2
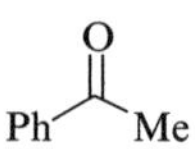	1.2×10^7	9.4	13.4
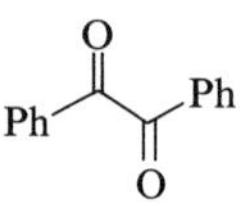	3.3×10^8	9.3	4.2
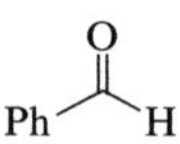	4.1×10^8		
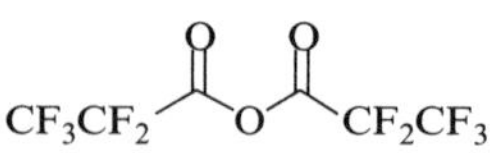	5.7×10^8	8.9	0.8
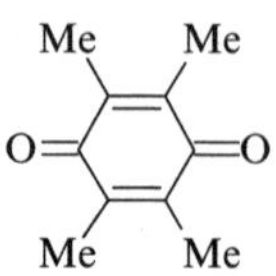	2.5×10^9		

5.3.2 HYDROSILYLATION OF CARBONYL GROUPS

The reduction of ketones with silicon hydrides has been occasionally performed by radical chemistry for a synthetic purpose. The radical adduct is stabilized by the α-silyloxyl substituent and for R_3Si (R = alkyl and/or phenyl) the hydrogen abstraction from the parent silane is much slower than a primary alkyl radical (cf. Chapter 3). On the other hand, $(TMS)_3SiH$ undergoes synthetically useful addition to the carbonyl group and the reactions with dialkyl ketones afford yields $\geq$ 70 % under standard experimental conditions, i.e., AIBN, 80–85 °C [45,51]. Reaction (5.25) shows as an example the reduction of 4-*tert*-butyl-cyclohexanone that affords mainly the *trans* isomer indicating that the axial H abstraction is favoured [51].

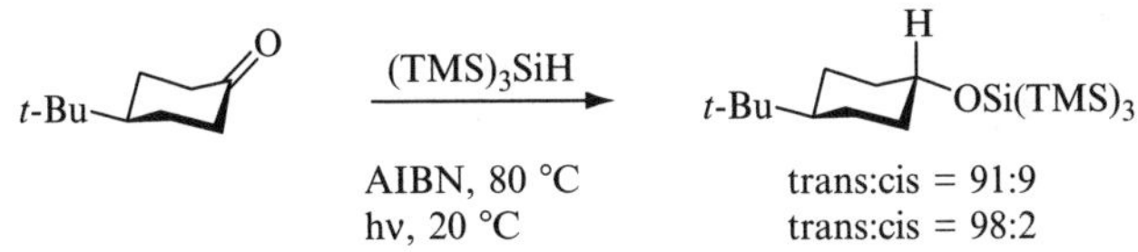

(5.25)

The stereochemical outcome of $(TMS)_3SiH$ addition to chiral ketones can be predicted by the Felkin–Ahn model. This model, usually applied to the nucleophilic addition to a carbonyl group, has been extended to the case of the H abstraction step involving the α-silyloxyalkyl radical [52–54]. Indeed, the hydrosilylation of ketone **29** with this silane leads predominatly to **30** via the intermediate radical **32**, which is bent in the ground state in accord with EPR measurements and theoretical calculations [53]. Scheme 5.6 shows for comparison the classic experiment by Cram for the reduction of ketone by $LiAlH_4$ and the radical counterpart with $(TMS)_3SiH$ in the presence of thiol as the catalyst, which furnishes the hydrogen atom. The *syn*:*anti* product ratios obtained from hydride ion (**34**:**35**) and hydrogen atom abstraction (**37**:**38**) are very similar [52]. In summary, the 1,2-stereoinduction in carbon-centred radicals bearing an α-silyloxy substituent follows the Felkin–Anh rule (*syn* product).

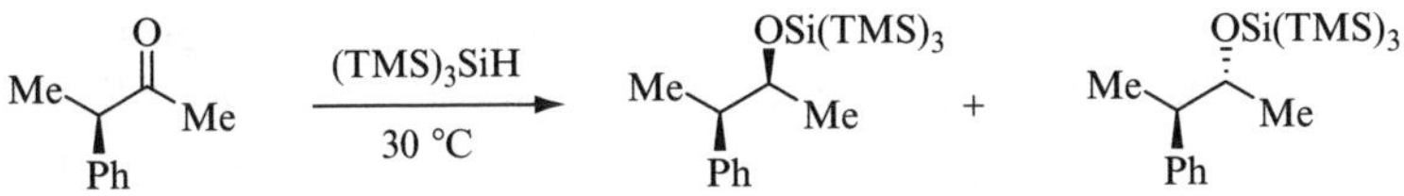

(5.26)

29 **30** **31**

64%, **30**:**31** = 4:1

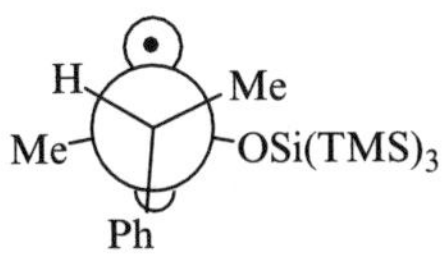

32

It has been reported that $(TMS)_3SiCl$ can be used for the protection of primary and secondary alcohols [55]. Tris(trimethylsilyl)silyl ethers are stable to the usual conditions employed in organic synthesis for the deprotection of other silyl groups and can be deprotected using photolysis at 254 nm, in yields ranging from 62 to 95%. Combining this fact with the hydrosilylation of ketones and aldehydes, a radical pathway can be drawn, which is formally equivalent to the ionic reduction of carbonyl moieties to the corresponding alcohols.

Reactions (5.27) and (5.28) provide good evidence for the participation of radical intermediates in the initial hydrosilylation process [51,56]. Indeed, the

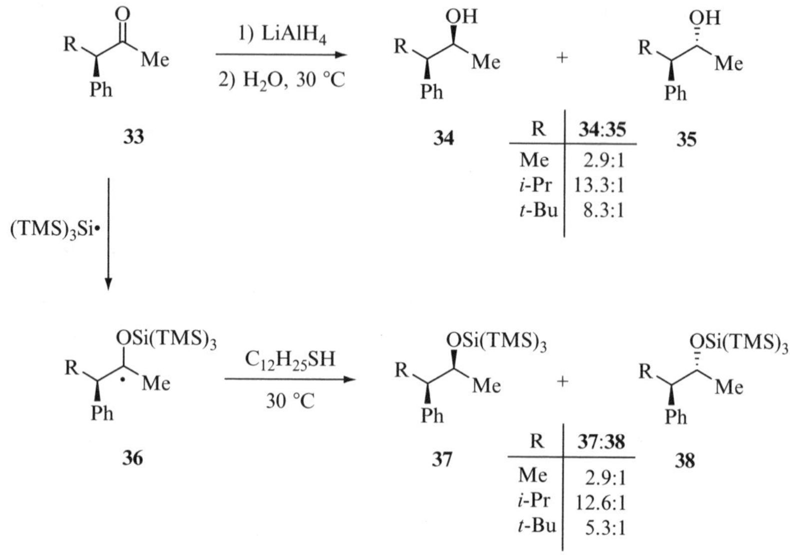

33 34 35

R	34:35
Me	2.9:1
i-Pr	13.3:1
t-Bu	8.3:1

36 37 38

R	37:38
Me	2.9:1
i-Pr	12.6:1
t-Bu	5.3:1

Scheme 5.6 Cram's rule for radical reactions

products derived from a fast ring opening of the intermediate α-silyloxyl radicals prior to hydrogen abstractions. Carbaldehyde **39** has been used as a probe for discrimination between radical and cationic mechanisms [57]. Indeed the ring-opening reaction depended on whether a radical or cationic intermediate develops at the carbonyl carbon, with consequent cleavage at the C1—C2 or C1—C3 bond, respectively [58]. For example, the reaction with $(TMS)_3SiH$ under standard radical conditions afforded a mixture of (*Z*)- and (*E*)-alkene **40** from regioselective ring opening of the C1—C2 bond (Scheme 5.7). Interestingly, the reaction of carbaldehyde **39** with tetramesityldisilene **41** at −60 °C yielded the cyclic compound **42** as the only product providing unequivocal evidence for the presence of radical intermediates in the reaction of carbonyl compounds with disilene [57].

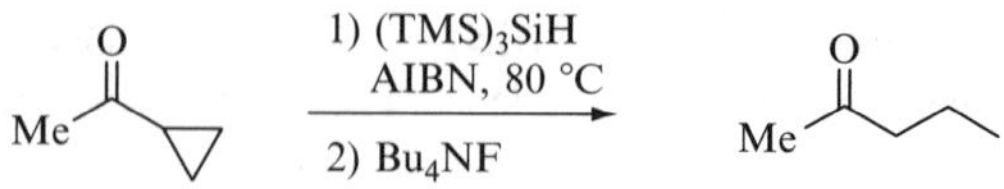

75% (5.27)

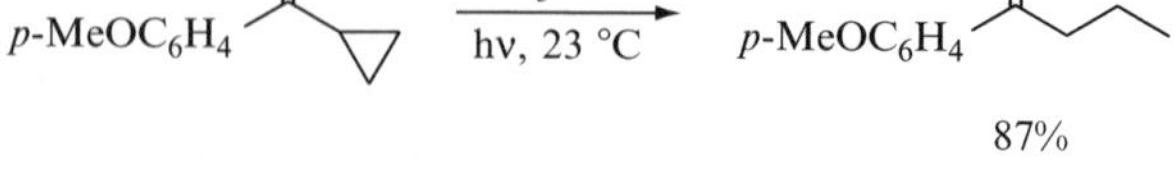

87% (5.28)

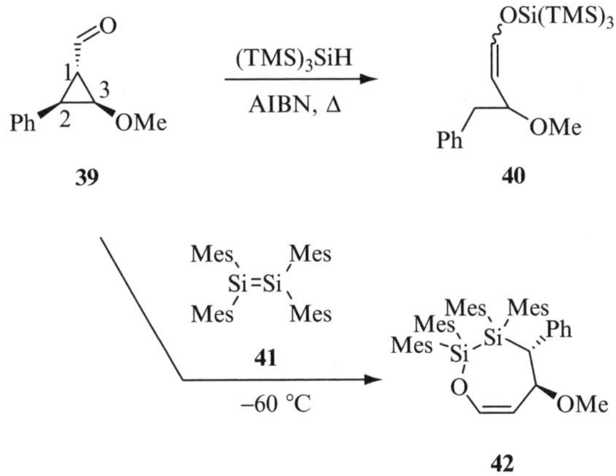

Scheme 5.7 The use of carbaldehyde **39** as a mechanistic probe

The reduction of α-hydroxy ketones **43** to 1,2-diols **44** was achieved in good yield and high diastereoselectivity in favour of the *anti* product by using Cl_3SiH and photolytic conditions [56]. In the proposed mechanism of Scheme 5.8, the addition proceeded via the intermediate adduct **45**, which participated in the stereo-controlled hydrogen donation, and via the α-silylether **46**, affording 1,2-diols in a preferential *anti* conformation.

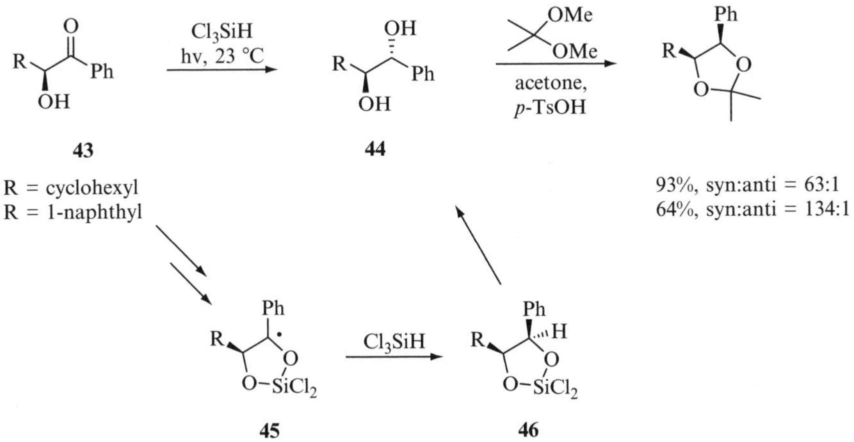

Scheme 5.8 Reduction of α-hydroxy ketones using Cl_3SiH

A variety of carbonyl compounds react with $PhSeSiMe_3$ in the presence $Bu_3SnH/AIBN$ to afford the corresponding hydrosilylation derivatives [59,60]. Generally good yields are obtained only for aromatic substituted aldehydes or ketones. Reactions (5.28) and (5.29) show this case for a few aldehydes and α-dicarbonyl compounds, respectively. The proposed reaction mechanism is summarized in Scheme 5.9 and involves the addition of $Me_3Si\bullet$ radical to the carbonyl group and subsequent hydrogen abstraction from Bu_3SnH. In turn, the $Bu_3Sn\bullet$ radical displaced the $Me_3Si\bullet$ radical from $PhSeSiMe_3$ and completes the radical reaction cycle (see Section 1.1).

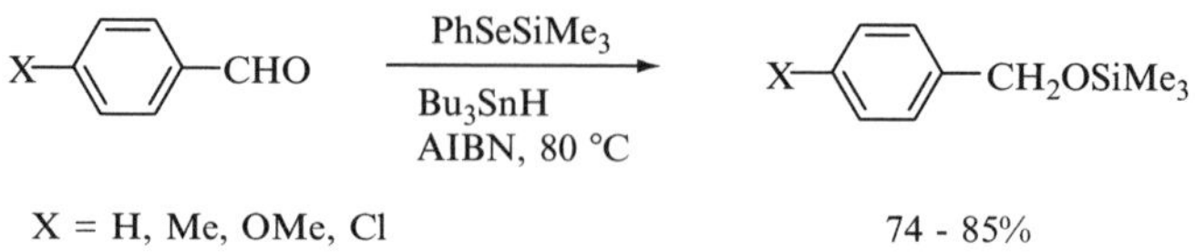

(5.29)

(5.30)

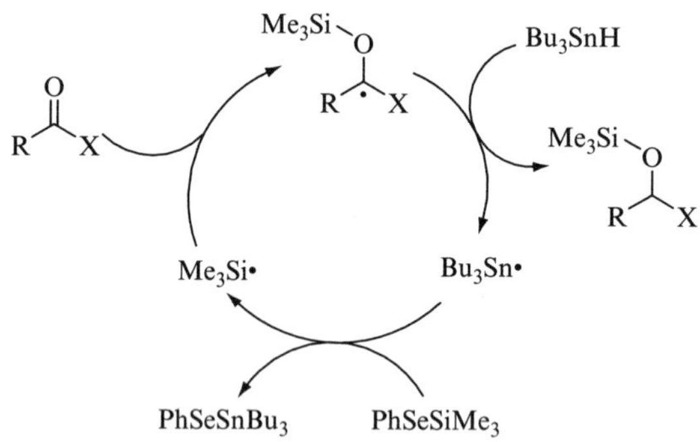

Scheme 5.9 Propagation steps for hydrosilylation of carbonyl compounds using $PhSeSiMe_3/Bu_3SnH$ system

5.3.3 RADICAL BROOK REARRANGEMENT

The 1,2 migration of the silyl group in Reaction (5.31) somehow recalls the addition of silyl radicals to carbonyl moieties. The initial α-silyl alkoxyl radical **47** could be formally one of the two possible adducts of silyl radical to the carbonyl group, although evidence has never been found for such a reaction. On the other hand, the rearranged carbon-centred radical **48** is identical to the adduct of silyl radical with the carbonyl group. In this section, we briefly describe this rearrangement, which is also called radical Brook rearrangement because it resembles the well known analogous ionic reaction [61].

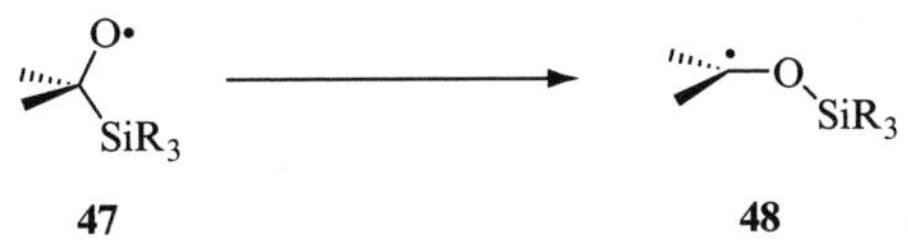

(5.31)

EPR studies showed that 1,2 migration of the silyl group in the prototype $Me_3SiCH_2O\bullet$ radical is a facile process proceeding even at −83 °C [62]. However, the reduction of bromide **49** with Bu_3SnH under standard radical conditions afforded compound **50** as the only product (Reaction 5.32), which suggests that the ring opening of the intermediate α-silyl alkoxyl radical **51** is faster than the radical Brook rearrangement [63].

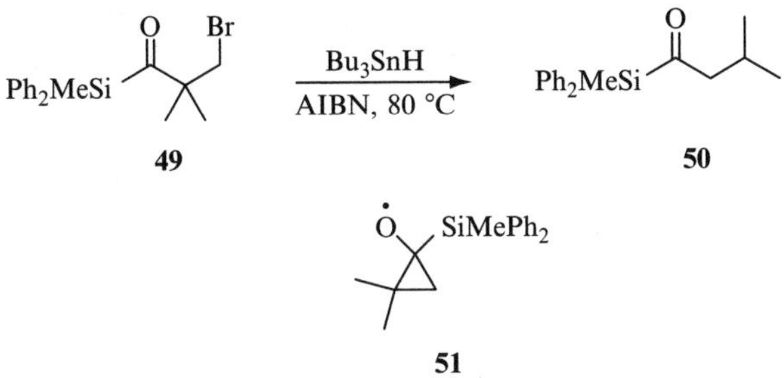

(5.32)

It is worth mentioning that in a few cases the β-elimination of the silyl radical from the α-silyl alkoxyl radical (**47**) with the formation of corresponding carbonyl derivative was observed [63,64]. Evidently the fate of α-silyl alkoxyl radical depends on the method of radical generation and/or the nature of the substrate. Two examples that delineate the potentialities of this rearrangements are reported in Reactions (5.33) and (5.34). In the former, the 5-*exo* cyclization of secondary alkyl radical on the carbonyl moiety followed by the radical Brook rearrangement afforded the cyclopentyl silyl ether [65], whereas Reaction (5.34) shows the treatment of an α-silyl alcohol with lead tetracetate to afford the mixed acetyl–silyl acetal under mild conditions [63].

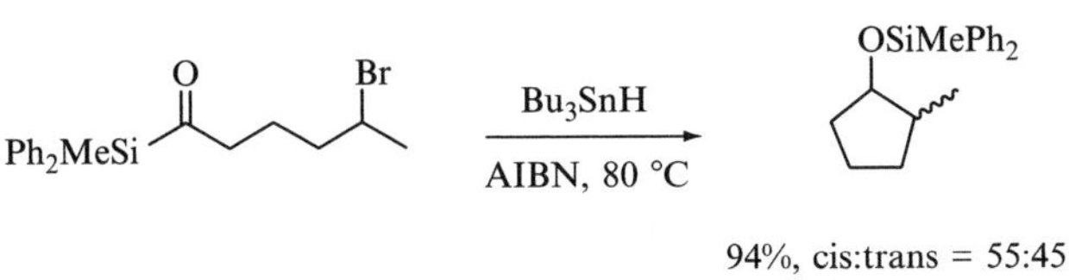

(5.33)

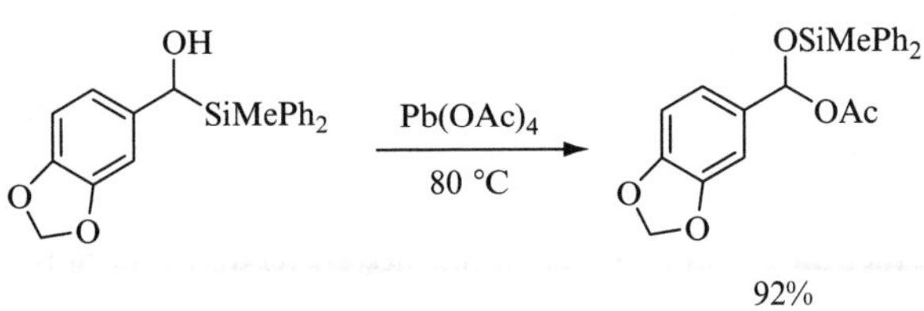

(5.34)

5.4 OTHER CARBON–HETEROATOM MULTIPLE BONDS

Addition of silyl radicals to carbon–nitrogen multiple bonds has mainly been investigated by EPR spectroscopy [9,66].

Studies on the addition of silyl radicals to compounds containing C=N bonds are quite extensive [66]. Silicon-centred radicals add to the C=N moiety either at the nitrogen or at the carbon atom depending on the nature of the substituents (Scheme 5.10). In the majority of cases, the addition at the nitrogen atom is the preferred one as is expected thermodynamically. Furthermore, it has been shown by EPR that 1,2-migration of the Me_3Si group from carbon to nitrogen in the $Me_3SiCH_2N(\bullet)R$ occurs readily (for R = H, the rate constant is estimated to be $3 \times 10^3\ s^{-1}$ at 27 °C), and it is sensitive to the presence of sterically large groups at the nitrogen atom (for R = *t*-Bu, the rate constant is estimated to be $3 \times 10^1\ s^{-1}$ at 27 °C) [67].

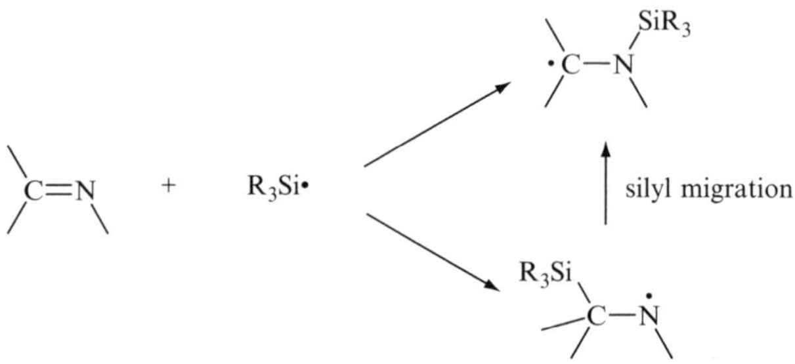

Scheme 5.10 Reaction paths for the addition of silyl radicals to C=N double bonds

The adduct of silyl radicals to 4-substituted pyridines and pyrazine monitored by EPR results from the attack at the nitrogen atom to give radicals **52** and **53**, respectively [68,69]. The rate constant for the addition of $Et_3Si\bullet$ radical to pyridine is about three times faster than for benzene (Table 5.3) [24].

52 **53**

The addition of the $Et_3Si\bullet$ radical to the C=N bond of nitrones also occurs readily. A rate constant of $7.1 \times 10^7\ M^{-1}\,s^{-1}$ at 27 °C has been obtained for Reaction (5.35), whereas information on the structure of the adduct radicals has been obtained by EPR spectroscopy [13,70].

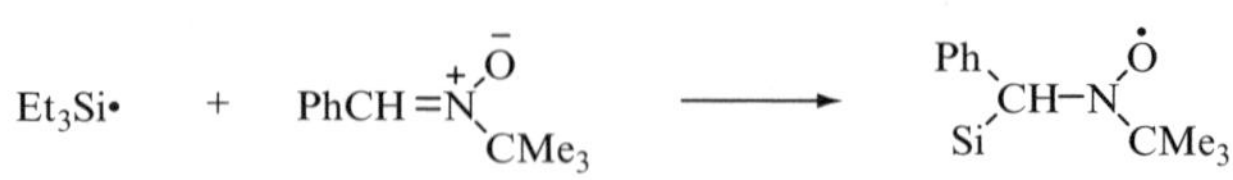

(5.35)

The EPR technique has also been employed to investigate the adducts of the reaction of silyl radicals with various nitrile-*N*-oxides [71]. As an example, the addition of $Et_3Si^\bullet$ and $Ph_3Si^\bullet$ radicals to *p*-chlorobenzonitrile *N*-oxide gives the corresponding persistent iminoxyl radicals in a *syn:anti* ratio that depends strongly on the nature of the silyl group (Reaction 5.36). Moreover, the presence of silyl groups produces a stabilization of *syn* isomer with respect to the corresponding species where the silyl group is replaced by a hydrogen atom.

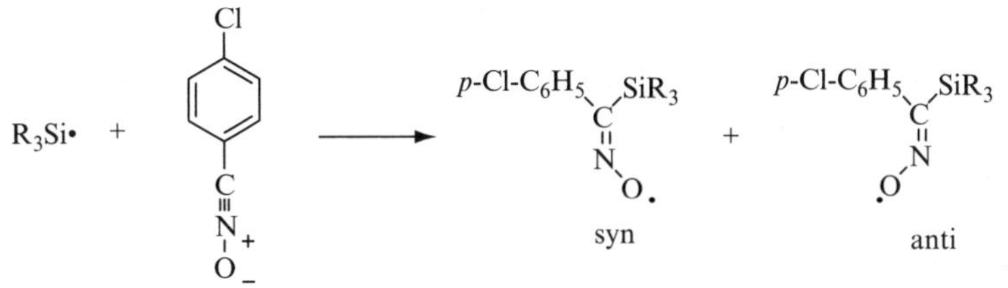

(5.36)

R = Et syn:anti = 1:1.7
R = Ph syn:anti = 1:0.7

The addition of silyl radicals to thiocarbonyl derivatives is a facile process leading to α-silylthio adducts (Reaction 5.37). This elementary reaction is the initial step of the radical chain deoxygenation of alcohols or Barton–McCombie reaction (see Section 4.3.3 for more details). However, rate constants for the formation of these adducts are limited to the value for the reaction of $(TMS)_3Si^\bullet$ radical with the xanthate c-$C_6H_{11}OC(S)SMe$ (Table 5.3), a reaction that is also found to be reversible [15]. Structural information on the α-silylthio adducts as well as some kinetic data for the decay reactions of these species have been obtained by EPR spectroscopy [9,72].

Table 5.3 Kinetic data for the addition of silyl radicals to heteroatom-containing multiple bonds

Substrate	Silyl radical	T(°C)	k/M^{-1} s^{-1}	Reference
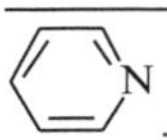	$Et_3Si^\bullet$	27	1.3×10^6	13
$PhCH{=}\overset{+}{N}(O^-)CMe_3$	$Et_3Si^\bullet$	27	7.1×10^7	13
$(RO)(MeS)C{=}S$	$(TMS)_3Si^\bullet$	21	1.1×10^9	15
t-Bu—N=C=O	$Et_3Si^\bullet$	27	5.5×10^6	13
Me—N=C=O	$Me_3Si^\bullet$	−109	2.3×10^6	77
O=C=O	$Me_3Si^\bullet$	−109	3.2×10^4	77
$Me{-}\overset{+}{N}(O^-){=}O$	$Et_3Si^\bullet$	27	4.3×10^7	13
t-Bu$-\overset{+}{N}(O^-){=}O$	$(TMS)_3Si^\bullet$	21	1.2×10^7	15

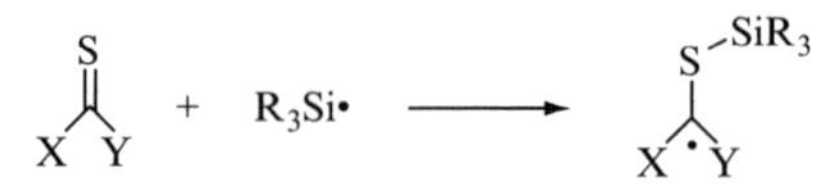

(5.37)

5.5 CUMULENES AND HETERO-CUMULENES

Addition of silyl radicals to cumulenes and their isoelectronic derivatives has mainly been studied by EPR spectroscopy. The adducts of Me_3Si• radical with the two substituted allenes **54** and **55** have been recorded [73,74]. The attack occurs at the central atom affording unconjugated allyl-type radicals. In particular the adduct radical with **55** has been described as a very persistent 'perpendicular' allyl radical [74].

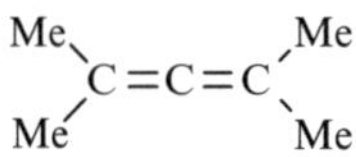

54

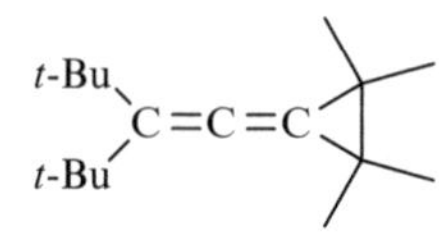

55

A theoretical study of the reactivity of prototype ketene $CH_2{=}C{=}O$ with several radicals including H_3Si• has been reported. Scheme 5.11 shows that three adducts are possible with the corresponding energy barriers for their formation calculated at B3LYP/6-31G* level of theory [75]. However, all the levels of theory used in this study predicted that silyl radical prefers to add to the carbon terminus of ketene.

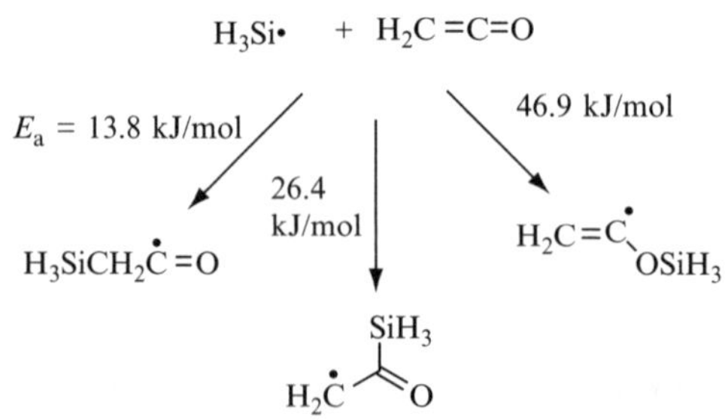

Scheme 5.11 Theoretical study of the reactivity of ketene

Trialkylsilyl radicals add to alkyl isocyanate to form imidoyl radicals **56** (Reaction 5.38). Detailed EPR studies established intermediates **56** to be strongly bent at the carbon bearing the unpaired electron [76]. The absolute rate constant for the reaction of Et_3Si• radical with *tert*-butyl isocyanate was found to be $5.5 \times 10^6\ M^{-1}\,s^{-1}$ at 27 °C [13], whereas the relative rate of the addition of Me_3Si• radicals to alkyl isocyanates was found to decrease in the order R = Me > Et > *i*-Pr > *t*-Bu at −109 °C, and this trend was proposed to be steric in origin [76]. In a similar manner, silyl radicals were found to add to carbon dioxide to give the radical adduct **57** (Reaction 5.39) [77]. The rate constant for the addition of Me_3Si• radical to MeN=C=O is two orders of magnitude faster than the addition to O=C=O at −109 °C (see Table 5.3).

$R_3Si\bullet$ + R'N=C=O ⟶ 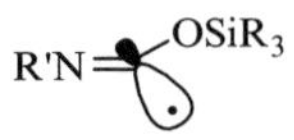 (5.38)

56

$R_3Si\bullet$ + O=C=O ⟶ 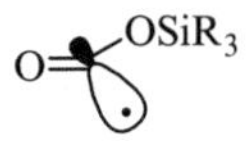 (5.39)

57

The addition of trisubstituted silanes to carbonyl sulfide has been applied to the synthesis of the corresponding silanethiol derivatives (Reaction 5.40) [78]. In Scheme 5.12 the mechanism is depicted, starting from the silyl radical attack to the sulfur atom of O=C=S and ejection of carbon monoxide. The resulting silanethiyl radical abstracts hydrogen from the starting silane, to give the silanethiol and to generate 'fresh' silyl radical (see Section 3.4).

$$R_3SiH + S{=}C{=}O \xrightarrow[60\ ^\circ C]{t\text{-BuONNOBu-}t} R_3SiSH \qquad (5.40)$$

R = alkyl and/or aryl > 60%

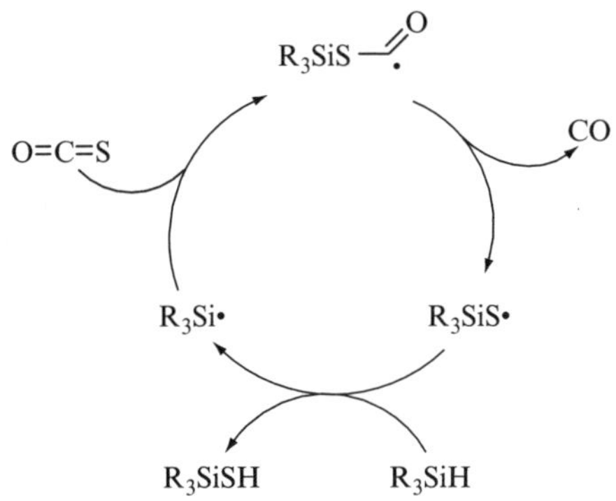

Scheme 5.12 Propagation steps for the formation of silanethiols

5.6 HETEROATOM–HETEROATOM MULTIPLE BONDS

The addition of silyl radicals to nitrogen–nitrogen and nitrogen–heteroatom double bonds has mainly been investigated by EPR spectroscopy.

The adducts of azodicarboxylate with $Et_3Si\bullet$ or $Ph_3Si\bullet$ radicals [79], have been identified as hydrazyl radicals (Reaction 5.41). The addition of $R_3Si\bullet$ radical to alkyl azides takes place at the terminal nitrogen (Reaction 5.42) affording the triazenyl σ-type radical in which the SOMO is in the NNN plane (only one canonical structure is shown) [80,81].

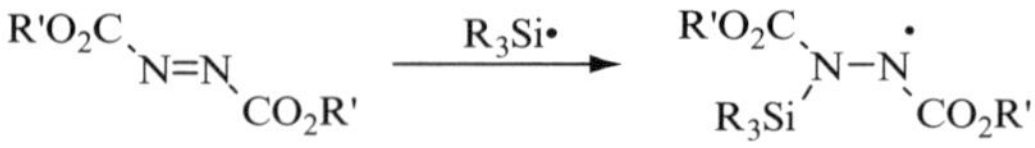

(5.41)

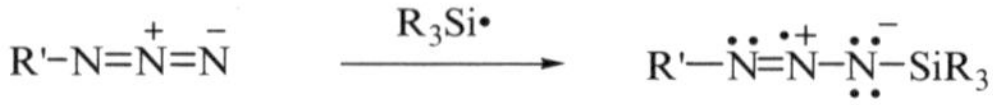

(5.42)

The adducts derived from the reaction of a variety of alkyl or alkoxyl substituted silyl radicals with aminoboranes have also been recorded [82]. The silyl radical addition takes place at the boron site to give the aminyl-borane radical (Reaction 5.43) and structural information for this class of radicals has been obtained.

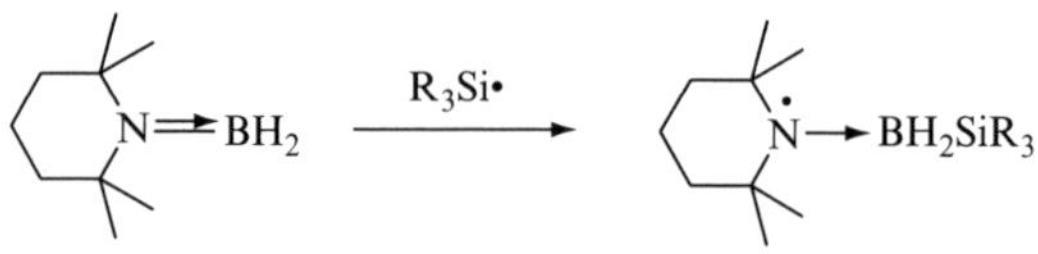

(5.43)

The addition of silyl radicals to the N═O bond is very well documented. Indeed, silyl radicals add rapidly to aromatic, heteroaromatic and aliphatic nitro compounds to give persistent trialkylsilyloxy nitroxide radicals [9]. Rate constants for the addition of $Et_3Si\bullet$ radical to $MeNO_2$ [13] and of $(TMS)_3Si\bullet$ radical to t-$BuNO_2$ [15] are in the range of 10^7–10^8 $M^{-1}\,s^{-1}$, respectively, at room temperature (Table 5.3). Structural information about a large number of adduct radicals as well as some kinetic information associated with these species have been obtained by EPR spectroscopy [83,84]. For example, $Et_3Si\bullet$ radicals add to 2-nitrothiophene to give the silyloxynitroxide that was detected both in the *cis* and *trans* arrangements (Reaction 5.44). For the interconversion of the two conformers, an activation energy of 49.3 kJ/mol was obtained [85].

(5.44)

Detailed kinetic studies as a function of temperature have been carried out on the fate of radicals **58** and **59**, derived from the addition of $Ph_3Si\bullet$ and $(TMS)_3Si\bullet$ to t-$BuNO_2$ and $MeNO_2$, respectively [86,87]. Activation parameters for the first-order decay of radical **58** ($\log A/s^{-1} = 11.1$ and $E_a = 78.7$ kJ/mol) and radical **59** ($\log A/s^{-1} = 13.5$ and $E_a = 80.5$ kJ/mol) were obtained that correspond to rate constants of 0.002 and $0.2\,s^{-1}$, respectively, at room temperature. Evidence that nitroxide **58** fragments preferentially at the nitrogen–carbon bond [86] and nitroxide **59** fragments at the nitrogen–oxygen bond [87] are based on the detection of secondary adducts.

58 **59**

An Arrhenius expression for the reaction of $Me_3Si\bullet$ radical with molecular oxygen is available from the gas-phase kinetics, from which a rate constant of ca $1 \times 10^{10}\,M^{-1}\,s^{-1}$ at room temperature can be calculated [88].

The reaction of molecular oxygen with a variety of silyl radicals ($Me_3Si\bullet$, $Et_3Si\bullet$, n-$Bu_3Si\bullet$, t-$Bu_3Si\bullet$, $Ph_2MeSi\bullet$, and $Ph_3Si\bullet$) has been investigated by EPR spectroscopy [89]. Based on ^{17}O-labelling experiments, it was found that the structure of t-$Bu_3SiO_2\bullet$ resembles the structure of alkylperoxyl radicals, i.e., the two oxygen nuclei are magnetically nonequivalent (Reaction 5.45), and this radical has more π spin density on the terminal oxygen than alkylperoxyl radicals. Furthermore, trialkylsilylperoxyl radicals exist in equilibrium with a tetraoxide (Reaction 5.46) at temperatures below $-40\,°C$ with $\Delta H^\circ = -46 \pm 8\,kJ/mol$ and $\Delta S^\circ < -30\,cal\,K^{-1}\,mol^{-1}$.

$$R_3Si\bullet + O_2 \longrightarrow R_3Si{-}O{-}O^\bullet \qquad (5.45)$$

$$2\,R_3Si{-}O{-}O^\bullet \rightleftharpoons R_3Si{-}O{-}O{-}O{-}O{-}SiR_3 \qquad (5.46)$$

t-$BuMe_2SiH$ was found to undergo a radical-initiated oxidation to the corresponding hydroperoxide and this represents the first case of a silane oxidation that resembles a hydrocarbon oxidation [90]. Kinetic studies carried out by following the oxygen uptake allowed the oxidizability of t-$BuMe_2SiH$ to be obtained in the temperature range of 295–350 K. The reaction mechanism is shown in Scheme 5.13.

A variety of alkyl and/or phenyl substituted silicon hydrides are found to be oxidized to the corresponding silanols in very good yields in the presence of molecular oxygen and N-hydroxyphthalimide (**60**) and Co(II) catalysis (Reaction 5.47) [91, 92]. It is worth mentioning that n-Bu_3SiH is essentially inert under conditions by which Et_3SiH affords silanol in an 87 % yield and steric effects are invoked to play an important role. From a mechanistic point of view, it was suggested that silylperoxyl radicals, derived from the addition of silyl radicals to oxygen, abstract rapidly hydrogen from **60** ($k \sim 7 \times 10^3\,M^{-1}\,s^{-1}$ at room temperature) to give the corresponding nitroxide, which in turn abstracts hydrogen from the starting silane and completes the chain (Reaction 5.48). Since the DH(O—H) in **60** is 369 kJ/mol [92], Reaction (5.48) is expected to be nearly thermoneutral for Ph_3SiH and strongly endothermic ($\Delta H_r = -29\,kJ/mol$)

Initiation steps:

$$\text{AIBN} \xrightarrow{\Delta,\ O_2} \text{ROO}\bullet \xrightarrow{t\text{-BuMe}_2\text{SiH}} t\text{-BuMe}_2\text{Si}\bullet$$

Propagation steps:

$$t\text{-BuMe}_2\text{Si}\bullet + O_2 \longrightarrow t\text{-BuMe}_2\text{SiOO}\bullet$$

$$t\text{-BuMe}_2\text{SiOO}\bullet + t\text{-BuMe}_2\text{SiH} \longrightarrow t\text{-BuMe}_2\text{SiOOH} + t\text{-BuMe}_2\text{Si}\bullet$$

Termination steps:

$$2\ t\text{-BuMe}_2\text{SiOO}\bullet \longrightarrow \text{non-radical products}$$

Scheme 5.13 Reaction mechanism of the oxidation of t-$BuMe_2SiH$

for trialkylsilyl hydrides (cf. Chapter 2). Finally, the resulting silanol is suggested to derive from the reaction of silylhydroperoxide with Co(II).

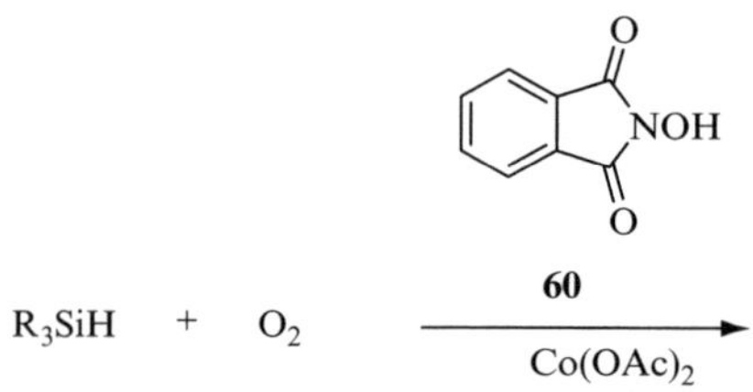

(5.47)

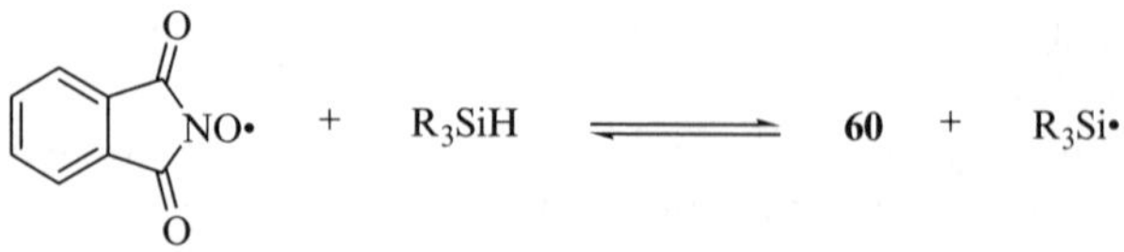

(5.48)

Hydroxysilylation of alkenes with R_3SiH under molecular oxygen was also successfully achieved using **60** as catalyst [93]. Reaction (5.49) shows two examples using Et_3SiH where the conversion is ca. 60% and selectivity higher than 90%. Similar results were obtained with Ph_3SiH whereas $(TMS)_3SiH$ afforded only hydrosilylation products. It was suggested that Reaction (5.48) is one of the propagation steps. Consecutive additions of silyl radical to alkene and of the adduct radical to oxygen afford a peroxyl radical that it expected to be the precursor of the isolated product.

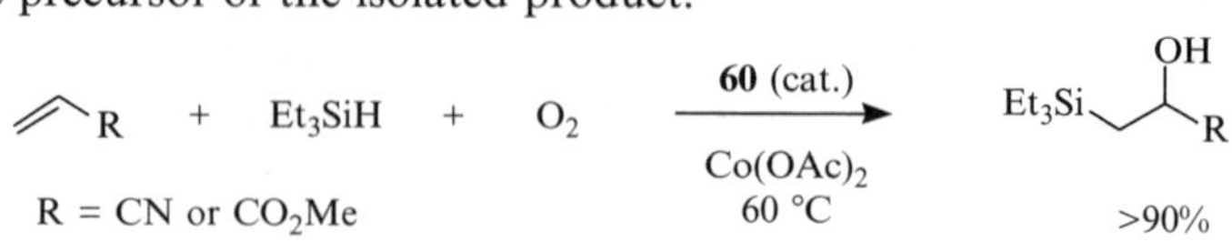

(5.49)

In connection with the above procedure, Co(III)-alkyl and Co(III)-alkylperoxo complexes were found to catalyse triethylsilylperoxidation of alkenes with O_2 and Et_3SiH (Reaction 5.50). Although the proposed radical chain mechanism does not involve Et_3SiOO• radical, this species is believed to play a role in the initiation step [94].

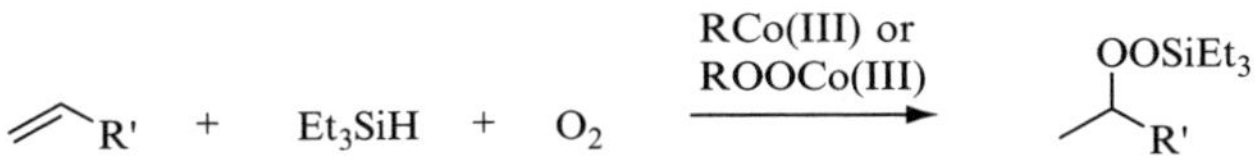

(5.50)

$(TMS)_3SiH$ and poly(hydrosilane)s show autoxidation to form polysiloxanes. ^{18}O-Labelling experiments with $(TMS)_3SiH$ and $(TMS)_2MeSiH$ allowed for envisaging an operating mechanism that involves consecutive rearrangements (see Sections 6.5 and 8.2 for details).

The addition of silyl radicals to P=S and P=Se bonds have been used for synthetic purposes. A variety of silicon hydrides react with $Ph_3P{=}S$ by a radical chain mechanism to give the corresponding silanethiols in good yields (Reaction 5.51) [95]. The reaction mechanism is similar to the reaction of silanes with O=C=S reported in Scheme 5.12. The intermediate phosphoranyl adduct **61** was identified by EPR and the data obtained suggest that the unpaired electron reside in the PS σ^* orbital. $(TMS)_3SiH$ reacts with phosphine sulfides and phosphine selenides under free-radical conditions to give the corresponding phosphines in good yields [96]. The initial step of these reactions is the addition of $(TMS)_3Si$• radical to P=S and P=Se bonds. For details and stereochemical studies on P-chiral compounds, see Section 4.3.4.

$$Ph_3P{=}S + R_3SiH \longrightarrow Ph_3P + R_3SiSH \qquad (5.51)$$

$$Ph_3\dot{P}{-}S{-}Si(OCMe_3)_3$$

61

5.7 REFERENCES

1. Sommer, L.H., Pietrusza, E.W., and Whitmore, F.C., *J. Am. Chem. Soc.*, 1947, **69**, 188. Burkhard, C.A., and Krieble, R.H., *J. Am. Chem. Soc.*, 1947, **69**, 2687. Barry, A.J., De Pree, L., Gilkey, J.W., and Hook, D.E., *J. Am. Chem. Soc.*, 1947, **69**, 2916.
2. *Comprehensive Handbook on Hydrosilylation*, B. Marciniec, (Ed.), Pergamon, Oxford, 1992.
3. El-Durini, N.M.K., and Jackson, R.A., *J. Organomet. Chem.*, 1982, **232**, 117.
4. Speier, J.L., *Adv. Organomet. Chem.*, 1979, **17**, 407.
5. Ojima, I. In *The Chemistry of Organic Silicon Compounds*, S. Patai and Z. Rappoport (Eds), Wiley, Chichester, 1989, pp. 1479–1526.
6. Ojima, I., Li, Z., and Zhu, J. In *The Chemistry of Organic Silicon Compounds*, Volume 2, Z. Rappoport and Y. Apeloig (Eds), Wiley, Chichester, 1998, pp. 1687–1792.
7. Marciniec, B., *Silicon Chem.*, 2002, **1**, 155.

8. Sakurai, H. In *Free Radicals*, Volume II, J.K. Kochi (Ed.), Wiley, New York, 1973, pp. 741–808.
9. Alberti, A., and Pedulli, G.F., *Rev. Chem. Intermed.*, 1987, **8**, 207.
10. Dohmaru, T. In *Chemical Kinetics of Small Organic Radicals*, Volume III, Z.B. Alfassi (Ed.), CRC Press, Boca Raton, FL, 1988, pp. 165–195.
11. Sakurai, H., Kyushin, S., Nakadaira, Y., and Kira, M., *J. Phys. Org. Chem.*, 1988, **1**, 197.
12. Guerra, M., *J. Am. Chem. Soc.*, 1992, **114**, 2077.
13. Chatgilialoglu, C., Ingold, K.U., and Scaiano, J.C., *J. Am. Chem. Soc.*, 1983, **105**, 3292.
14. Lusztyk, J., Maillard, B., and Ingold, K.U., *J. Org. Chem.*, 1986, **51**, 2457.
15. Ballestri, M., Chatgilialoglu, C., Clark, K.B., Griller, D., Giese, B., and Kopping, B., *J. Org. Chem.*, 1991, **56**, 678.
16. Arthur, N.L., and Potzinger, P., *Organometallics*, 2002, **21**, 2874.
17. Lozano, A.E., Alonso, A., Catalina, F., and Peinado, C., *Macromol. Theory Simul.*, 1999, **8**, 93.
18. Jackson, R.A., *J. Chem. Soc., Chem. Commun.*, 1974, 573.
19. Chatgilialoglu, C., Ballestri, M., Ferreri, C., and Vecchi, D., *J. Org. Chem.*, 1995, **60**, 3826.
20. Chatgilialoglu, C., Altieri, A., and Fischer, H., *J. Am. Chem. Soc.*, 2002, **124**, 12816.
21. Griller, D., Marriott, P.R., Nonhebel, D.C., Perkins, M.J., and Wong, P.C., *J. Am. Chem. Soc.*, 1981, **103**, 7761.
22. Kira, M., Sugiyama, H., and Sakurai, H., *J. Am. Chem. Soc.*, 1983, **105**, 6436.
23. Lunazzi, L., Placucci, G., and Grossi, L., *J. Chem. Soc., Perkin Trans. 2,* 1982, 875.
24. Chatgilialoglu, C., Ingold, K.U., Lusztyk, J., Nazran, A.S., and Scaiano, J.C., *Organometallics*, 1983, **2**, 1332.
25. Kopping, B., Chatgilialoglu, C., Zehnder M., and Giese, B., *J. Org. Chem.*, 1992, **57**, 3994.
26. Smadja, W., Zahouily, M., and Malacria, M., *Tetrahedron Lett.*, 1992, **33**, 5511.
27. Roberts, B.P., *Chem. Soc. Rev.*, 1999, **28**, 25.
28. Dang, H.-S., and Roberts, B.P., *Tetrahedron Lett.*, 1995, **36**, 2875.
29. Smadja, W., Zahouily, M., Journet, M., and Malacria, M., *Tetrahedron Lett.*, 1991, **32**, 3683.
30. Haque, M.B., and Roberts, B.P., *Tetrahedron Lett.*, 1996, **37**, 9123.
31. Haque, M.B., Roberts, B.P., and Tocher, D.A. *J. Chem. Soc., Perkin Trans. 1*, 1998, 2881.
32. Du, W., Kaskar, B., Blumbergs, P., Subramanian, P.-K., and Curran, D.P., *Bioorg. Med. Chem.*, 2003, **11**, 451.
33. Chatgilialoglu, C., Ballestri, M., Vecchi, D., and Curran, D.P., *Tetrahedron Lett.*, 1996, **37**, 6383.
34. Amrein, S., and Studer, A., *Helv. Chim. Acta*, 2002, **85**, 3559.
35. Matsumoto, A., and Ito, Y., *J. Org. Chem.*, 2000, **65**, 5707.
36. Wiberg, K.B., and Waddell S.T., *J. Am. Chem. Soc.*, 1990, **112**, 2184.
37. Scaiano, J.C., and McGarry, P.F., *Can. J. Chem.*, 1998, **76**, 1474.
38. Chatgilialoglu, C., and Ferreri, C. In *Supplement C2: The Chemistry of Triple-Bonded Functional Groups*, S. Patai (Ed.), Wiley, Chichester, 1994, pp. 917–944.
39. Miura, K., Oshima, K., and Utimoto, K., *Bull. Chem. Soc. Jpn.*, 1993, **66**, 2356.
40. Gilman, H., and Wittenberg, D., *J. Org. Chem.*, 1958, **23**, 501.
41. Hudson, A., and Jackson, R.A., *Chem. Commun.*, 1969, 1323.
42. Fischer, H., and Paul, H. In *Magnetic Properties of Free Radicals*, H. Fischer and K.-H. Hellwege (Eds) Landolt–Börnstein New Series, Volume II/9b, Springer, Berlin, 1977.

43. Neugebauer, F.A. In *Magnetic Properties of Free Radicals*, H. Fischer (Ed.), Landolt–Börnstein New Series, Volume II/17b, Springer, Berlin, 1987.
44. Alberti, A., and Hudson, A., *Chem. Phys. Lett.*, 1977, **48**, 331.
45. Alberti, A., and Chatgilialoglu, C., *Tetrahedron*, 1990, **46**, 3963.
46. Prokofev, A.I., Prokofeva, T.I., Belostotskaya, I.S., Bubnov, N.N., Solodovnikov, S.P., Irshov, V.V., and Kabachnik, M.I., *Tetrahedron*, 1979, **35**, 2471.
47. Alberti, A., Hudson, A., and Pedulli, G.F., *J. Chem. Soc., Faraday 2*, 1980, **76**, 948.
48. Alberti, A., Chatgilialoglu, C., Pedulli, G.F., and Zanirato, P., *J. Am. Chem. Soc.*, 1986, **108**, 4993.
49. Chatgilialoglu, C., Ingold, K.U., and Scaiano, J.C., *J. Am. Chem. Soc.*, 1982, **104**, 5119.
50. Cooper, J., Hudson, A., and Jackson, R.A. *J. Chem. Soc., Perkin Trans. 2*, 1973, 1933.
51. Kulicke, K.J., and Giese, B., *Synlett*, 1990, 91.
52. Giese, B., Damm, W., Dickhaut, J., Wetterich, F., Sun, S., and Curran, D.P., *Tetrahedron Lett.*, 1991, **32**, 6097.
53. Damm, W., Dickhaut, J., Wetterich, F., and Giese, B., *Tetrahedron Lett.*, 1993, **34**, 431.
54. Giese, B., Bulliard, M., Dickhaut, J., Halbach, R., Hassler, C., Hoffmann, U., Hinzen, B., and Senn, M., *Synlett*, 1995, 116.
55. Brook, M.A., Balduzzi, S., Mohamed, M., and Gottardo, C., *Tetrahedron*, 1999, **55**, 10027.
56. Enholm, E.J., and Schulte II, J.P., *J. Org. Chem.*, 1999, **64**, 2610.
57. Samuel, M.S., Jenkins, H.A., Hughes, D.W., and Baines, K.M., *Organometallics*, 2003, **22**, 1603.
58. Newcomb, M., and Toy, P.H., *Acc. Chem. Res.*, 2000, **33**, 449.
59. Nishiyama, Y., Kajimoto, H., Kotani, K., and Sonoda, N., *Org. Lett.*, 2001, **3**, 3087.
60. Nishiyama, Y., Kajimoto, H., Kotani, K., Nishida, T., and Sonoda, N., *J. Org. Chem.*, 2002, **67**, 5696.
61. Brook, A.G., *Acc. Chem. Res.*, 1974, **80**, 1866.
62. Harris, J.M., MacInnes, I., Walton, J.C., and Maillard, B., *J. Organomet. Chem.*, 1991, **403**, C25.
63. Paredes, M.D., and Alonso, R., *J. Org. Chem.*, 2000, **65**, 2292.
64. Robertson, J., and Burrows, J.N., *Tetrahedron Lett.*, 1994, **35**, 3777.
65. Chang, S.-Y., Jiaang, W.-T., Cherng, C.-D., Tang, K.-H., Huang, C.-H., and Tsai, Y.-M., *J. Org. Chem.*, 1997, **62**, 9089.
66. Chatgilialoglu, C., *Chem. Rev.*, 1995, **95**, 1229.
67. Harris, J.M., Walton, J.C., Maillard, B., Grelier, S., and Picard, J.-P., *J. Chem. Soc., Perkin Trans. 2*, 1993, 2119.
68. Alberti, A., and Pedulli, G.F., *Tetrahedron Lett.* 1978, 3283.
69. Alberti, A., and Pedulli, G.F., *J. Organomet. Chem.*, 1983, **248**, 261.
70. Haire, D.L., Oehler, U.M., Krygsman, P.H., and Janzen, E.G., *J. Org. Chem.*, 1988, **53**, 4535.
71. Alberti, A., Barbaro, G., Battaglia, A., Guerra, M., Bernardi, F., Dondoni, A., and Pedulli, G.F., *J. Org. Chem.*, 1981, **46**, 742.
72. Alberti, A., Benaglia, M., Bonini, B.F., and Pedulli, G. F., *J. Chem. Soc., Faraday Trans. 1*, 1988, **84**, 3358.
73. Davis, W.H., Jr., and Kochi, J.K., *Tetrahedron Lett.*, 1976, 1761.
74. Griller, D., Cooper, J.W., and Ingold, K.U. *J. Am. Chem. Soc.*, 1975, **97**, 4269.
75. Sung, K., and Tidwell, T.T. *J. Org. Chem.*, 1998, **63**, 9690.
76. Baban, J.A., Cook, M.D., and Roberts, B.P. *J. Chem. Soc., Perkin Trans. 2*, 1982, 1247.
77. Johnson, K.M., and Roberts, B.P. *J. Chem. Res. (S)*, 1989, 352.

78. Cai, Y., and Roberts, B.P. *Tetrahedron Lett.*, 2001, **42**, 763.
79. Roberts, B.P., and Winter, J.N. *Tetrahedron Lett.*, 1979, 3575.
80. Roberts, B.P., and Winter, J.N. *J. Chem. Soc., Perkin Trans. 2*, 1979, 1353.
81. Brand, J.C., and Roberts, B.P. *J. Chem. Soc., Perkin Trans. 2*, 1982, 1549.
82. Green, I.G., Johnson, K.M., and Roberts, B.P. *J. Chem. Soc., Perkin Trans. 2*, 1989, 1963.
83. Forrester, A.R. In *Magnetic Properties of Free Radicals*, H. Fischer and K.-H. Hellwege (Eds.), Landolt–Börnstein New Series, Volume II/9c1, Springer, Berlin, 1979.
84. Forrester, A.R. In *Magnetic Properties of Free Radicals*, H. Fischer (Ed.), Landolt–Börnstein New Series, Volume II/17d1, Springer, Berlin 1989.
85. Camaggi, C.M., Lunazzi, L., Pedulli, G.F., Placucci, G., and Tiecco, M. *J. Chem. Soc., Perkin Trans. 2*, 1974, 1226.
86. Lucarini, M., Pedulli, G.F., Alberti, A., and Benaglia, M. *J. Am. Chem. Soc.*, 1992, **114**, 9603.
87. Ballestri, M., Chatgilialoglu, C., Lucarini, and M., Pedulli, G.F. *J. Org. Chem.*, 1992, **57**, 948.
88. Niiranen, J.T., and Gutman, D. *J. Phys. Chem.*, 1993, **97**, 4106.
89. Howard, J.A., Tait, J.C., and Tong, S.B. *Can. J. Chem.*, 1979, **57**, 2761.
90. Timokhin, V.I., Lutsyk, D.A., Zaborovskii, A.B., Prystansky, R.E., and Chatgilialoglu, C. *Russ. J. Gen. Chem.*, 2003, in press.
91. Minisci, F., Recupero, F., Punta, C., Guidarini, C., Fontana, F., and Pedulli, G.F. *Synlett*, 2002, 1173.
92. Minisci, F., Recupero, F., Pedulli, G.F. and Lucarini, M., *J. Mol. Catal. A: Chem.*, 2003, **204/205**, 63.
93. Tayama, O., Iwahama, T., Sakaguchi, S., and Ishii, Y., *Eur. J. Org. Chem.*, 2003, 2286.
94. Tokuyasu, T., Kunikawa, S., Masuyama, A., and Nojima, M. *Org. Lett.* 2002, **4**, 3595.
95. Cai, Y., and Roberts, B.P. *Tetrahedron Lett.*, 2001, **42**, 4581.
96. Romeo, R., Wozniak, L.A., and Chatgilialoglu, C. *Tetrahedron Lett.*, 2000, **41**, 9899.

6 Unimolecular Reactions

6.1 CYCLIZATION REACTIONS OF SILYL RADICALS

The importance of carbon-centred radical cyclizations in organic chemistry has been documented in the large number of papers published each year and numerous reviews and books dealing with this subject. In Chapter 7 the reader can find a collection of such processes mediated by organosilanes. The silicon-centred radical cyclizations have instead received very little attention, although there has been a growing interest in silicon-containing compounds from a synthetic point of view, due to their versatility and applicability to material science. As we shall see, this area of research is very active and some recent examples show the potentiality of silyl radical cyclization in the construction of complex molecules.

Early work was focused to establish the preference for *exo*- vs *endo*-mode of cyclization. However, the absence of an effective method for generation of alkyl and/or aryl substituted silyl radicals made this task difficult. The reaction of prototype alkanesilane **1** with thermally generated *t*-BuO• radicals at 145 °C after 4 h afforded a 48 % yield of unreacted starting material and 19 % yield of a six-membered cyclic product (Scheme 6.1) [1]. Moreover, EPR studies of the same reaction recorded the spectra at temperatures between −30 and 0 °C, which were identified as the superimposition of two species having allylic-type (**2**) and six-membered ring (**3**) structures, respectively [2]. At higher temperatures radical **2** predominates; therefore, the low yield detected in the product studies could derive from the extensive *t*-BuO• attack on the allylic hydrogens.

On the other hand, product studies of volatile materials for the reaction of silane **4** (no allylic hydrogen available) with thermally generated *t*-BuO• radicals at 46 °C revealed the formation of cyclic silanes in a 46 % overall yield

Organosilanes in Radical Chemistry C. Chatgilialoglu
© 2004 John Wiley & Sons, Ltd ISBN: 0-471-49870-X

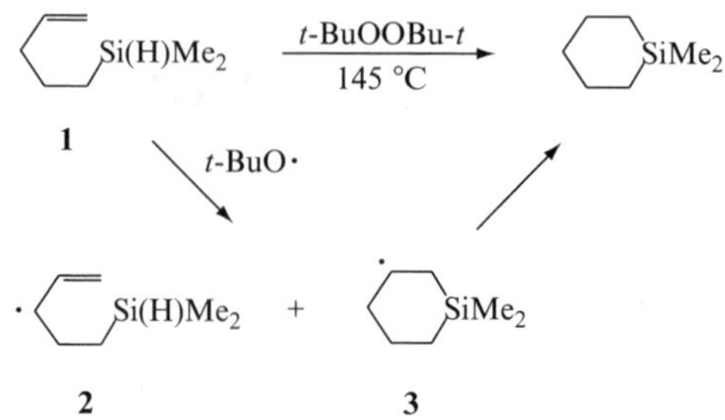

Scheme 6.1 Summary from product and EPR studies

(Scheme 6.2) [2]. Evidence for a five-membered ring product in 1% yield has also been obtained. Furthermore, EPR studies of the same reaction allowed for the identification of the silacyclohexyl radical **6** with two unequivalent β-hydrogen atoms in the temperature range between −73 and −25 °C, as the only observable intermediate. A kinetic investigation placed the rate constant for the cyclization of radical **5** between $10^7 < k < 10^9\ s^{-1}$ [2].

From the above description, it is clear that the cyclization of pent-4-ene substituted silyl radicals is a fast reaction and that formation of the six-membered ring is very strongly favoured. This is of course in antithesis with the analogous carbon-centered radicals, where the kinetically favored product has the five-membered ring structure produced by an *exo*-cyclization (cf. Section 7.3). The reversal in the regioselectivity of the silyl radicals was initially explained based on structural factors, such as the longer C—Si bond length and pyramidal geometry of the silicon centre. Semi-empirical calculations (MINDO/3) suggested that such behaviour of silyl radicals is due to an enthalpy factor caused by electronic effects [3]. The six-membered ring formation was also observed in the reaction of analogous heterosubstituted derivatives **7** (Reaction 6.1) [1].

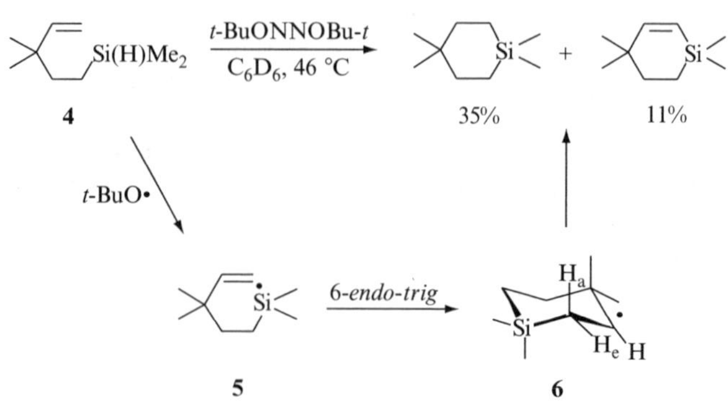

Scheme 6.2 Preference for 6-*endo-trig* cyclization of silyl radicals

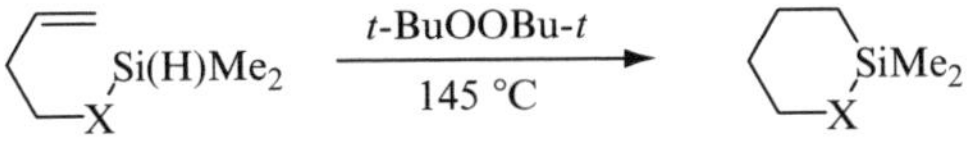

(6.1)

7

X = O 24%
X = $SiMe_2$ 23%

The *endo*-mode of cyclization is found to be the preferred path also in the lower homologues. Reaction (6.2) shows the reactions of two silanes (**8**) with thermally generated *t*-BuO• radicals to afford the five-membered ring in low yields via a 5-*endo-trig* cyclization [1]. EPR spectra recorded from these two silanes with photogenerated *t*-BuO• radicals are assigned to secondary alkyl radical intermediates formed by an intermolecular addition involving the expected silyl radical and the parent silane [2].

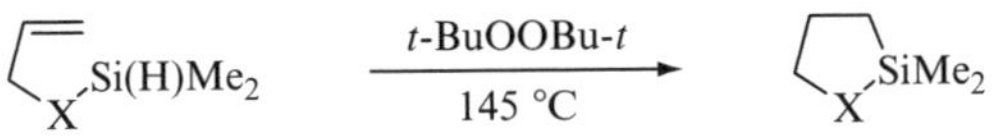

(6.2)

8

X = CH_2 18%
X = $SiMe_2$ 13%

Irradiation of trisilane **9** in cyclohexane with 254 nm light generates a very persistent radical, whose EPR spectra assigned to structure **11** (Scheme 6.3). The quenching of radical **11** with ethanethiol gave the expected silacyclobutene **12** as a single diastereoisomer, whereas the reaction with molecular oxygen afforded the lactone **13** [4]. It was suggested that upon homolytic cleavage of Si—$SiMe_2Ph$ in **9**, the initially formed silyl radical **10** undergoes a rearrangement to the more stable silacyclobutenyl radical **11**. The proposed rearrangement can be viewed as a 3-*exo-trig* cyclization of radical **10** followed by a specific fission of the internal C—C bond in the constrained bicycle intermediate to give radical **11**.

Allyloxysilanes (**14**) undergo radical chain cyclization in the presence of di-*tert*-butyl hyponitrite as radical initiator and thiol as a catalyst at ca 60 °C (Reaction 6.3) [5]. The thiol promotes the overall abstraction from the Si—H moiety as shown in Scheme 6.4 and the silyl radical undergoes a rapid 5-*endo-trig* cyclization. Indeed, EPR studies on the reaction of *t*-BuO• radical with silanes **14** detected only spectra from the cyclized radicals even at −100 °C, which implies that the rate constants for cyclization are $\geq 10^3\ s^{-1}$ at this temperature.

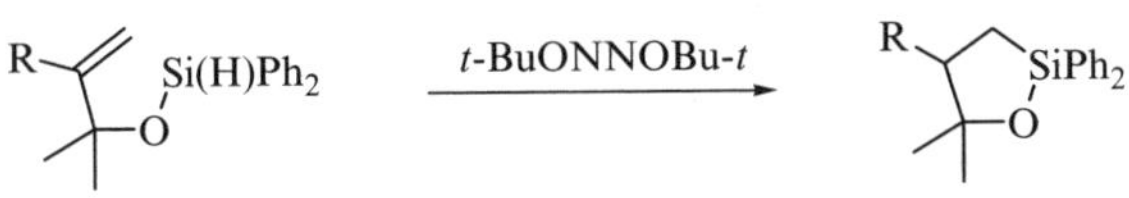

(6.3)

14

R = H t-$C_{12}H_{25}SH$ (cat.), hexane, 60 °C 95%
R = Me Ph_3SiSH (cat.), dioxane, 65 °C 95%

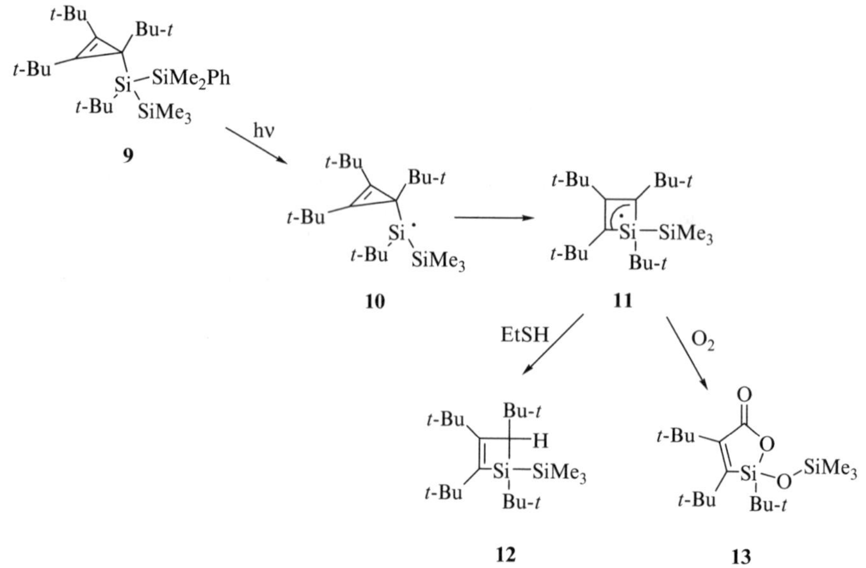

Scheme 6.3 Proposed reaction mechanism for the formation of radical **11** and its fate

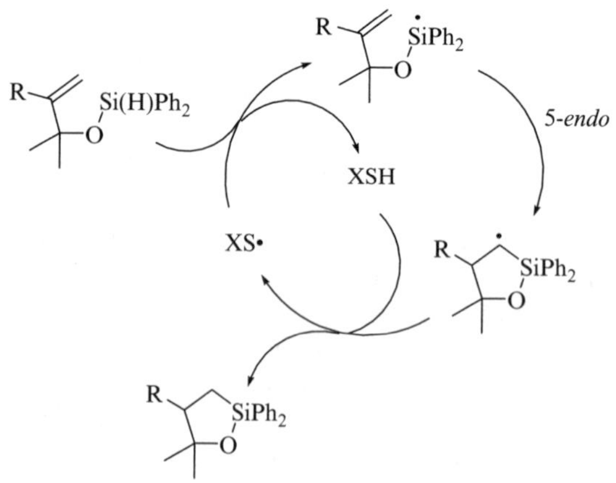

Scheme 6.4 Propagation steps for the intramolecular radical hydrosilylation of a carbon–carbon double bond catalysed by thiol

Intramolecular hydrosilylation of the higher homologues **15** and **20** under similar conditions gave also excellent yields of cyclized products [5]. The homoallyloxysilanes **15** afforded a mixture of six- and five-membered ring products in a ratio of 4.5:1 for R = Me and 2.5:1 for R = Ph in favour of the larger ring (Reaction 6.4). The EPR spectra obtained by the reaction of *t*-BuO• radical with silane **15** showed contemporaneously radicals **18** and **19** derived from the 6-*endo* and 5-*exo* cyclizations, respectively, in a 9:1 ratio for R = Me at −83 °C and a 4:1 ratio for R = Ph at −83 °C, ratios that parallel those of the final products **16** and **17** at 65 °C. For chair–chair interconversion of **18** (R = Ph), a rate constant of ~ $4 \times 10^7 s^{-1}$ at −85 °C was also obtained from a detailed lineshape analysis. On the other hand, silane **20** afforded exclusively a seven-membered heterocycle derived from a 7-*endo* cyclization (Reaction 6.5).

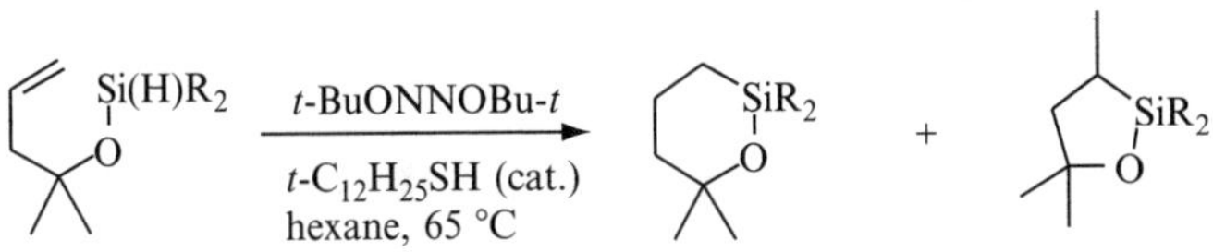

(6.4)

15 **16** **17**

R = Me 90%, **16**:**17** = 82:18
R = Ph 95%, **16**:**17** = 72:28

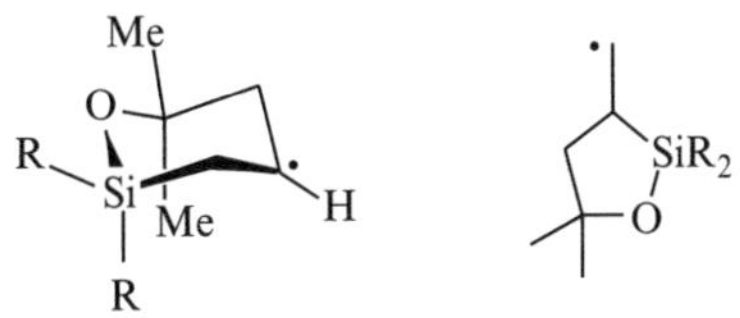

18 **19**

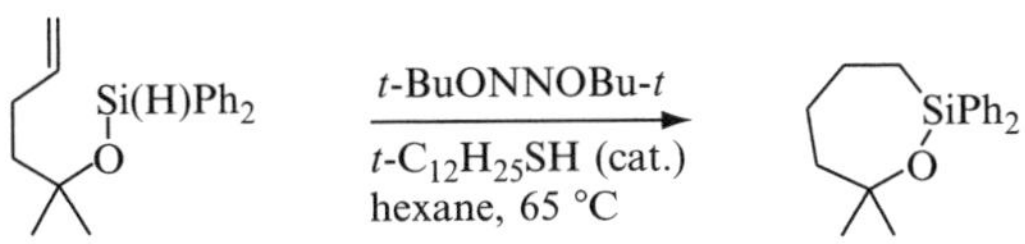

(6.5)

20 95%

Intramolecular hydrosilylation of alkenyloxysilyl radicals has also been investigated using silylated cyclohexadienes as the starting substrates [6]. Scheme 6.5 shows the reaction of **21** in hexane at 80–85 °C and in the presence of di-*tert*-butyl hyponitrite as radical initiator. The crude reaction mixture was treated with an excess of PhLi to provide alcohol **22** in moderate yields. Intermediates **23** and **24** are the expected species involved in the 5-*endo-trig* cyclization.

Analogously silylated cyclohexadiene **25** having propargyl alcohol as a pendant was used in the radical intramolecular hydrosilylation followed by ionic ring opening to provide alcohol **26** in 55 % yield (Scheme 6.6) [6]. The cyclization of silyl radical **27** to radical **28** represents an example of a 5-*endo-dig* process.

Sequences (or cascades) of radical reactions involving the 5-*endo-trig* cyclization of silyl radical to an allyloxy-type substituent as the key step have been developed and applied to the synthesis of natural products [7–11]. The concept and the sequence of radical reactions is outlined in Scheme 6.7. A radical of

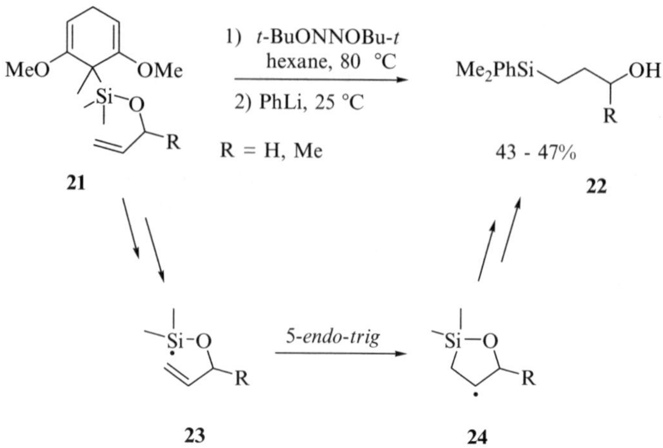

Scheme 6.5 Intramolecular radical hydrosilylation of a carbon–carbon double bond

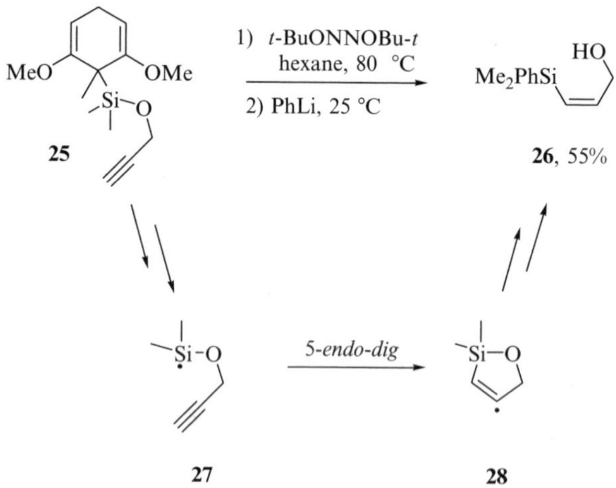

Scheme 6.6 Intramolecular radical hydrosilylation of a carbon–carbon triple bond

general formula **29** undergoes a 5-*exo-dig* cyclization and the resulting vinyl radical **30** abstracts hydrogen from the silane moiety intramolecularly to afford the silicon-centred radical **31**. This radical closes in a 5-*endo-trig* mode to generate radical **32**, whose fate (either reduction or β-elimination of the R group) depends on the nature of substituent R. The linked X—Y—Z groups could be a variety of atoms and substituents. The initial species **29** is a carbon-centred radical of alkyl, α-substituted alkyl, acyl or α-substituted acyl radical types and its precursor preferably is a phenylseleno derivative, although bromides and thiocarbonates have also been used [7–11].

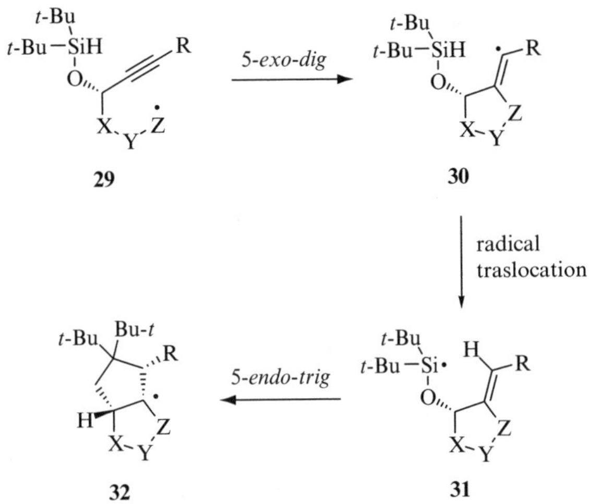

Scheme 6.7 Sequential radical reactions including 5-*endo-trig* cyclization of silyl radical

A few examples are chosen in order to illustrate the potentialities of this remarkable methodology. In Reaction (6.6) the sequence is initiated by the removal of the PhSe group and the formation of a carbamoyl radical. It is worth mentioning that the stereochemical outcome of these cascade reactions is controlled by the stereochemistry of the oxygen-bearing asymmetric carbon in **29**. Indeed, Reactions (6.6) and (6.7) show clearly the stereochemical control. On the other hand, Reactions (6.7) and (6.8) illustrate the role of R which is carried as a terminal group in the acetylenic moiety. For R = Ph the last step is the hydrogen abstraction, whereas for R = $SnBu_3$, the last step is the ejection of Bu_3Sn• radical (cf. Scheme 6.7).

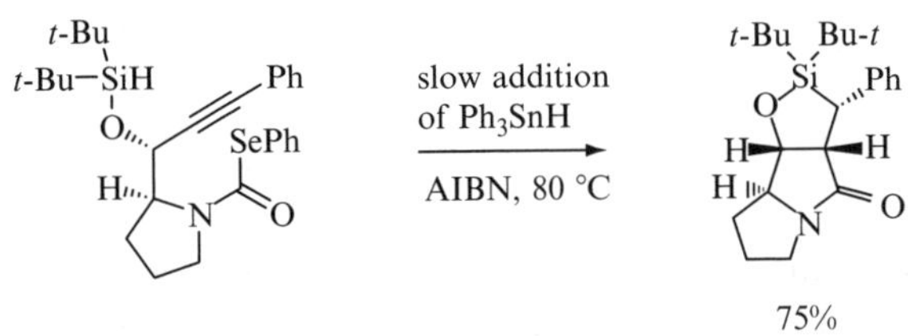

(6.6)

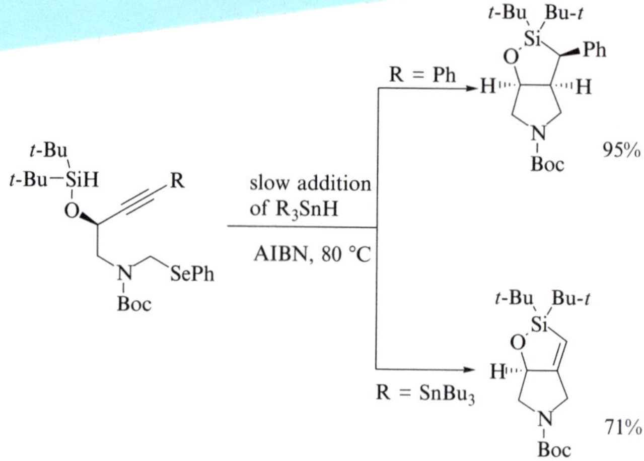

(6.7)

(6.8)

6.1.1 FIVE-MEMBERED RING EXPANSION

The above series of intramolecular hydrosilylations of alkoxysilanes allowed for the formation of cyclic alkoxysilanes, which are very useful intermediates. For example, they can be treated with fluoride ion under the conditions of Tamao oxidation to yield diols [12]. Other established procedures for the preparation of cyclic alkoxysilanes are the intramolecular hydrosilylation of alkoxysilanes using transition metal catalysis [13], as well as the radical cyclization of 3-oxa-4-silahexenyl radicals of type **32** [14–16] that has been successfully applied to the synthesis of biologically important branched nucleosides [17,18] and *C*-glycosides [19]. Interestingly, the cyclization of radical **32** afforded cyclic alkoxysilanes **35** and/or **36**, whose relative percentages strongly depend on the concentration of the reducing agent (Scheme 6.8). For example, at high concentration of Bu_3SnH the five-membered **35** is formed, whereas at a low Bu_3SnH concentration the six-membered **36** predominates. The independent formation of radical **33** showed that the six-membered ring formation is an authentic ring enlargement of radical **33** and not a direct cyclization of radical **32** [16].

Scheme 6.9 shows the radical clock methodology approach used for obtaining the rate constant of the ring expansion (k_{re}). Radical **37** was obtained by the reaction of the corresponding phenylseleno derivative with $(TMS)_3SiH$. A relative rate constant of $k_H/k_{re} = 20.2\ M^{-1}$ was obtained at 80 °C under first-order kinetics. Taking $k_H = 1.2 \times 10^6\ M^{-1}\ s^{-1}$ at 80 °C, the k_{re} value for the ring expansion **37** → **38** was calculated to be $6.1 \times 10^4\ s^{-1}$ at 80 °C [15].

Extensive mechanistic investigation of the ring expansion **33** → **34** was performed in order to differentiate between a ring-opening reaction to give a silyl radical **39** (path a), followed by the 6-*endo* cyclization, or a pentavalent silicon transition state **40** (path b). It was clearly demonstrated that the ring expansion proceeds via a pentavalent silicon transition state (Scheme 6.10) [16].

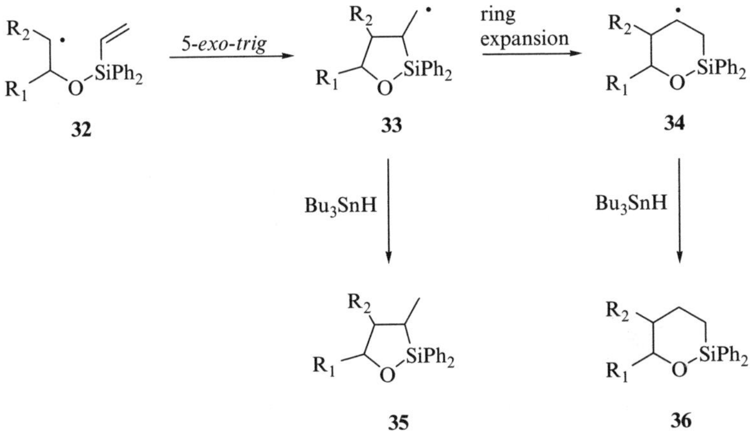

Scheme 6.8 Preferred mode of radical **32** cyclization and subsequent ring expansion

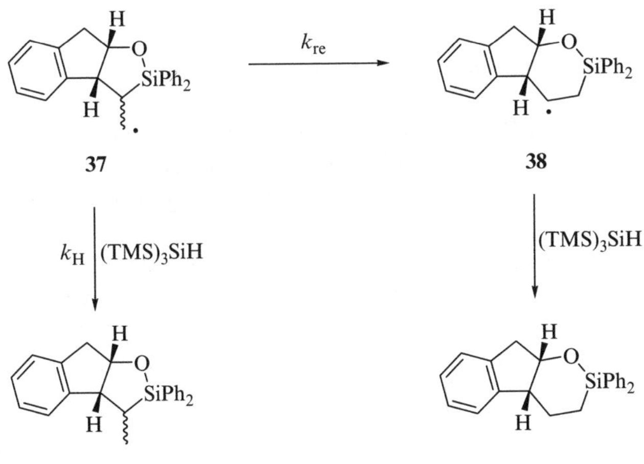

Scheme 6.9 Radical clock methodology approach for measuring the rate constant of ring expansion (cf. Section 3.1.1)

The above-described ring expansion reaction could play an important role in the outcome of silyl radical cyclization. Indeed, a cyclic product could derive first from the *exo*-mode and then transformed to the *endo*-mode product via the ring-expansion pathway (Scheme 6.11). For example, the products reported in Schemes 6.1 and 6.2 could be interpreted by this cyclization–expansion process, because of the experimental conditions (the relatively high temperature and absence of a hydrogen donor). On the other hand, products from both modes of cyclization are observed in the cyclization of silane **15** (Reaction 6.4). In this case, although thiols are very good hydrogen atom donors, the catalytic amount used could have not been enough to suppress completely the ring expansion

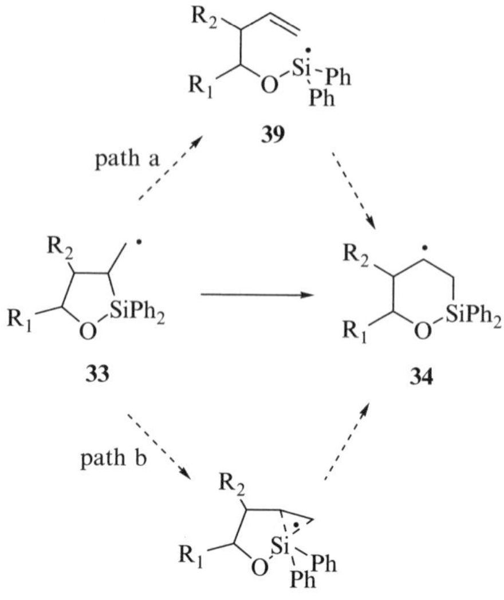

Scheme 6.10 Possible reaction paths for the ring expansion

process. However, EPR ratios of the observed intermediate radicals for Reaction (6.4) as well as for the reactions reported in Scheme 6.2 showed them to be parallel to those of the final products, thus confirming the occurrence of the two cyclization modes.

Undoubtedly, further work will be necessary in order to get a solid background for silyl radical cyclizations. In particular, further investigations on the nature of the pendant alkenyl substituent as well as on the ring size and substituent effects on the silicon in the ring expansion can be expected in the future.

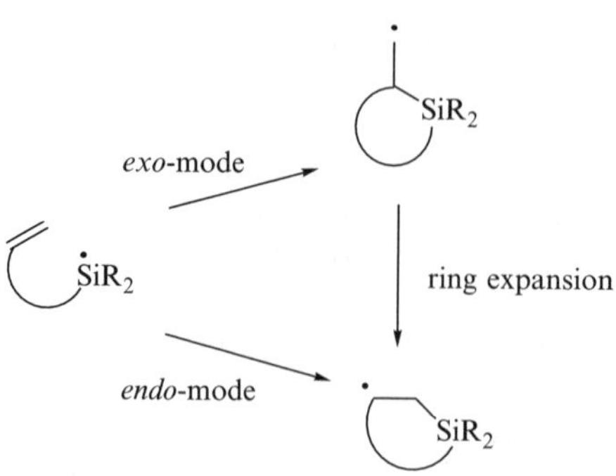

Scheme 6.11 Two possible pathways for the formation of a radical product having a larger ring

6.2 ARYL MIGRATION

The intramolecular addition of silyl radicals to aromatic rings has also attracted some attention. Early work on the silyl radical obtained by the reaction of silanes **41** with thermally generated *t*-BuO• radicals at 135 °C showed the formation of rearranged products only for $n = 3$ or 4, whereas for $n = 1, 2, 5$, and 6 no rearrangement took place [20].

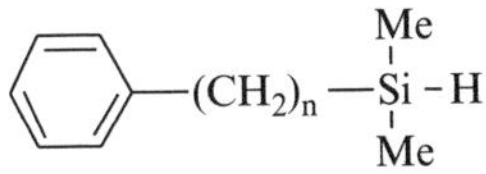

41 (n=1, 2, 3, 4, 5, 6)

Scheme 6.12 summarizes the experimental findings for $n = 4$, in the case of **41** and of two of its methylated derivatives [20]. The isolated products **45** and **47** derived from quenching of the radical **44** or by removal of the hydrogen atom from **46**, respectively. The silyl radical **42** undergoes either an *ortho*-attack to afford the cyclohexadienyl radical **46** or an *ipso*-attack to give the spiro intermediate **43**, which then rearomatizes to form the carbon-centred radical **44**. The reaction **42** → **44** is formally a 1,5-phenyl migration from carbon to silicon and the reverse step a 1,5-phenyl migration from silicon to carbon, via a common spiro intermediate **43**. Indeed, radical **44** generated as a primary species during the reduction of corresponding halides with Bu_3SnH, gave as products the two silanes derived from the quenching of radicals **42** and **44**, confirming the reversibility of these paths [21].

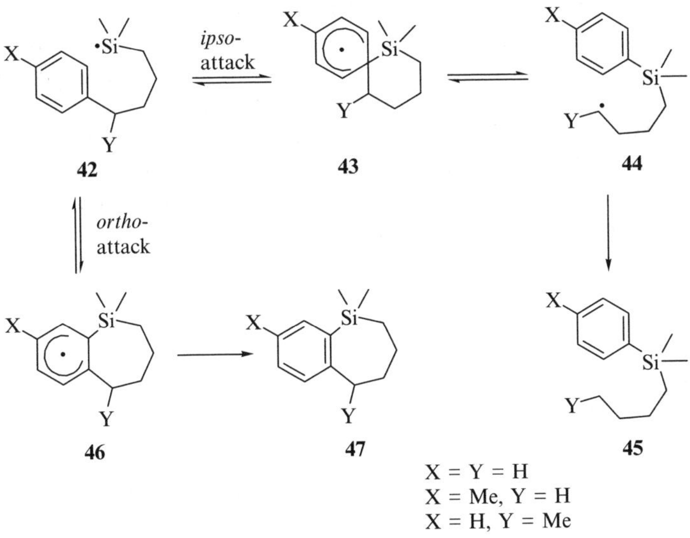

Scheme 6.12 Intramolecular addition of silyl radicals to aromatic rings: *ortho*- vs *ipso*-attack

From the reaction of photogenerated *t*-BuO• radicals with silanes **48**, in the cavity of the EPR spectrometer, the corresponding spiro cyclohexadienyl radicals are the only detectable species [22]. For $n = 3$, the intermediate radical **49** showed a spectral feature that suggests a chair-like arrangement and a chair–chair interconversion barrier of 18.4 kJ/mol was obtained from a detailed line-shape analysis.

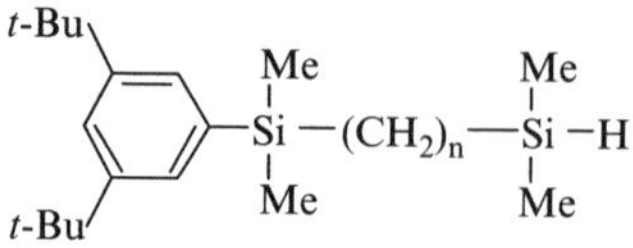

48 (n=2, 3, 4)

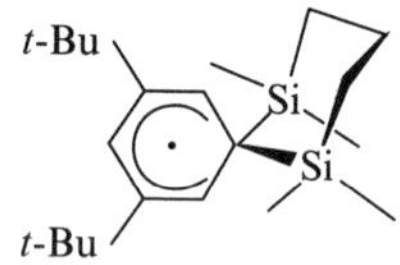

49

From synthetic point of view, the 1,5-phenyl migration discussed above (Reaction **42** ⇌ **44** in Scheme 6.12) has been successfully applied to the synthesis of 3-phenylated alcohols [23]. Few examples are given in Reaction (6.9) where an alkyl bromide or iodide having the diphenyl(trimethylsilyl)silyl ether in the γ-position reacted with Bu_3SnH under free-radical conditions to give the desired alcohols in good diastereoselectivity, after treatment with methyllithium. In the mechanistic paths shown in Reaction (6.10), the secondary alkyl radical **52** is initially formed by halogen abstraction, then undertakes an intramolecular *ipso*-attack to the migrating phenyl group to form the cyclohexadienyl radical **53**. The observed 1,3 selectivity has been explained by assuming a transition state in which both substituents R and Me prefer the equatorial positions (as shown in radical **53**). Rearomatization of spiro intermediate **53** then leads either to silyl radical **54** or returns back to radical **52**. Silyl radical **54** abstracts the halogen atom from the starting halide and completes the chain. Evidence that the reversible reaction **54** → **53** is unimportant under the experimental conditions is also obtained. This process was shown to proceed well also for 1,4-phenyl migration and even with a higher selectivity (Reaction 6.11) [23]. The stereochemical outcome of the 1,4-phenyl transfer reaction has been explained by assuming a transition state that resembles the chair arrangement shown in **58**, with the substituents R and Me in pseudoequatorial positions.

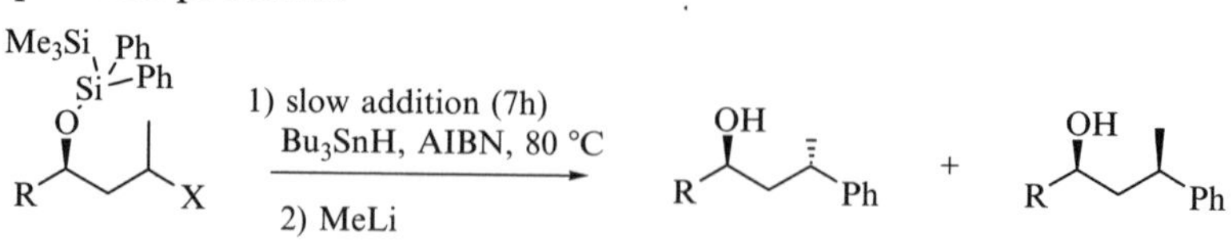

(6.9)

X = I, Br

50 **51**

R = Me	70%, **50**:**51** = 10:1
R = *i*-Pr	82%, **50**:**51** = 13:1
R = *t*-Bu	93%, **50**:**51** = 10:1
R = Ph	62%, **50**:**51** = 11:1

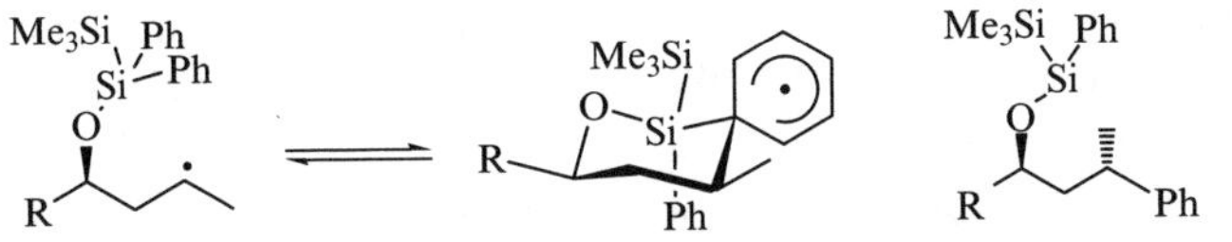

52 **53** **54** (6.10)

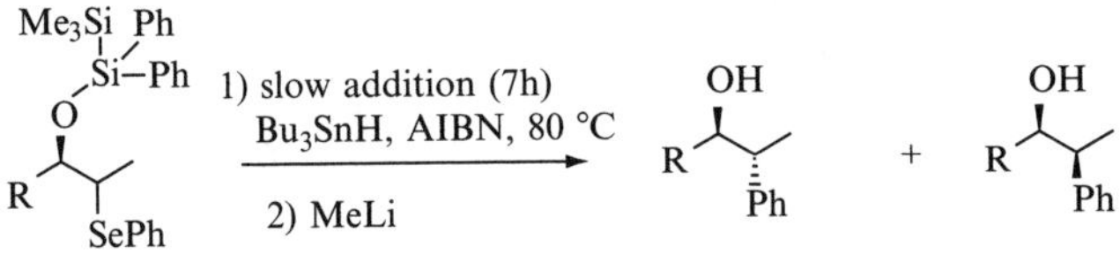

55 **56** **57** (6.11)

R = Me	81%, **56**:**57** = 14:1
R = Ph	74%, **56**:**57** = 18:1

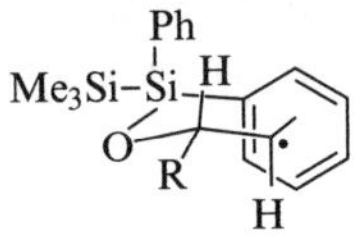

58

The same methodological approach has been extended to the synthesis of biaryls [24]. Reaction (6.12) shows a few examples in which the transfers of functionalized aryl groups from silicon to aryl radicals are successful. A mechanistic scheme similar to that reported in Reaction (6.10) has been proposed.

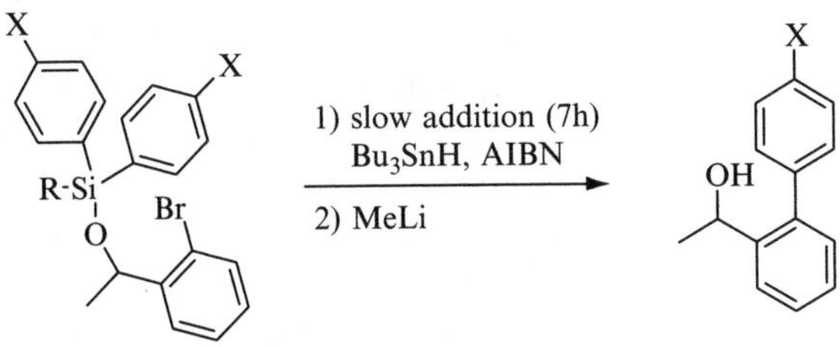

(6.12)

R = Ph	X = H	71%
R = $SiMe_2Ph$	X = F	49%
R = Me	X = OMe	77%

6.3 ACYLOXY MIGRATION

The 1,2 migration of an acyloxy group in β-(acyloxy)alkyl radicals has been the subject of numerous synthetic and mechanistic studies [25]. These rearrangements generally proceed in a nondissociative manner through either a five-centre/five-electron shift (**59**) or three-centre/three-electron shift (**60**), the latter being a more polarized transition state. All collected data are inconsistent with a stepwise pathway via a five-membered cyclic intermediate radical [25].

59 **60**

Analogous silicon radical reactions are limited to two reports [26,27]. The reaction of photogenerated *t*-BuO• radical with silane **61** in hexadecane as solvent containing CCl_4 in varying concentrations, provided the two chlorides shown in Reaction (6.13) [26]. Although no detailed kinetic studies are reported, the fact that alkyl chloride was the major product even in the presence of 5 M CCl_4 suggests a very fast 1,2 migration of the acetyloxy group.

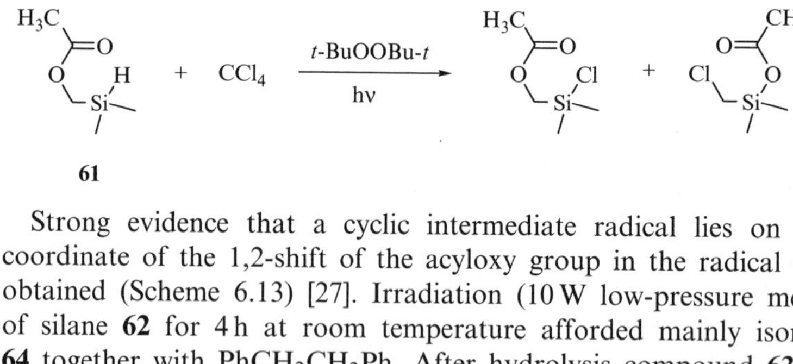

61

(6.13)

Strong evidence that a cyclic intermediate radical lies on the reaction coordinate of the 1,2-shift of the acyloxy group in the radical **65**, has been obtained (Scheme 6.13) [27]. Irradiation (10 W low-pressure mercury lamp) of silane **62** for 4 h at room temperature afforded mainly isomers **63** and **64** together with $PhCH_2CH_2Ph$. After hydrolysis compound **63** is quantitatively transformed in $RC(O)CH_2Ph$. These results are consistent with Scheme 6.13, where the initial formation of silyl radical **65** is followed by radical cyclization to radical **66**. In turn radical **66** either can be trapped by benzyl radical to give **63** or further rearranges to radical **67**, prior to the benzyl radical trapping. Deuterium-labelled experiments indicated that the formation of **63** occurs in the solvent cage whereas **64** arises as both cage and escape products. EPR spectra recorded during the photolysis of silane **62** at around −100 °C consisted of the superimposition of radicals **65** and **67**, which suggests a rate constant in the range of 10^2–$10^3\ s^{-1}$ for the 1,2 migration of an acyloxy group at this temperature. Therefore, the intramolecular addition of silyl radical to the ester moiety is a fast process and a few orders of magnitude faster than the intermolecular addition due to entropic effects (cf. Chapter 5).

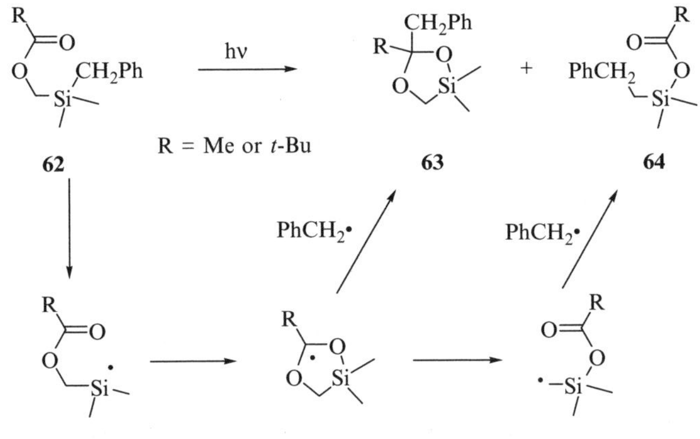

Scheme 6.13 A stepwise pathway via five-membered cyclic intermediate radical **66** in the 1,2 migration of an acyloxy group

6.4 INTRAMOLECULAR HOMOLYTIC SUBSTITUTION AT SILICON

It has been found theoretically that bimolecular homolytic substitution (S_H2) reaction of the methyl radical at the silicon atom in disilane can proceed via both *backside* and *frontside* attack mechanism (Scheme 6.14) [28]. The transition state for the backside attack **(68)** is predicted to adopt an exactly collinear arrangement of the attacking methyl radical and the leaving silyl radical, as it is generally believed for the majority of S_H2 reactions. The transition state for the frontside attack **(69)** is predicted to involve an attack angle of around 80 °. At the highest level of theory, the energy barriers of the two paths are very similar being 47.4 and 48.6 kJ/mol for the backside and frontside attack, respectively. This information can be easily extrapolated to the intramolecular homolytic substitution (S_Hi) reactions for mechanistic consideration, by imagining links between the attacking carbon-centred radical and one of the two silicon centres.

Reflux of bromide **70** in benzene and in the presence of small amounts of $(TMS)_3SiH$ and AIBN afforded the silabicycle **71** in a 88% yield (Reaction 6.14) [29]. On the other hand, the reduction of iodide **72** with $(TMS)_3SiH$ under standard experimental conditions gave the silacyclopentane **74** in a 24% yield, along with the reduced product **73** (Reaction 6.15) [30]. Scheme 6.15 shows the propagation steps for Reaction (6.14). The key step for this transformation is the S_Hi reaction at the central silicon atom via a five-membered transition state which occurred with a rate constant of $2.4 \times 10^5\ s^{-1}$ at 80 °C. Similar reactions are observed in the hydrosilylation of 1,6-dienes with $(TMS)_3SiH$ when the concentration of reducing agent is kept low by a slow addition in order to avoid the trapping of a carbon-centred radical [29,30]. Reaction (6.16) shows an

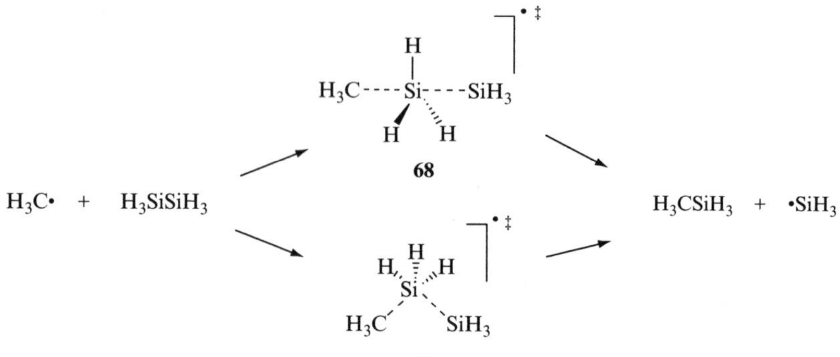

Scheme 6.14 Backside and frontside substitution reactions of methyl radical with disilane

example of addition of $(TMS)_3SiH$ to the 1,7-diene **75**, which affords the bicyclized product **76** in an isomeric mixture (1.7:1) of *cis*- and *trans*-fused compounds [30].

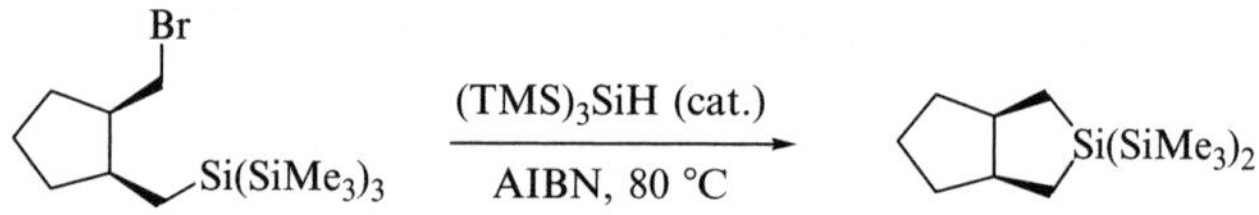

(6.14)

70 **71**, 88%

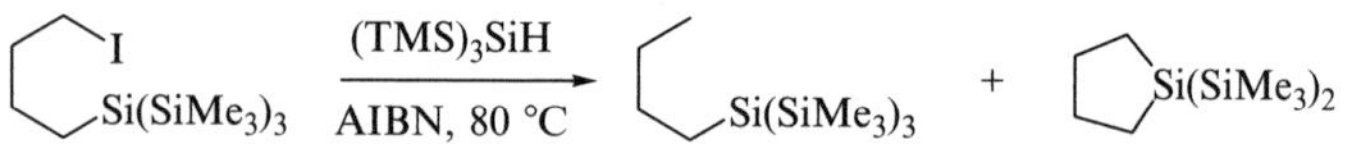

(6.15)

72 **73**, 63% **74**, 24%

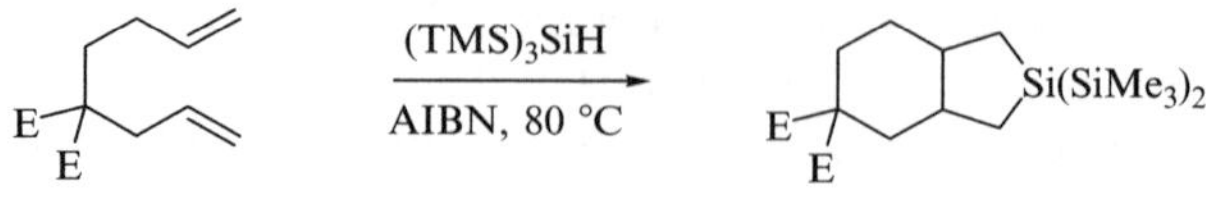

(6.16)

75, E = CO_2Me **76**, 52%

The reaction has also been extended to the analogous vinyl bromides [30]. Indeed, the alkenyl bromide **77** under normal reduction conditions gave the bicyclic compound **78** in good yield by an S_Hi reaction given by the vinyl radical (Reaction 6.17). Under these conditions, the reduction products could not be observed which suggests a very fast unimolecular reaction.

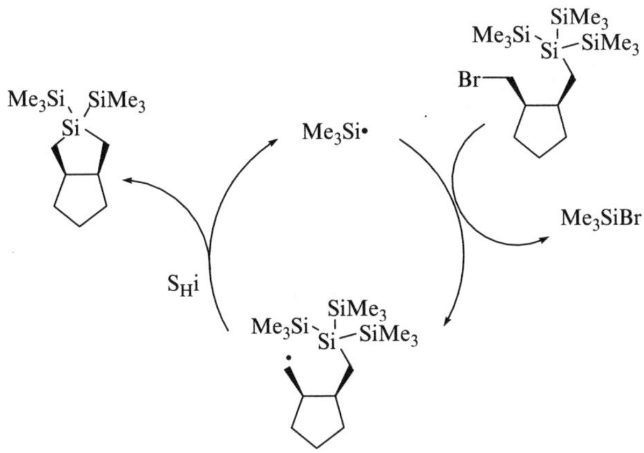

Scheme 6.15 Propagation steps of Reaction (6.14) involving an S_Hi step via a five-membered transition state

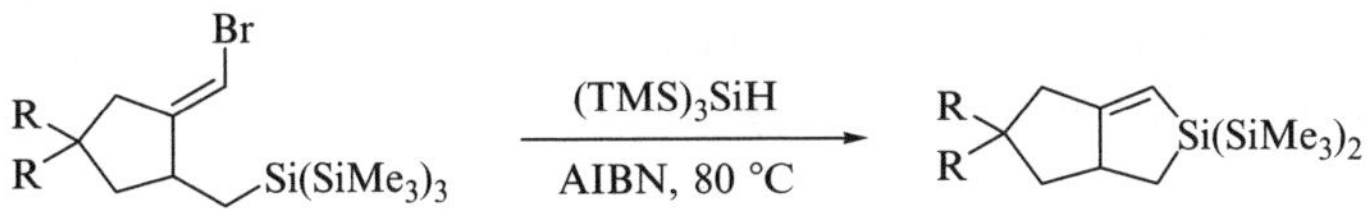

(6.17)

77, R = CO_2Me **78**, 75%

The S_Hi reaction at silicon positioning a carbon-centred radical in the γ-position to a silyl ether group, has been introduced as an alternative method for the preparation of cyclic alkoxysilanes (Scheme 6.16) [31,32]. Reaction (6.18) shows the extension of the chemistry reported above by using aryl radicals, which afforded the cyclic alkoxysilane in good yield. However, by replacing the two $SiMe_3$ groups by phenyls the yield decreased considerably (Reaction 6.19) in favour of the product derived from 1,5-phenyl migration, which was obtained in a 55% yield (cf. Reactions 6.9 and 6.10). Indeed, a better leaving group like $SnMe_3$ is necessary for increasing the efficiency of cyclic alkoxysilane. Kinetic data on the S_Hi reaction at silicon that parallels the efficiencies observed in Reaction (6.19) have also been obtained.

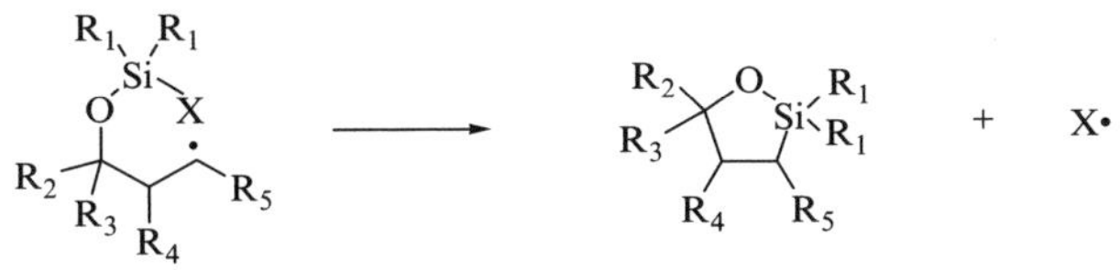

Scheme 6.16 S_Hi reaction at silicon: preparation of cyclic alkoxysilanes

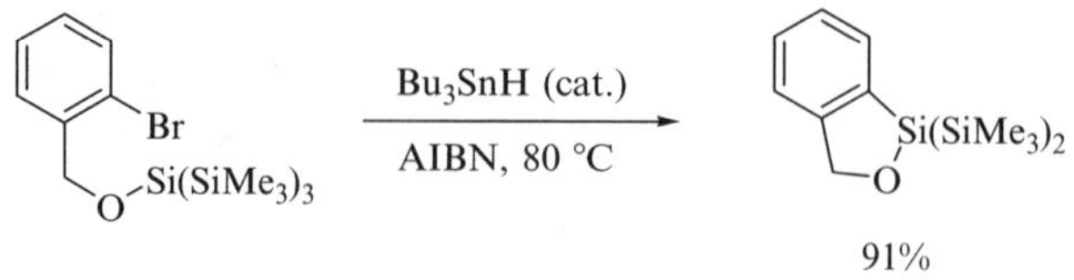

(6.18)

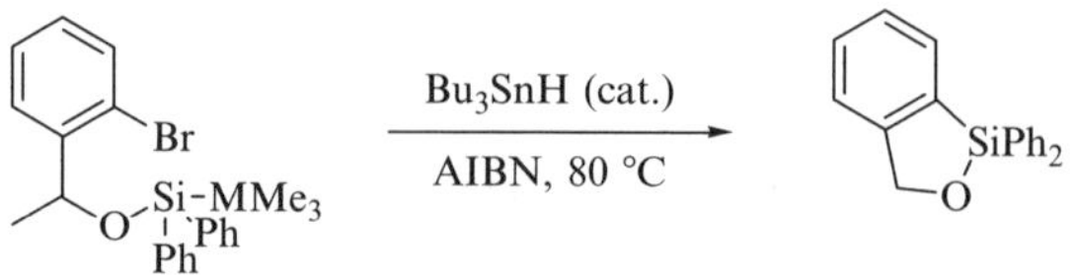

(6.19)

M = Si	28%
M = Ge	39%
M = Sn	84%

Interestingly, the corresponding reaction with the silylether group in the β-position afforded completely different products [23]. Indeed, the reduction of **79** gave the *ortho*-silylated phenol **80** in a 50% yield (Scheme 6.17). The key step is the S_Hi reaction of aryl radical **81** at Me_3Si silicon atom with formation of silyl radical **82**, confirming the preference for a five-membered transition state.

Photochemical behaviour of compounds **83–86** [33] in the gas phase has been reported, in order to distinguish between silyl radical and silylene formation. Photolysis of the noncyclic precursors **83** and **84** gave products derived from silyl radicals, which come from a direct Si—Si bond homolysis, with a little evidence of silylene formation. In contrast, dimethylsilylene (Me_2Si:) was observed as a direct photoproduct from the cyclic precursors **85** and **86**. The reaction sequence including a S_Hi step shown in Scheme 6.18 for the formation of dimethylsilylene was proposed to explain the different observations for cyclic and noncyclic systems.

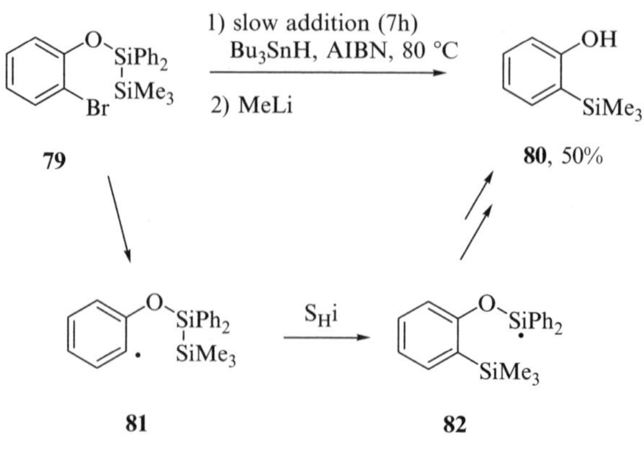

Scheme 6.17 S_Hi reaction at silicon: the preference for a five-membered transition state

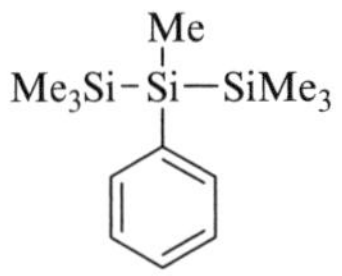

83

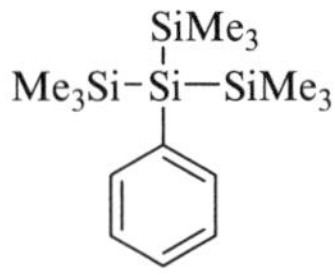

84

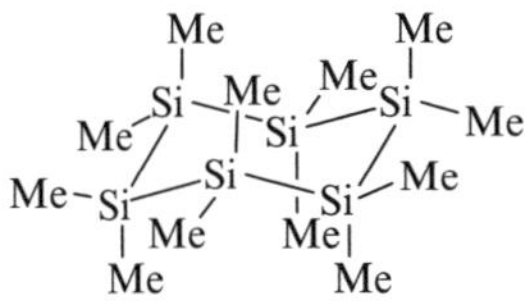

85

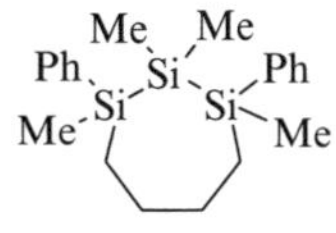

86

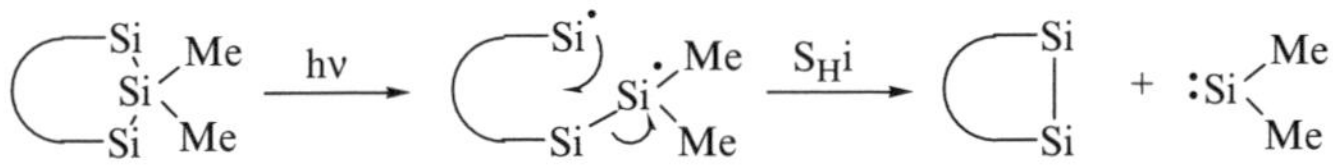

85 – **86**

Scheme 6.18 Reaction mechanism for the photolysis of cyclic compounds **85** and **86**

6.5 HOMOLYTIC ORGANOSILICON GROUP TRANSFER

As we showed previously in this chapter, silicon easily expands its valence shell to become a penta-coordinated species. Therefore, it is expected that an intramolecular transfer of organosilicon group from X to Y may occur via a radical mechanism, provided that the R substituents on the silicon atom are not good leaving groups (Reaction 6.20).

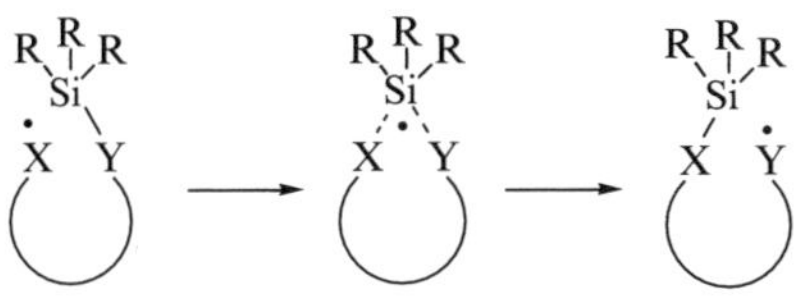

(6.20)

Radical rearrangements that involve a homolytic 1,2 transfer of organosilicon group between similar or different atoms are well documented. Scheme 6.19 shows the classes of reaction that involve 1,2 silyl group migration between atoms of the second row of the periodic table. Examples of 1,2 organosilicon group migration from carbon to carbon can be found in the expansion of five- to six-membered rings, previously discussed in this chapter (see Section 6.1.1). An example of 1,2 organosilicon group transfer from carbon to oxygen is given by the radical Brook rearrangement discussed in Section 5.3.3. Examples of migration from carbon to nitrogen are also discussed in Chapter 5 in connection with

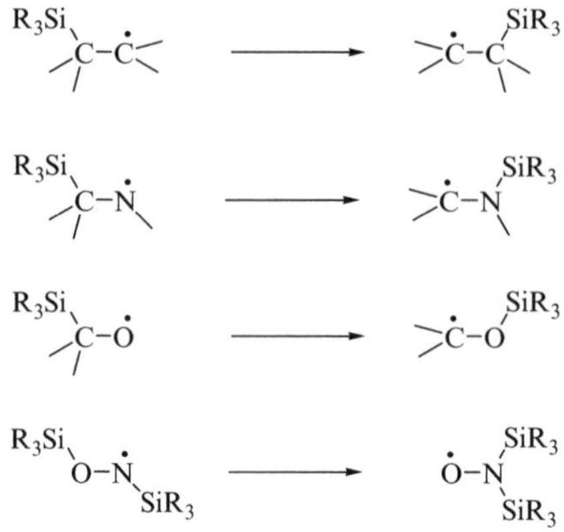

Scheme 6.19 Radical 1,2 silyl migrations

the addition of silyl radicals to C=N bonds (see Section 5.4). Similar 1,2 silyl shifts from oxygen to nitrogen were observed in solution (Scheme 6.19) [34]. In the majority of cases, only a penta-coordinated transition state, derived from the frontside attack as in Reaction (6.20), is possible.

Examples of 1,2 migration of silyl group from silicon to other atoms are also well documented. Reaction (6.21) that involves a transfer from silicon to carbon has been observed in the gas phase at elevated temperatures [35]. From the measured Arrhenius parameters, i.e. $\log A/s^{-1} = 12.3$ and $E_a = 90\,kJ/mol$, it is clear that the reaction is unimportant around room temperature. However, this type of reaction is probably the essential step in the formation of silicon–carbon fibres by pyrolysis of methylpolysilanes.

(6.21)

Silanethiols **87** readily undergo a radical-induced skeletal rearrangement to give **88** (Reaction 6.22) [36,37]. The key step for these reactions is the 1,2 migration of silyl group from silicon to sulfur (Reaction 6.23). Based on this reactivity, silanethiols **87** have been used as reducing agents for a fast hydrogen donation (see Section 4.6).

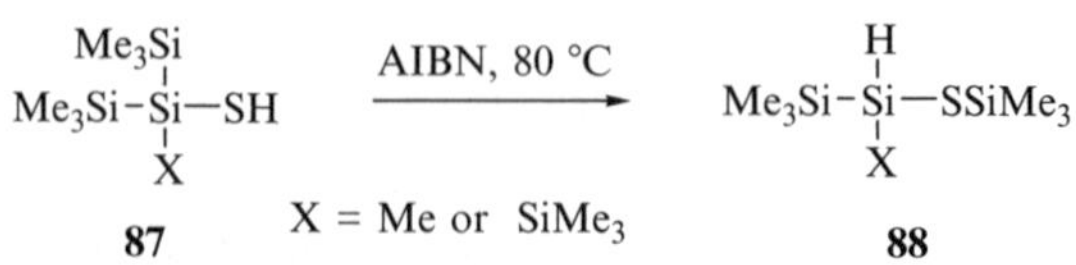

87 **88** (6.22)

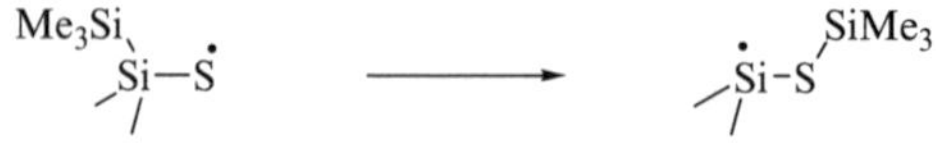

(6.23)

Evidence that the 1,2 transfer of the Me_3Si group from silicon to oxygen is an effective process (Reaction 6.24) has been obtained from a mechanistic investigation of the reaction of $(TMS)_3SiH$ with nitroalkanes and nitroxides [38,39]. A lower limit rate constant of $10^7\,s^{-1}$ at room temperature was obtained for Reaction (6.24).

(6.24)

The 1,2 transfer of the organosilicon group from silicon to oxygen has also been invoked in the mechanism of the autoxidation of $(Me_3Si)_3SiH$, $(Me_3Si)_2MeSiH$ and poly(hydrosilane)s in general (see Section 8.2) [40,41]. Details of the radical chain autoxidation of $(Me_3Si)_3SiH$ and $(Me_3Si)_2MeSiH$, proved by oxygen-labelling experiments, are shown in Scheme 6.20. Silylperoxyl radical **89** rearranges to **90** by means of an unusual mechanism followed by a 1,2 transfer of a Me_3Si group to give **91**.

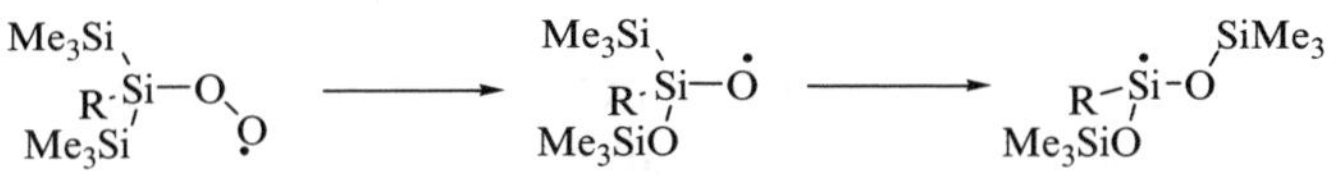

89 **90** **91**

Scheme 6.20 Reaction intermediates involved in the autoxidation of silyl substituted silicon hydrides (R = Me or Me_3Si)

The intramolecular migration of the organosilicon groups observed in the silyl radical adducts of α-diones and *ortho*-quinones (see Section 5.3.1) are examples of the 1,4 transfer of a silyl group from oxygen to oxygen. Another example is shown in Reaction (6.25) for the adduct of 3,6-di-*tert*-butyl-*ortho*-benzoquinone with Me_3Si• radical. The activation parameters for this symmetrical fluxional motion are $\log A/s^{-1} = 13.2$ and $E_a = 38.9\,kJ/mol$, which correspond to a rate constant of $1.8 \times 10^6\,s^{-1}$ at 20°C [42]. It is also observed that for unsymmetrical α-carbonyl derivatives, where the two adducts are not equivalent, the equilibrium lies towards the more stable adduct radical [43].

Homolytic 1,5 transfer of an organosilicon group from enoxy oxygen to alkoxy oxygen has also been observed [44]. As shown in Scheme 6.21, radical reaction of compound **92** with Bu_3SnH afforded compound **93** in a 92% yield.

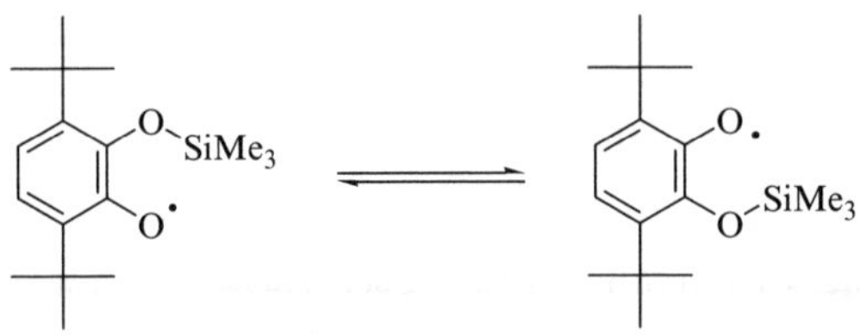

(6.25)

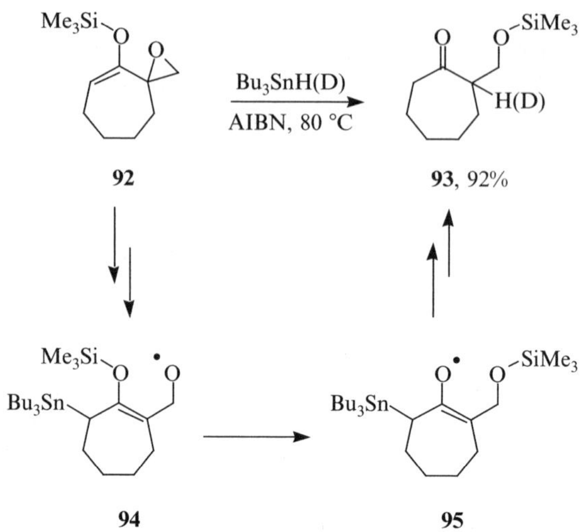

Scheme 6.21 Homolytic 1,5 transfer of Me_3Si group from an enoxy oxygen to alkoxy oxygen

When Bu_3SnH was replaced with Bu_3SnD, the deuterated compound **93** was isolated in 86 % yield. The intermediate alkoxy radical **94** derived by addition of Bu_3Sn• radical, following by ring opening of the epoxide moiety, undergoes a fast 1,5 silyl migration to give **95** prior to hydrogen abstraction. A lower limit rate constant of $10^9\,s^{-1}$ is estimated for such a unimolecular process. By replacing Me_3Si with t-$BuMe_2Si$ group the 1,5 transfer occurred more slowly since the corresponding alkoxy radical **94** was partially quenched. An analogous reaction with a chiral organosilicon moiety proceeded with retention of configuration most likely through a frontside attack mechanism [45]. The 1,5 translocation of SiH_3 group between two alkyl carbon atoms or two alkoxy oxygen atoms (cf. Reaction 6.20) has also been investigated theoretically [46]. Concerted frontside transition states were located with much lower barriers for the shift between the two oxygens.

6.6 REFERENCES

1. Barton, T.J., and Revis, A., *J. Am. Chem. Soc.*, 1984, **106**, 3802.
2. Chatgilialoglu, C., Woynar, H., Ingold, K.U., and Davies, A.G., *J. Chem. Soc., Perkin Trans. 2*, 1983, 555.
3. Sarasa, J.P., Igual, J., and Poblet, J.M., *J. Chem. Soc., Perkin Trans. 2*, 1986, 861.
4. Maier, G., Kratt, A., Schick, A., Reisenauer, H.P., Barbosa, F., and Gescheidt, G., *Eur. J. Org. Chem.*, 2000, 1107.
5. Cai, Y., and Roberts, B.P., *J. Chem. Soc., Perkin Trans. 1*, 1998, 467.
6. Amrein, S., and Studer, A., *Chem. Commun.*, 2002, 1592.
7. Clive, D.L.J., and Cantin, M., *J. Chem. Soc., Chem. Commun.*, 1995, 319.
8. Clive, D.L.J., and Yang, W., *J. Chem. Soc., Chem. Commun.*, 1996, 1606.

9. Sannigrahi, M., Mayhew, D.L., and Clive, D.L.J., *J. Org. Chem.*, 1999, **64**, 2776.
10. Clive, D.L.J., Yang, W., MacDonald, A.C., Wang, Z., and Cantin, M., *J. Org. Chem.*, 2001, **66**, 1966.
11. Clive, D.L.J., and Ardelean, E.-S., *J. Org. Chem.*, 2001, **66**, 4841.
12. Jones, G.R., and Landais, Y., *Tetrahedron*, 1996, **52**, 7599.
13. Bergens, S.H., Noheda, P., Whelan, J., and Bosnich, B., *J. Am. Chem. Soc.*, 1992, **114**, 2121 and 2128.
14. Shuto, S., Kanazaki, M., Ichikawa, S., and Matsuda, A., *J. Org. Chem.*, 1997, **62**, 5676.
15. Sugimoto, I., Shuto, S., and Matsuda, A., *Synlett*, 1999, 1766.
16. Shuto, S., Sugimoto, I., Abe, H., and Matsuda, A., *J. Am. Chem. Soc.*, 2000, **122**, 1343.
17. Shuto, S., Kanazaki, M., Ichikawa, S., Minakawa, N., and Matsuda, A., *J. Org. Chem.*, 1998, **63**, 746.
18. Sukeda, M., Shuto, S., Sugimoto, I., Ichikawa, S., and Matsuda, A., *J. Org. Chem.*, 2000, **65**, 8988.
19. Shuto, S., Yahiro, Y., Ichikawa, S., and Matsuda, A., *J. Org. Chem.*, 2000, **65**, 5547.
20. Sakurai, H., and Hosomi, A., *J. Am. Chem. Soc.*, 1970, **92**, 7507.
21. Wilt, J.W. In *Free Radicals*, Volume I, J.K. Kochi (Ed.), Wiley, New York, 1973, pp. 333–501.
22. Sakurai, H., Nozue, I., and Hosomi, A., *J. Am. Chem. Soc.*, 1976, **98**, 8279.
23. Amrein, S., Bossart, M., Vasella, T., and Studer, A., *J. Org. Chem.*, 2000, **65**, 4281.
24. Studer, A., Bossart, M., and Vasella, T., *Org. Lett.*, 2000, **2**, 985.
25. Beckwith, A.L.J., Crich, D., Duggan, P.J., and Yao, Q., *Chem. Rev.*, 1997, **97**, 3273.
26. Wilt, J.W., and Keller, S.M., *J. Am. Chem. Soc.*, 1983, **105**, 1395.
27. Kira, M., Yoshida, H., and Sakurai, H., *J. Am. Chem. Soc.*, 1985, **107**, 7767.
28. Matsubara, H., Horvat, S.M., and Schiesser, C.H., *Org. Biomol. Chem.*, 2003, **1**, 1199.
29. Kulicke, K.J., Chatgilialoglu, C., Kopping, B., and Giese, B., *Helv. Chim. Acta*, 1992, **75**, 935.
30. Miura, K., Oshima, K., and Utimoto, K., *Bull. Chem. Soc. Jpn.*, 1993, **66**, 2348.
31. Studer, A., *Angew. Chem. Int. Ed.*, 1998, **37**, 462.
32. Studer, A., and Steen, H., *Chem. Eur. J.*, 1999, **5**, 759.
33. Borthwick, I., Baldwin, L.C., Sulkes, M., and Fink, M.J., *Organometallics*, 2000, **19**, 139.
34. West, R., and Boudjouk, P., *J. Am. Chem. Soc.*, 1973, **95**, 3983.
35. Davidson, I.M.T., Potzinger, P., and Reiman, B., *Ber. Bunsenges. Phys. Chem.*, 1982, **86**, 13.
36. Pitt, C.G., and Fowler, M.S., *J. Am. Chem. Soc.*, 1968, **90**, 1928.
37. Ballestri, M., Chatgilialoglu, C., and Seconi, G., *J. Organomet. Chem.*, 1991, **408**, C1.
38. Ballestri, M., Chatgilialoglu, C., Lucarini, M., and Pedulli, G.F., *J. Org. Chem.*, 1992, **57**, 948.
39. Lucarini, M., Marchesi, E., Pedulli, G.F., and Chatgilialoglu, C., *J. Org. Chem.*, 1998, **63**, 1687.
40. Chatgilialoglu, C., Guarini, A., Guerrini, A., and Seconi, G., *J. Org. Chem.*, 1992, **57**, 2208.
41. Chatgilialoglu, C., Guerrini, A., Lucarini, M., Pedulli, G.F., Carrozza, P., Da Roit, G., Borzatta, V., and Lucchini, V., *Organometallics*, 1998, **17**, 2169.
42. Prokofev, A.I., Prokofeva, T.I., Belostotskaya, I.S., Bubnov, N.N., Solodovnikov, S.P., Irshov, V.V., and Kabachnik, M.I., *Tetrahedron*, 1979, **35**, 2471.

43. Alberti, A., Martelli, G., Hudson, A., Pedulli, G.F., Tiecco, M, and Ciminale, J., *J. Organomet. Chem.*, 1979, **182**, 333.
44. Kim, S., Do, J.Y., and Lim, K.M., *J. Chem. Soc., Perkin Trans. 1*, 1994, 2517.
45. Horvat, S.M., Kim, S., and Schiesser, C.H., *Chem. Commun.*, 2003, 1182.
46. Kim, S., Horvat, S.M., and Schiesser, C.H., *Aust. J. Chem.*, 2002, **55**, 753.

7 Consecutive Radical Reactions

The use of free-radical reactions in multi-step synthesis has steadily increased with time. Indeed, synthetic strategies based on radical reactions have become more and more popular among chemists since a wide selection of functional groups are now available to generate carbon-centred radicals [1]. The knowledge of radical reactivity has increased to such a level as to aid in making the necessary predictions for performing sequential transformations. The predictability has extended to include also the formation of new stereogenic centres, so to render radical reactions of special interest across the field of asymmetric synthesis [1–4]. Stereoselectivities can be dictated by nearby stereocentres, by chiral additives, and by chiral catalysts.

As far as the use of silanes as mediators in consecutive radical reactions is concerned, the knowledge of their hydrogen donor abilities coupled with the steric hindrance given by the silicon substituents has contributed substantially in this area, with interesting results in terms of reactivity and stereoselectivity.

7.1 BASIC CONCEPTS OF CARBON–CARBON BOND FORMATION

The carbon-centred radical R• resulting from the initial atom (or group) removal by a silyl radical or by addition of a silyl radical to an unsaturated bond can be designed to undergo a number of consecutive reactions prior to H atom transfer. The key step in these consecutive reactions generally involves the inter- or intramolecular addition of R• to a multiple-bonded carbon acceptor. Care has to be taken in order to ensure that the effective rate of the radical addition is higher than the rate of H atom transfer. Standard synthetic planning can be based on the knowledge of rate constants, and coupled with reaction

Organosilanes in Radical Chemistry C. Chatgilialoglu
© 2004 John Wiley & Sons, Ltd ISBN: 0-471-49870-X

conditions, in order to control the concentration of the reducing agent (i.e., a slow addition by syringe pump) or, in the case of intermolecular addition reactions, to have a large excess of the radical acceptor.

As an example, the propagation steps for the reductive alkylation of alkenes are shown in Scheme 7.1. For an efficient chain process, it is important (i) that the $R'_3Si\bullet$ radical reacts faster with RZ (the precursor of radical R•) than with the alkene, and (ii) that the alkyl radical reacts faster with the alkene (to form the adduct radical) than with the silicon hydride. In other words, the intermediates must be *disciplined*, a term introduced by D. H. R. Barton to indicate the control of radical reactivity [5]. Therefore, a synthetic plan must include the task of considering kinetic data or substituent influence on the selectivity of radicals. The reader should note that the hydrogen donation step controls the radical sequence and that the concentration of silicon hydride often serves as the variable by which the product distribution can be influenced.

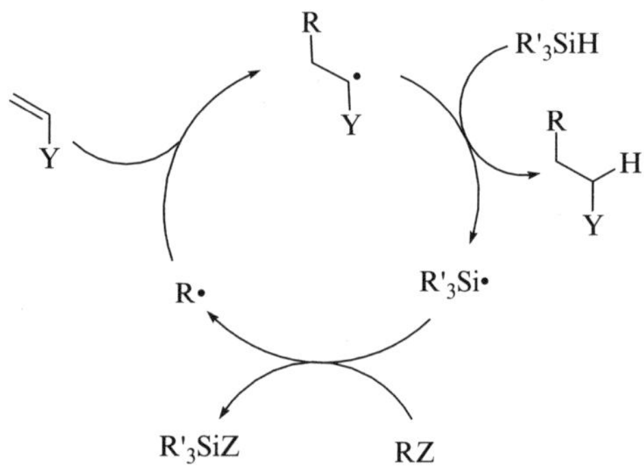

Scheme 7.1 Propagation steps for the intermolecular carbon–carbon bond formation

The majority of sequential radical reactions using silanes as mediators for the C—C bond formation deal with $(TMS)_3SiH$. Nevertheless, as we shall see there are some interesting applications using other silanes, too.

The general concepts of stereoselectivity in radical reactions have been illustrated in a number of recent books. Readers are referred to those books for a thorough treatment [1,2]. The following sections deal with a collection of applications where silanes act as mediators for smooth and selective radical strategies, based on consecutive reactions.

7.2 INTERMOLECULAR FORMATION OF CARBON–CARBON BONDS

In the initial work, the reaction of cyclohexyl iodide or isocyanide with a variety of alkenes mediated by $(TMS)_3SiH$ was tested, in order to find out the similarities with tin reagents (Reactions 7.1 and 7.2) [6,7]. For example, using a 1:1 ratio of cyclohexyl iodide and electron-poor terminal alkenes together with 1.2 equiv of $(TMS)_3SiH$ the desired product was obtained in a good yield. It is worth pointing out that with iodides the reaction yields are comparable with those obtained by the tin method, whereas with $(TMS)_3SiH$ the use of alkyl isocyanides as radical precursors has been possible.

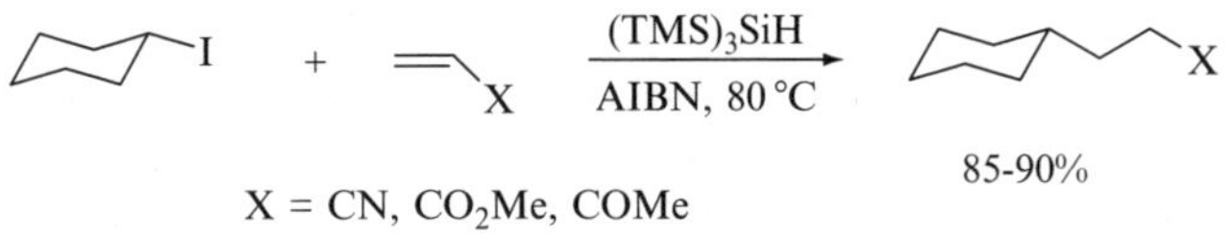

(7.1)

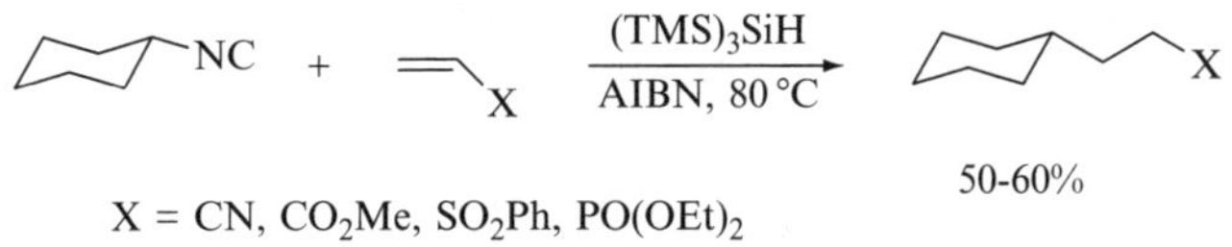

(7.2)

The intermolecular C—C bond formation mediated by $(TMS)_3SiH$ has been the subject of several synthetically useful investigations, with the scope of biological testing. A simple example is given in Reaction (7.3) for the synthesis of bishomolithocholic acid from the corresponding iodo derivative (**1**), which needed a modification on the steroid side chain to be introduced [8]. The ease of isolation and purification of this steroidal analogue allowed a smooth procedure to the subsequent biological assays. The effect of the bulky $(TMS)_3SiH$ can be appreciated in the reaction of β- or γ-substituted α-methylenebutyrolactones with *n*-BuI (Reaction 7.4) [9]. The formation of α, β- or α, γ-disubstituted lactones was obtained in good yield and a good diastereoselection is observed when one of the substituents is a phenyl ring. The prediction of stereochemistry is simplified by considering the conformation of the cyclic radical intermediate involved in the reduction step of the reaction sequence. The *anti* rule can be successfully applied, as in the majority of the cases involving cyclic radicals, that is, H donation preferentially occurs in *anti* fashion to the substituents, in order to reduce steric interference with bulky silicon hydride. As a matter of fact, for β- or γ-alkyl substituted substrates the same reaction was found to be less selective.

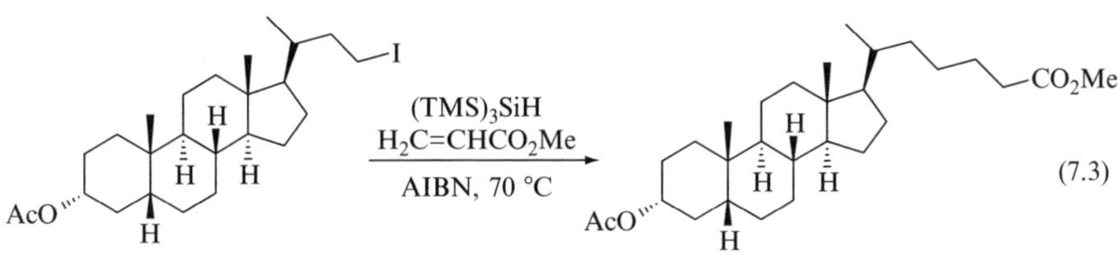

(7.3)

1 64%

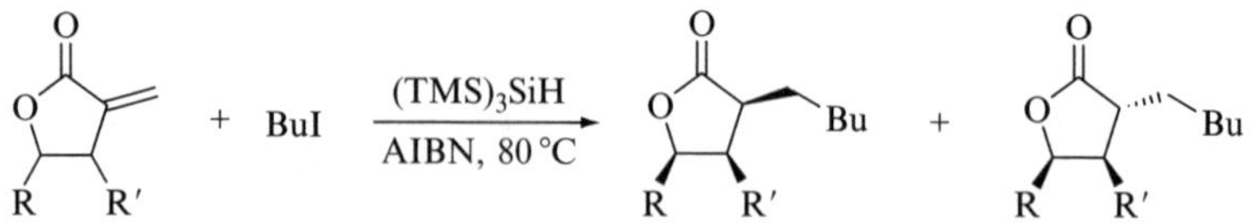

(7.4)

R = Ph, R′ = H 60%, cis:trans = 98:2

R = H, R′ = Ph 60%, cis:trans = 94:6

Stereoelectronic effects can be invoked for the radical reaction at anomeric centre of carbohydrates. The high stereoselective preparation of α-substituted C-glycosyl phosphonates in a α:β ratio of 98:2 was achieved by reductive addition of bromide **2** to α-phosphonoacrylate (Reaction 7.5) [10]. Yields (in parentheses) depend on the sugar configuration: D-galacto (80%), D-manno (47%), D-gluco (30%) and L-fuco (62%).

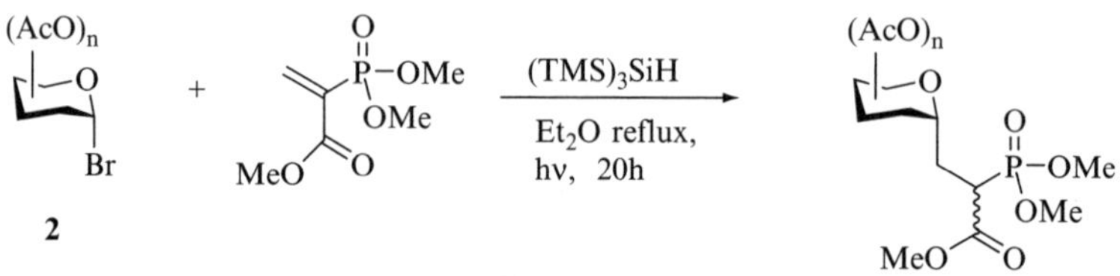

2

(7.5)

30 - 80%

Other silanes have been used as mediators in the intermolecular C—C bond formation. They can be used alone, as in the following example of the disilane $(Ph_2SiH)_2$ shown in Reaction (7.6), for the reductive addition of a bromide or a xanthate to phenyl vinyl sulfone [11,12].

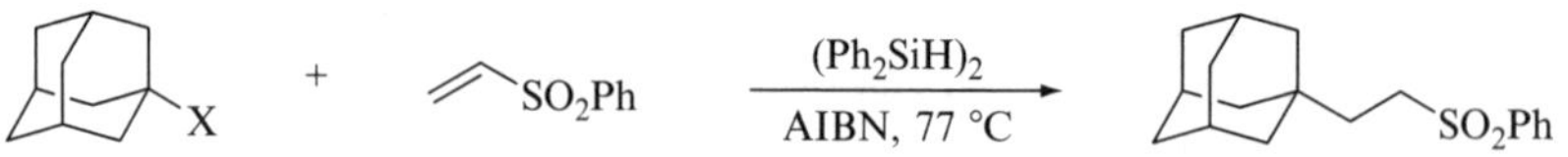

(7.6)

X = Br 88%

X = OC(S)SMe 71%

Alternatively, the couple Ph_3SiH/RSH has been used for the reductive alkylation of electron-rich terminal alkenes as shown in Reaction (7.7) [13]. In this example, 5 mol% of $MeOC(O)CH_2SH$ or Ph_3SiSH is used as the catalyst with a slight excess of Ph_3SiH. Silyl radicals obtain the electrophilic carbon-centred radical through the halogen removal. Subsequently, the thiol effects the hydrogen atom transfer to the adduct radical and the derived RS• radical regenerates the catalyst and the silyl radical (cf. Section 4.5). The enantioselective version of this reaction utilized prochiral alkenes, with Ph_3SiH in the presence of homochiral thiol catalysts (such as 1-thio-β-D-glucopyranose tetraacetate), and gave products of moderate enantiomeric purity.

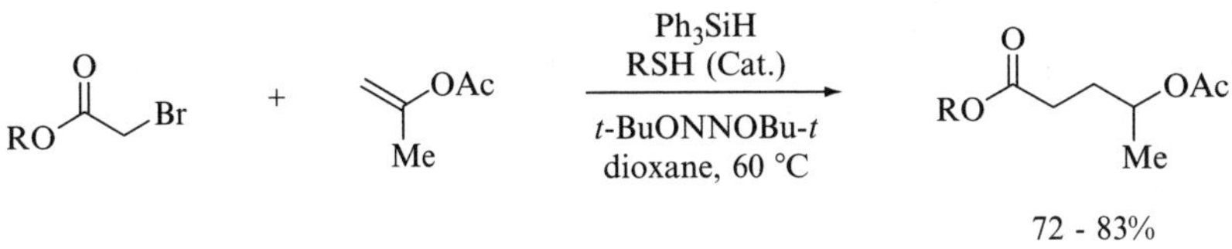

(7.7)

RSH = $MeOC(O)CH_2SH$ or Ph_3SiSH

The hydrogen abstraction by alkyl radicals from the Si—H moiety can successfully occur intramolecularly, thus allowing interesting strategies to be envisaged based on the use of silicon substituents both as protecting groups and as the H-donating moieties [14,15]. An example is given in Scheme 7.2 where the reaction of silane **3** with the bromide affords the compound **4** in a 71 % yield. In particular, the initially generated alkyl radical adds to silane **3**, providing radical **5** which undergoes an intramolecular hydrogen transfer reaction to give the silyl radical **6**. Bromine abstraction completes the cycle of radical chain reactions and the silyl group is easily removed under standard conditions, thus affording the final product.

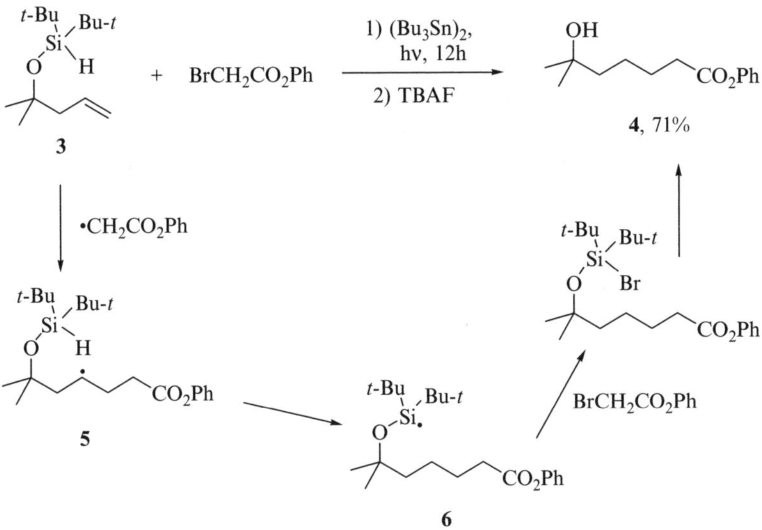

Scheme 7.2 Unimolecular chain transfer reaction

Carbonylation procedures have been successfully used for C—C bond forming radical strategies. Alkyl halides could be carbonylated under moderate pressure of CO (15–30 atm) in benzene at 80 °C in the presence of $(TMS)_3SiH$ and AIBN [16]. Reaction (7.8) shows the effect of the CO pressure on the carbonylation of a primary alkyl bromide. These radical chain reactions proceed by the addition of an alkyl radical onto carbon monoxide, which generates an acyl radical intermediate that, in turn, abstracts hydrogen from the silane to give the corresponding aldehyde. The successful application of carbonylation procedures to tandem strategies will be reported in Section 7.7.

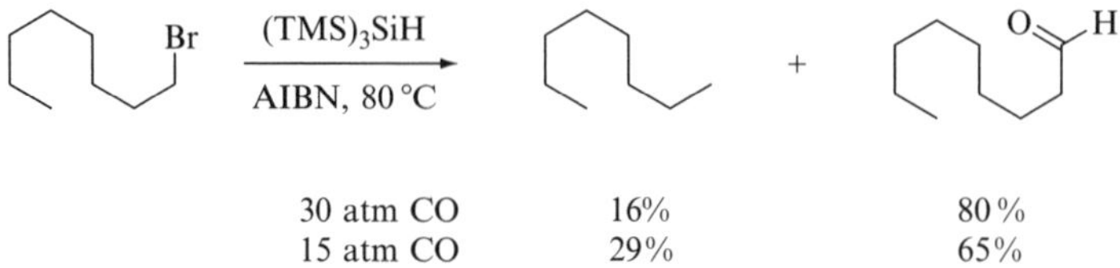

(7.8)

$(TMS)_3SiH$ has also been used as the mediator of C—C bond formation between an acyl radical and an α, β-unsaturated lactam ester (Reaction 7.9). The resulting ketone can be envisaged as potentially useful for the synthesis of 2-acylindole alkaloids [17]. Here, the effects of both H-donating ability and steric hindrance given by the silicon hydride can be seen.

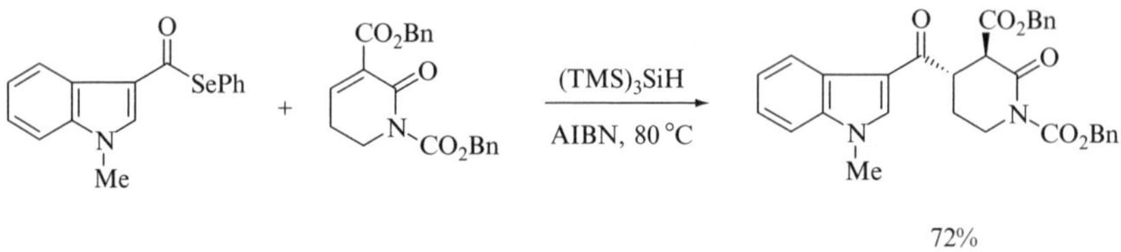

(7.9)

Intermolecular formation of C—C bond can be also coupled with β-elimination process. In this case, the reaction partners are planned in order to generate the alkyl radicals R•, which give the addition to unsaturated bonds. The radical adducts obtained so far do not terminate by H donation from the silane, but give the β-elimination process with loss of an appropriately designed group X•. Radical X• in turn abstracts hydrogen from silicon hydride in order to complete the radical chain. This strategy is carried out with success by the proper choice of the X group (that can give a stabilized radical species) and the hydrogen donor. Examples are given in Reactions (7.10) and (7.11), which formally represent an allylation and a cyanation protocol, respectively. Radical allylation with $(TMS)_3SiH$ as the mediator and 2-functionalized allyl phenyl sulfones, have from moderate to good yields, depending on the nature of the starting materials [18]. Two examples are given in Reaction (7.10). The reaction proceeded via addition of adamantyl radical to the double bond giving rise to an intermediate that undergoes β-scission to form $PhSO_2$• radical, which abstracts hydrogen from the silane to regenerate $(TMS)_3Si$• radical. The glycosyl radical cyanation (Reaction 7.11) yielded α-cyanoglycosides in variable yields, depending on the configuration of sugar: D-galacto (73 %), D-manno (40 %), D-gluco (71 %) and L-fuco (25 %). Only α-cyano anomers were formed [19]. The key propagation steps for these transformations are: the addition of glycosyl radical to the carbon atom of the isocyanide moiety, followed by a fragmentation yielding a *tert*-butyl radical and the desired glycosyl cyanide. *tert*-Butyl radical abstracts hydrogen from the silane to complete the chain.

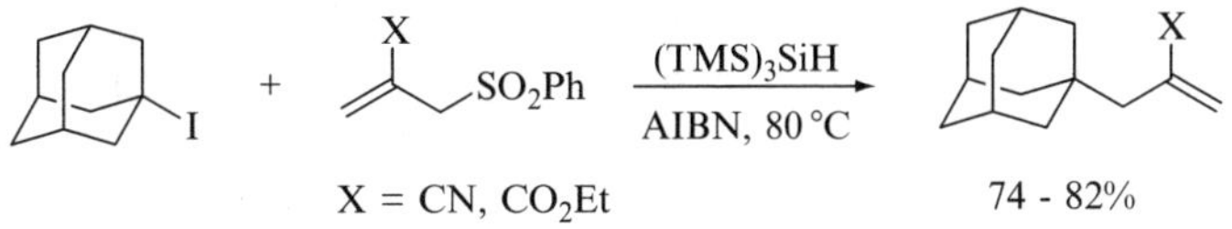

(7.10)

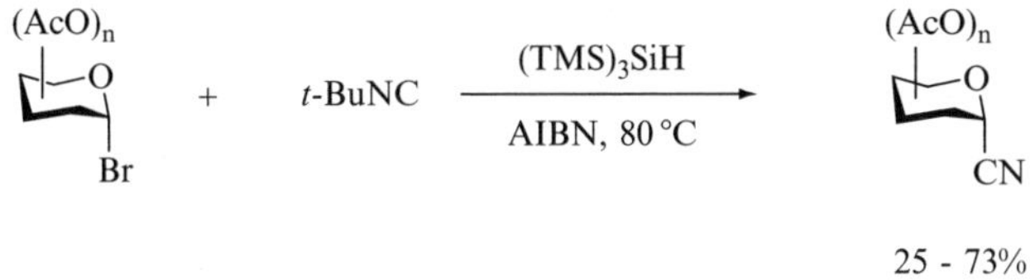

(7.11)

Heteroaromatic substitution can be successfully achieved by silicon hydrides or silanes used as the radical mediators with alkyl bromides or iodides [20,21]. Heteroaromatic bases activated by protonation with trifluoroacetic acid react with alkyl bromides or iodides and $(TMS)_3SiH$ under thermal conditions (in the presence of 1–2 equiv of AIBN in benzene), affording the desired product in moderate to good yields. An example is given in Reaction (7.12). Here, the stoichiometric quantity of AIBN ensured the rearomatization of the stabilized cyclohexadienyl-type radical intermediate, which reaches a stationary concentration suitable for intercepting the α-cyanoisopropyl radical, thus leading to the substitution product. At the same time, the electrophilicity of α-cyanoisopropyl radical prevents it from adding to the protonated heteroaromatic base, while completing the course of this nonchain process [20]. This method is the first report on the alkylation of heteroaromatic bases under non-oxidative conditions.

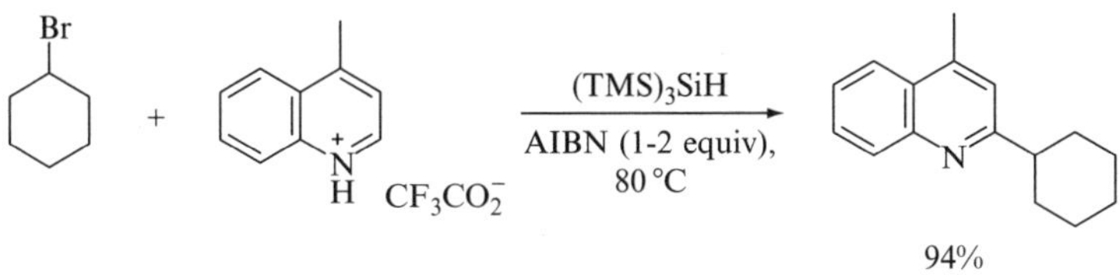

(7.12)

Similar results can be reached under photochemical conditions (irradiation with 400 W of visible light in CH_2Cl_2) using $(TMS)_3SiH$ or $(TMS)_4Si$ [21]. Other silanes are also successful mediators for alkylation of heteroaromatic compounds. 1,1,2,2-Tetraphenyldisilane also operates under thermal decomposition of AIBN [11], whereas $PhSiH_3$, Ph_2SiH_2 and Et_3SiH must be coupled with thermal decomposition of peroxides as radical initiators due to the stronger Si—H bonds [20].

7.3 INTRAMOLECULAR FORMATION OF CARBON–CARBON BONDS (CYCLIZATIONS)

Cyclization reactions have boosted the development of free-radical strategies in organic synthesis. The construction of five- and six-membered ring systems has dominated the field using Bu_3SnH as the mediator. The rate (and efficiency) of cyclization is strongly influenced by the presence of a heteroatom in the forming cycle, as well as by *geminal* disubstituents (Thorpe–Ingold effect). Guidelines for understanding the stereochemical outcome of radical cyclizations are well developed and have been summarized in various papers and reviews [3,22,23]. The so-called Beckwith–Houk rules for radical cyclization have been found to have an excellent predictive power allowing for the planning of radical steps in complex natural product synthetic strategies. The study of the 5-*exo* cyclization of methyl substituted hex-5-en-1-yl radicals has been the model for these rules (Scheme 7.3). The major isomers result from chair-like transition states, where the methyl substituent occupies an equatorial position, and the minor ones form boat-like transition states.

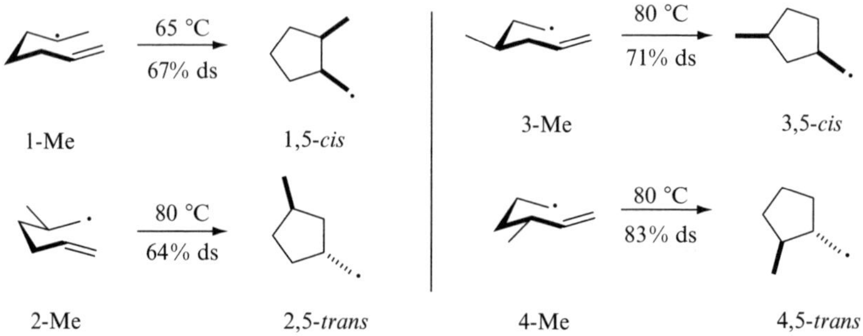

Scheme 7.3 Cyclization of methyl subsituted hex-5-en-1-yl radicals (chair-like arrangements)

Over the last few years, silicon hydrides have been introduced as the radical mediators and the intramolecular C—C bond formation (or cyclization) by these reagents has been the subject of numerous publications.

7.3.1 CONSTRUCTION OF CARBOCYCLES

Reaction (7.13) shows the simplest and most popular type of hex-5-en-1-yl radical cyclization mediated by $(TMS)_3SiH$. Thus, a 50 mmol solution of 5-hexenyl bromide and silane in the reported conditions leads to a 24:1 ratio of cyclized vs uncyclized products [6]. Similar results were obtained by replacing the bromide with isocyanide [7].

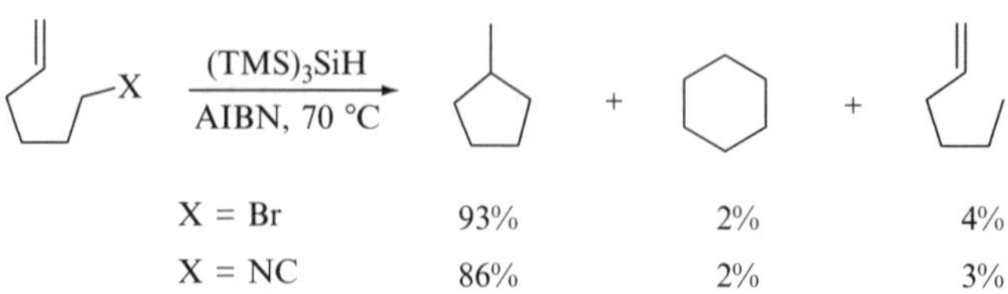

(7.13)

Using this procedure for the secondary isocyanide **7** (Reaction 7.14), the desired product was isolated after workup in 78% yield as a 4.6:1 mixture of *cis* and *trans* isomers [7].

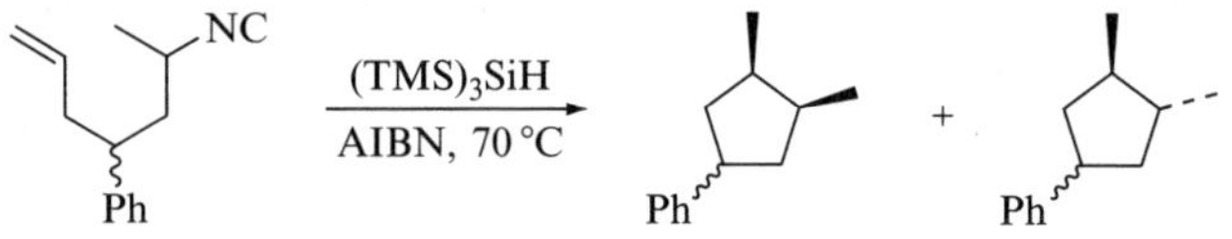

(7.14)

7 78% (4.6:1)

The behaviour of a series of methyl substituted hept-6-en-1-yl radicals, generated from the corresponding iodides with $(TMS)_3SiH$ and AIBN in benzene at 80 °C, has been investigated too. The low regioselectivity and poor stereoselectivities observed for these reactions are not unexpected given the conformational flexibility inherent in the hept-6-en-1-yl radical [24].

Five-membered ring formation has been used for preparing fused cyclic compounds, such as functionalized diquinanes [25] by the reaction of **8** with $(TMS)_3SiH$ under normal conditions. The expected product **9** was isolated in 80% yield and in a α: β ratio of 82:18, as the result of a kinetic controlled reaction. In Reaction (7.15), silyl radicals effect the PhSe group removal and generate the tertiary alkyl radical, which gives cyclization by adding to the double bond. The β-elimination of PhS• radical and the silane trapping of the thiyl radical complete the radical chain.

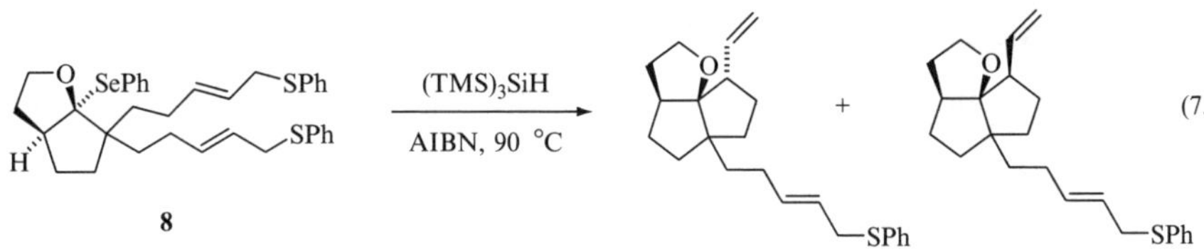

(7.15)

8

9, 80% (α:β = 82:18)

A convenient route to triquinanes is based on a strategy of silyl radical addition to conjugated dienes to form allylic type radicals and their subsequent intramolecular addition to C=C double bonds. By exposure of **10** to $(TMS)_3SiH$ and AIBN at 80 °C (Reaction 7.16) the triquinane **11** is obtained with an unoptimized 51% yield [26].

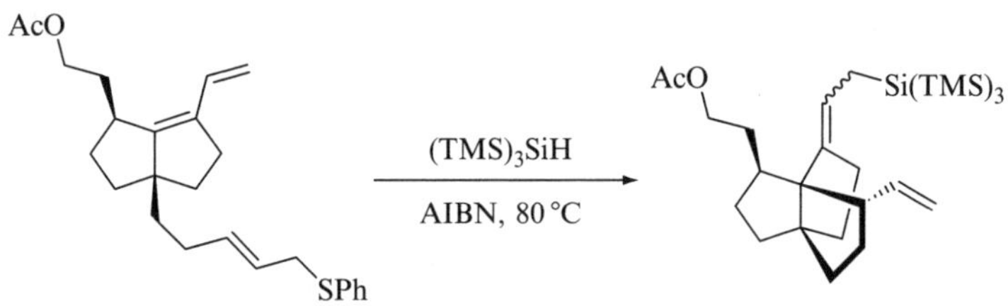

(7.16)

10 **11**, 51%

The cyclization of δ, ε-unsaturated acyl radicals has been the research subject of several groups [27]. The propagation steps for the prototype reaction are illustrated in Scheme 7.4. The 5-*exo*:6-*endo* product ratio varies with the change of the silane concentration due to the competition of hydrogen abstraction from the silane with the ring expansion path.

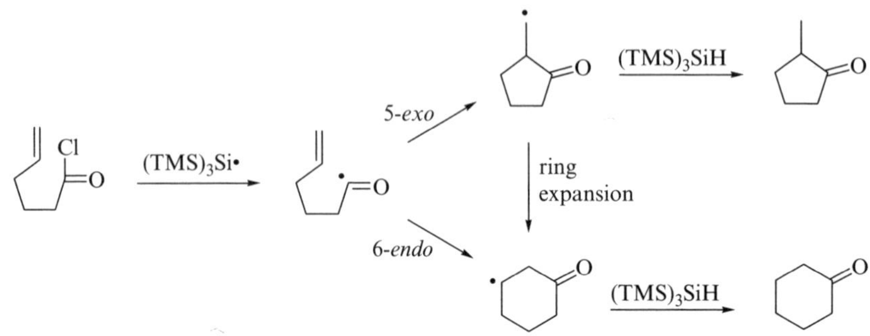

Scheme 7.4 Propagation steps involving the cyclization of acyl radicals

Cyclization of secondary alkyl radicals can occur with α, β-alkynyl esters, such as **12**, and proceeds with high stereoselectivity to give predominantly (*Z*)-exocyclic alkenes at low temperature upon reaction with $(TMS)_3SiH$ (Reaction 7.17) [28].

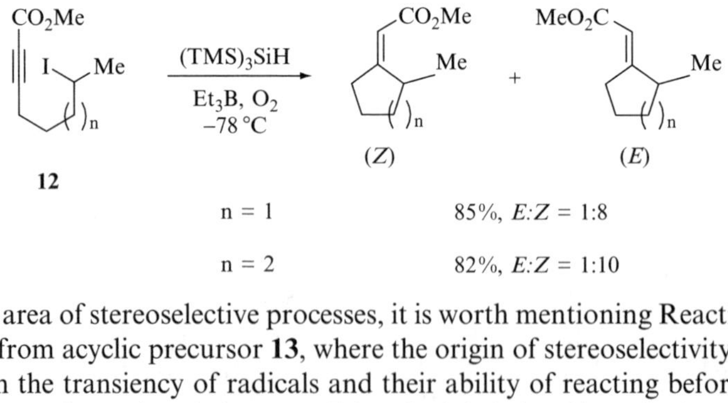

(7.17)

12

n = 1 85%, *E*:*Z* = 1:8

n = 2 82%, *E*:*Z* = 1:10

In the area of stereoselective processes, it is worth mentioning Reaction (7.18) starting from acyclic precursor **13**, where the origin of stereoselectivity could be found on the transiency of radicals and their ability of reacting before racemization or conformational changes. This principle is based on the knowledge of lifetime and reactivity of radicals and is called 'stereoselection at the steady state' [29].

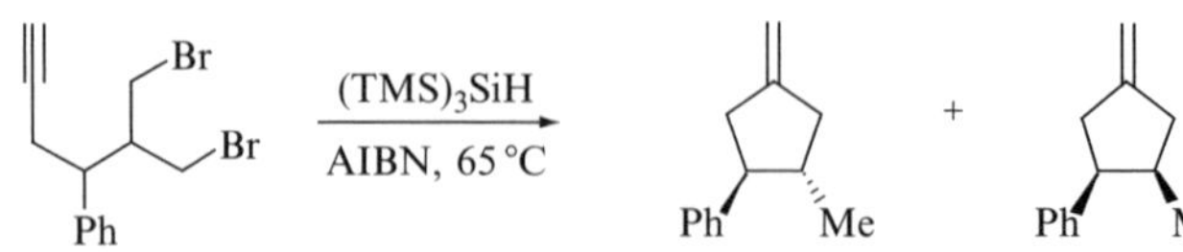

(7.18)

13

99%, trans:cis = 1.4:1

Taking advantage of the high reactivity of alkoxyl radicals towards organosilanes, which is 2–3 orders of magnitude higher than that of primary alkyl radicals (see Chapter 3), cyclization can start from halocarbonyl compounds using $PhSiH_3$ (Reaction 7.19) or $(TMS)_3SiH$ [30]. Both 6-*exo-trig* and 5-*exo-trig* cyclizations of alkyl radicals to the carbonyl moieties can be accomplished, with aldehydes working better than ketones.

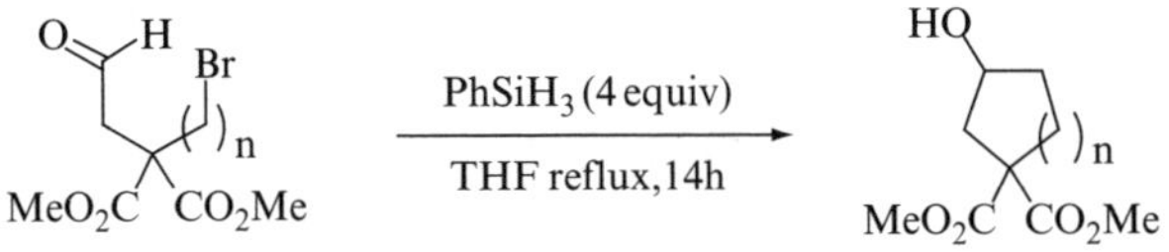

n = 1, 2 85-93% (7.19)

In Reaction (7.20) is reported the cyclization of a thermally unstable propargyl bromide cobalt complex mediated by Ph_2SiH_2 at room temperature and Et_3B/O_2 as the radical initiator. However, a mixture of reduced and bromine atom-transfer products (1:1.8 ratio) are isolated due to the low hydrogen donation of the employed silane [31].

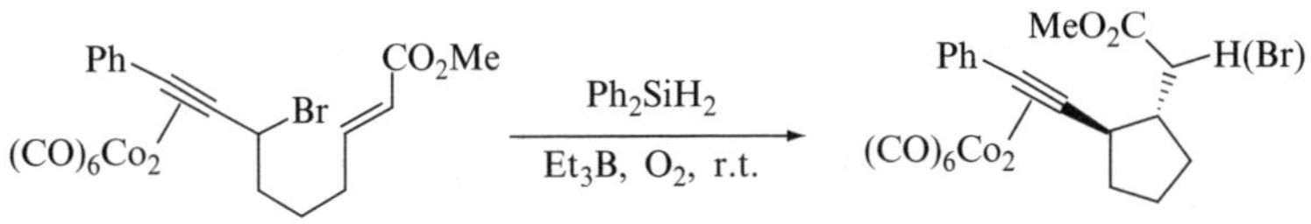

70% (7.20)

Silyl radical adds across the double or triple bonds to generate carbon-centred radicals ready for cyclization reactions. Two examples of hydrosilylation/cyclization of dienes are given by using Et_3SiH or the silylated cyclohexadienes **14** as reagents. Reaction (7.21) shows an example of the bridged bicyclic formation by initial attack of $Et_3Si•$ radical on the terminal alkene [32]. Reaction (7.22) shows the cyclization of a 1,6-diene and how flexible is this methodology to introduce any silyl group in the reaction products [33] (cf. Section 4.7). Yields are moderate to good and the *cis*:*trans* ratio of ca 4:1 does not change substantially with the nature of silylating group as expected. Another potentiality for the construction of carbocycles is based on the addition of silyl radicals to the carbonyl group followed by cyclization. Starting from ketoalkene **15**, the formation of a five-membered ring is obtained in good yield and discrete diastereoselectivity (Reaction 7.23) [34].

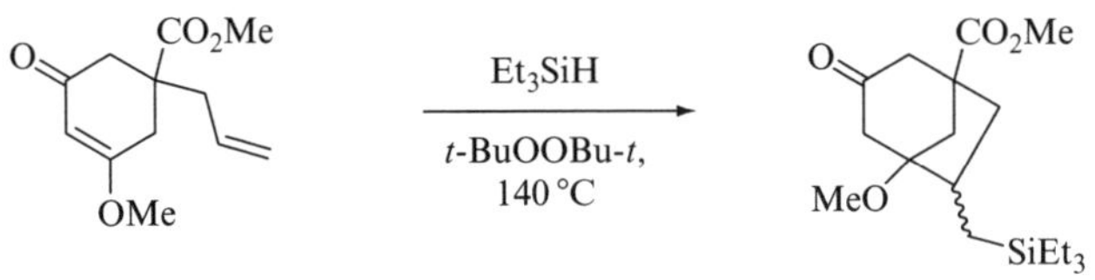

68% (7.21)

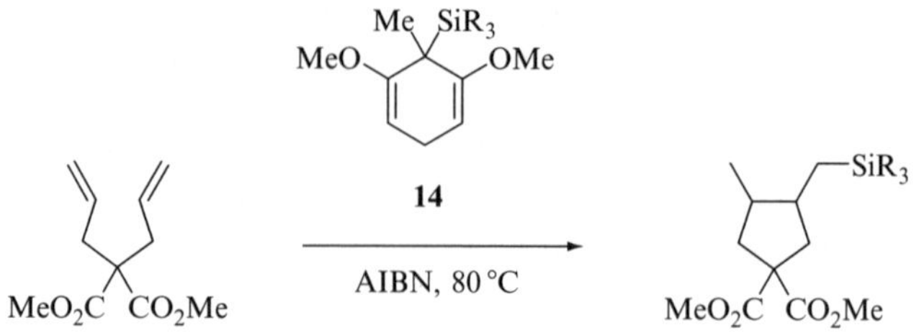

(7.22)

R_3Si = *i*-Pr_3Si, *t*-$BuMe_2Si$, $PhMe_2Si$ 47-84%

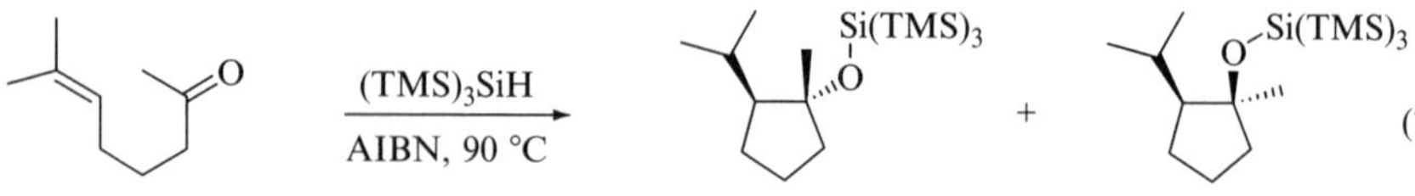

(7.23)

15 **16** **17**

67%, **16**:**17** = 70:30

7.3.2 CONSTRUCTION OF CYCLIC ETHERS AND LACTONES

Stereocontrolled synthesis of cyclic ether moieties gives access to a series of biologically important targets. Methods that utilize radicals as the reactive intermediates have gained considerable prominence in recent years, primarily due to the excellent stereocontrol that could be obtained. A thorough study of the formation of five- to seven-membered cyclic ethers has been performed by the cyclization of vinylogous sulfonates. An example is given in Reaction (7.24) [35]. Treatment of **18** with $(TMS)_3SiH$ and Et_3B at room temperature, in the presence of air, gave the cyclized product with a high *cis*-diastereoselectivity, due to the *exo* mode and with decreased yields depending on the ring size increases.

Several approaches dealt with the preparation of tetrahydrofuran derivatives, starting from optically active compounds. The sulfoxide **19** was debrominated by $(TMS)_3SiH$ and AIBN in refluxing benzene and behaved in a totally regio-selective 5-*exo* mode affording the ring closure. The diastereoselective outcome of the radical cyclization depended strongly on the configuration of the double bond, which reacted in the *s-trans* conformation during the cyclization, i.e., the attacking alkyl radical is in *anti* to the bulky *p*-tolyl group (Reaction 7.25) [36].

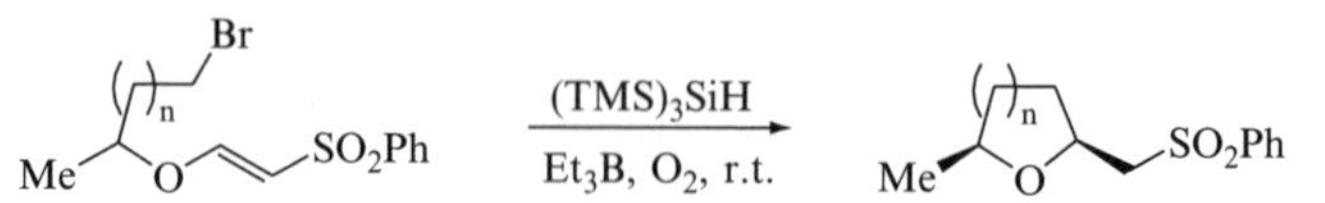

(7.24)

18

n = 1	99%, cis:trans ≥19:1
n = 2	90%, cis.trans ≥19:1
n = 3	34%, cis:trans ≥19:1

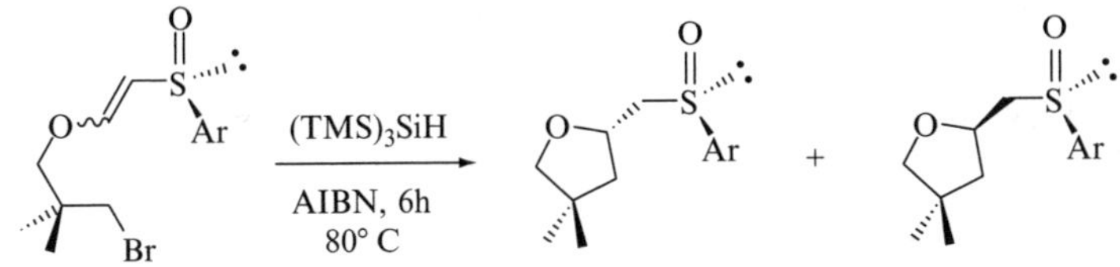

(7.25)

19 **20** (1*R*,3*S*) **21** (1*R*,3*R*)

19-(*E*)	93% (**20**:**21** = 82:18)
19-(*Z*)	90% (**20**:**21**≥ 2:98)

cis-2,5-Disubstituted tetrahydrofurans and *cis*-2,6-disubstituted tetrahydropyrans are conveniently formed by radical cyclization of β-alkoxyacrylates. A strategy using β-alkoxymethacrylate **22** led to the stereoselective preparation of the benzyl ether of (+)-methyl nonactate, resulting from exclusive formation of the threo product (Reaction 7.26) [37]. It is worth mentioning that in this strategy the 'cis-2,5' selectivity in the forming tetrahydrofuranyl ring is coupled with the 'threo' selectivity at the exocyclic α sites (**23**), and it should offer easy access to a number of natural products.

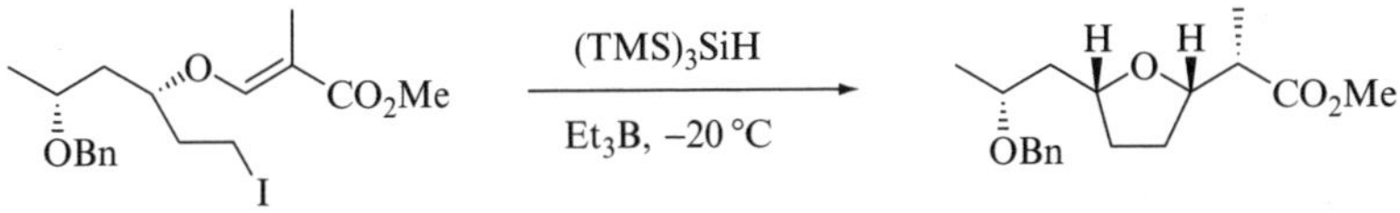

(7.26)

22 90%, de>96%

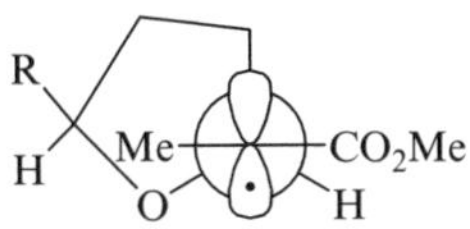

23

In the next example, the strategy for the formation of fused cyclic ethers utilized the formation of an intermediate α-heterosubstituted carbon radical, generated by the reaction of $(TMS)_3Si$• radical with *N*-(ethoxycarbonyl)-1,3-thiazolidine derivative [38]. This intermediate gives intramolecular C—C bond formation in the presence of proximate 1,2-disubstituted double bonds (Reaction 7.27). However, when terminal double bonds are used, the hydrosilylation of the double bond by $(TMS)_3SiH$ can compete with the reduction and prevent forming the desired C—C bond formation.

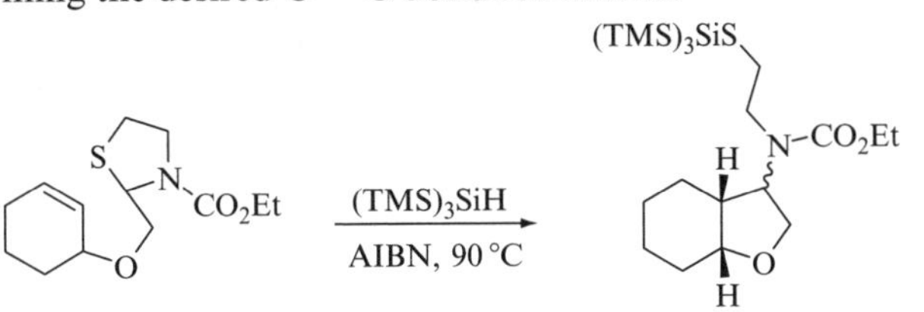

(7.27)

79%, *endo*:*exo* = 2.4:1

Fused cyclic ethers can be derived from appropriately substituted sugars. An example is given with the stereoselective 5-*exo* radical cyclization of allylic 2-bromo-2-deoxysugars, in the presence of 1,1,2,2-tetraphenyldisilane as the radical mediator and AIBN in refluxing ethyl acetate. The corresponding *cis*-fused bicyclic sugars have been prepared in moderate to good yields (Reaction 7.28) [39].

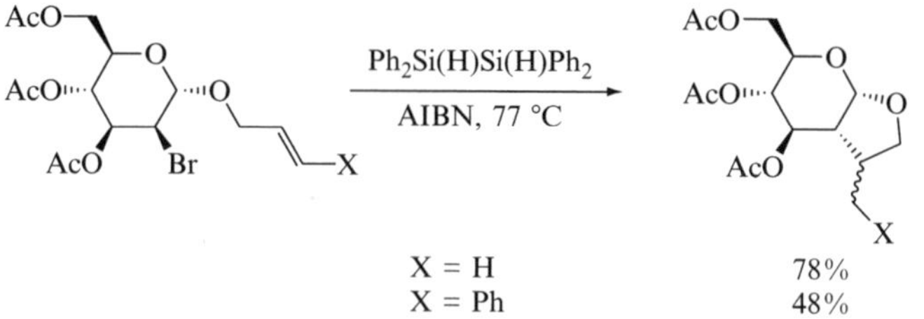

(7.28)

The chemoselective addition of silyl radical to the double bond of the β-alkenyloxyenone derivative **24** was instead planned in Reaction (7.29) and accompanied by a 5-*exo-trig* radical cyclization leading to the diastereomeric cyclic ether products [40].

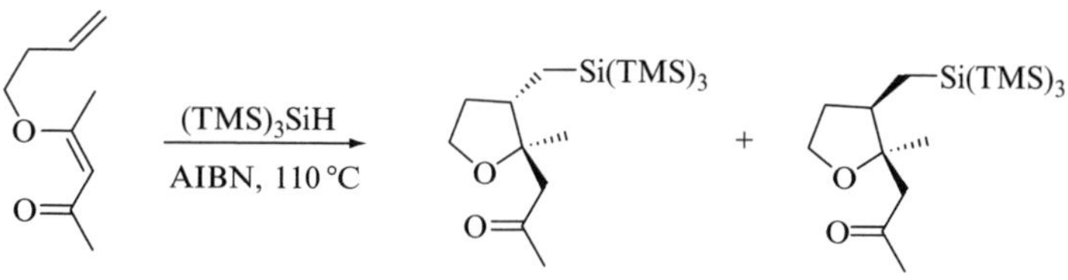

(7.29)

24 25% 58%

Acyl radical cyclization has been also used for the synthesis of oxygen-containing rings [27], using $(TMS)_3SiH$ as the mediator and phenylseleno esters as acyl radical precursors. Reaction (7.30) shows again the use of vinylogous sulfonates, such as **25**, which affords the *cis*-2,5-disubstituted cyclic derivatives in good yields on exposure at room temperature to $(TMS)_3SiH$ and Et_3B/O_2 combination as the initiator, following an *exo*-mode ring closure [35]. The selectivity is explained in the standard manner through a chair-like transition state. Related reactions in which CO_2Me replaced the SO_2Ph group behaved similarly [41]. Reaction (7.31) is relevant as the key step for the enantioselective synthesis of nonisoprenoid sequiterpene (−)-kumausallene [42]. The radical cyclization at low temperature furnished the cyclic ethers in a 32:1 mixture in favour of 2,5-*cis* diastereoisomer, whereas the analogous reaction at room temperature gave only modest diastereoselectivity (*cis*:*trans* = 8:1).

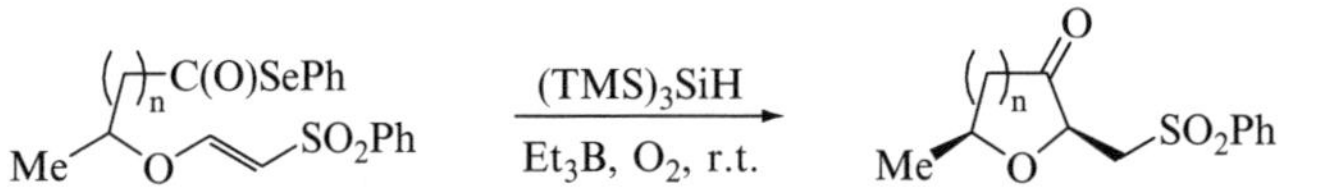

(7.30)

25

n = 1 87%, cis:trans = 17:1
n = 2 90%, cis.trans = 6:1
n = 3 63%, cis:trans = 17:1

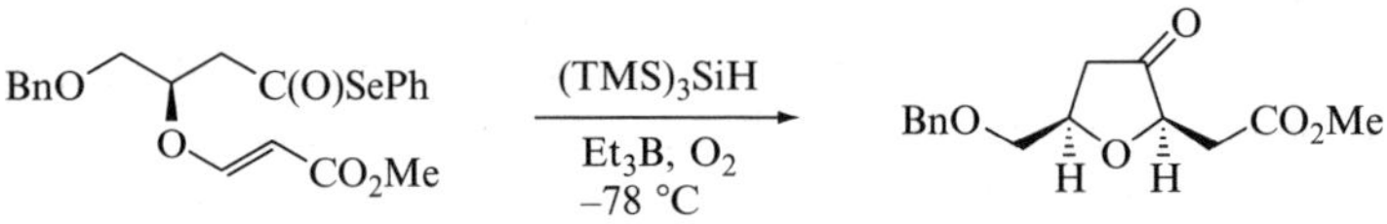

(7.31)

92 %, cis:trans = 32:1

In order to improve the poor diastereoselectivity of 6-*exo-trig* cyclizations, the configuration of a double bond was carefully considered (Reaction 7.32). Indeed, the *Z*-vinylogous sulfonate **26** under standard experimental conditions furnished the expected cyclic ethers with an excellent diastereoselectivity in favour of *cis*, thus confirming the relevance of the double bond geometry to the rotamer population. Indeed, the higher selectivity observed for **26**-(*Z*) is thought to arise from stronger 1,3-diaxial interactions in the chair-like transition state for the formation of the minor isomer [35]. This approach has been applied to cyclization of **27**, which afforded the *cis*-2,6-disubstituted tetrahydropyran-3-one in 81 % yield as a single diastereoisomer, with close structural requirements to C16–C26 fragment of the antitumour agent Mucocin (Reaction 7.33) [43]. Improved selectivity in 6-*exo-trig* cyclization was also found in substrates with a preexisting ring such as **28**, which gave high diastereoselectivity arisen from an increased rigidity imposed to the transition state (Reaction 7.34) [44].

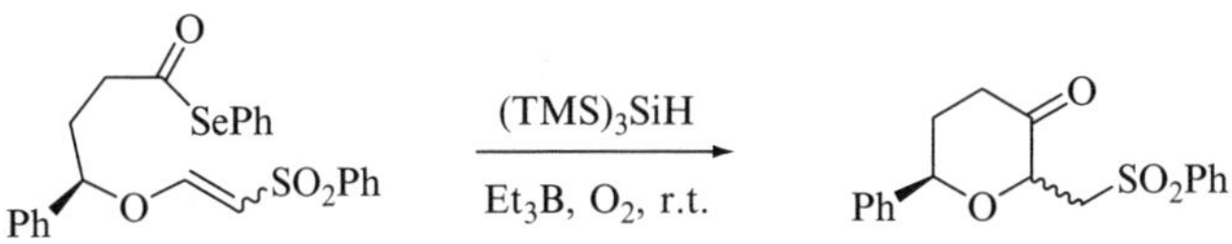

(7.32)

26-(*E*) 85%, cis:trans = 7:1
26-(*Z*) 84%, cis:trans = 35:1

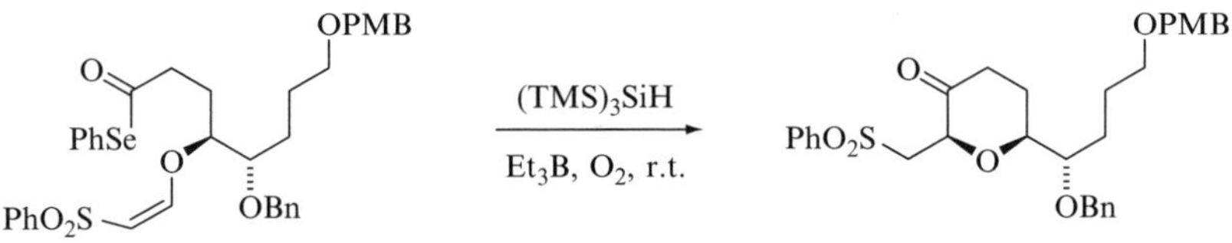

(7.33)

27 PMB = *p*-$MeOC_6H_4CH_2$ 81%

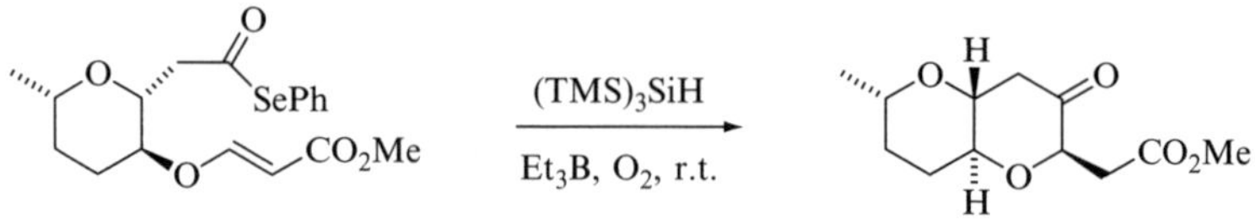

(7.34)

28 91%, ds>95:5

The use of $PhSeSiR_3$ coupled with a photoinduced electron transfer methodology has been used for the C—C bond formation via the group transfer radical reactions. Two examples are given in Reaction (7.35) starting either from a bromide or phenylselenide, which afford the same furanic compound [45,46]. The reaction mechanism of a certain complexity has been understood in some detail. Scheme 7.5 shows the mechanistic paths, concerning the construction of C—C bond (for the mechanistic paths of $PhSeSiR_3^{-\bullet}$ formation see Section 1.1). In particular, silyl radicals generated from the mesolysis of $PhSeSiR_3^{-\bullet}$ abstract Br or PhSe groups to generate alkyl radicals, while oxidative dimerization of the counteranion $PhSe^-$ to PhSeSePh functions as the radical terminator of the cyclized radical.

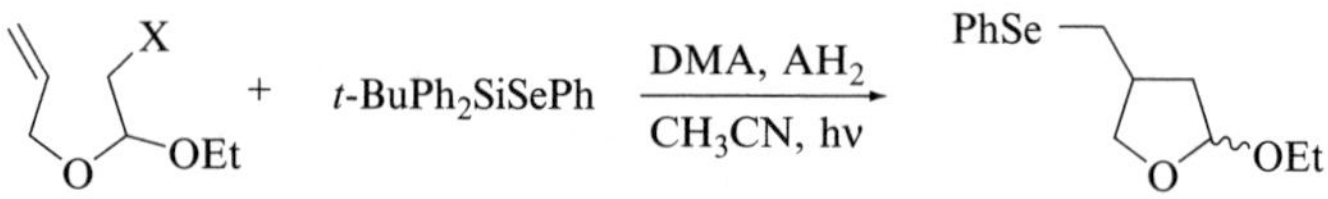

(7.35)

X = Br 81%, cis:trans = 16:84
X = SePh 75%

DMA = 9,10-dimethoxyanthracene
AH_2 = ascorbic acid

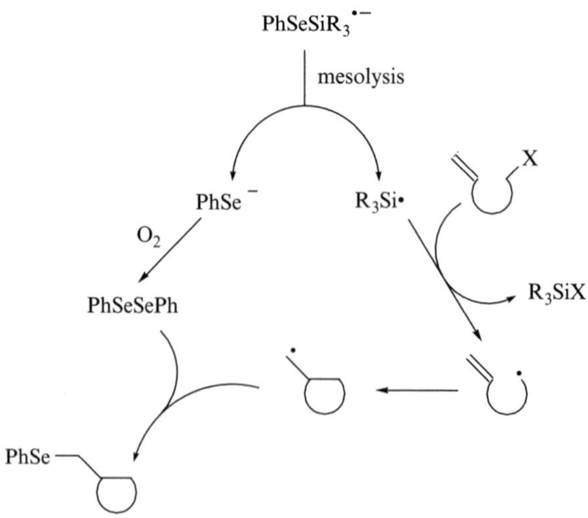

Scheme 7.5 Group transfer radical reation using a photoinduced electron transfer method for the generation of $PhSeSiR_3^{-\bullet}$ (see Scheme 1.1 in Chapter 1)

The synthesis of oxygen heterocycles in which cyclization onto a pendant alkyne is a key step has also been achieved. Reaction (7.36) shows an example of iodoacetal **29** cyclization at low temperature that afforded the expected furanic derivative in moderate *Z* selectivity [47]. A nice example of Lewis acid complexation which assists the radical cyclization is given by aluminium tris(2,6-diphenyl phenoxide) (ATPH) [48]. The β-iodoether **30** can be complexed by 2 equiv of ATPH, which has a very important template effect, facilitating the subsequent radical intramolecular addition and orienting the $(TMS)_3SiH$ approach from one face. The result is the formation of cyclization products with *Z* selectivity and in quantitative yield (Reaction 7.37).

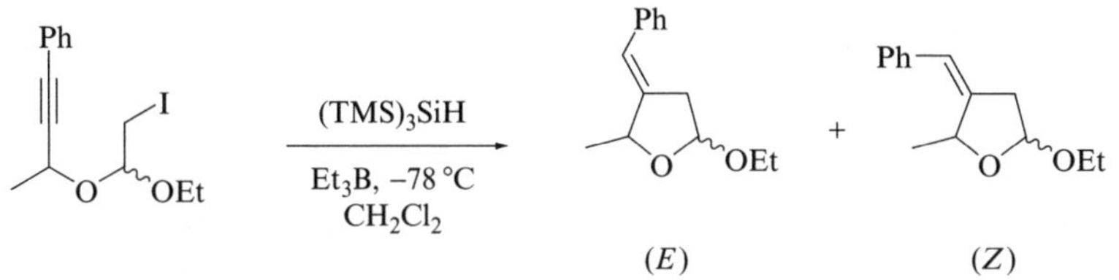

(7.36)

29

76%, *E*:*Z* = 1:6.7

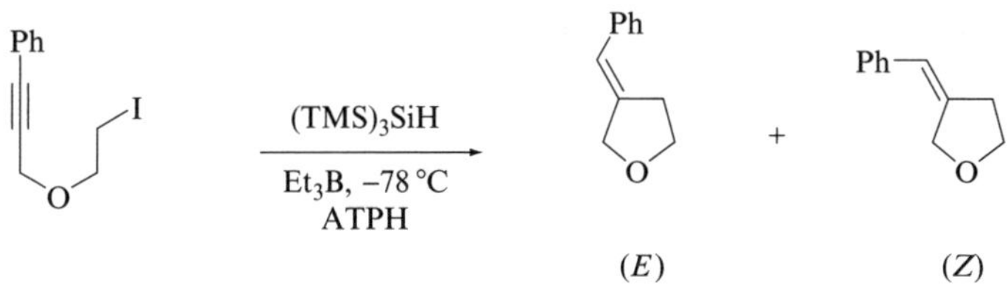

(7.37)

30

99%, *E*:*Z* = <1: >99

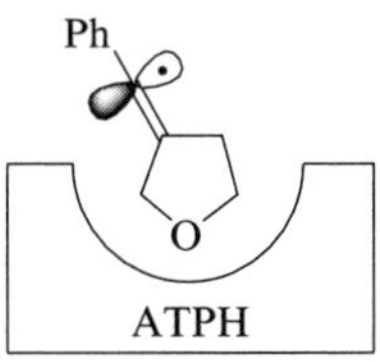

31

In analogy with the strategy of carbocyclic construction, 1,6-enynes containing an oxygen heteroatom in the carbon atom sequence have been used for 3,4-disubstituted tetrahydrofuran synthesis. The simplest example is given by the hydrosilylation of enyne at room temperature (Reaction 7.38) [49]. Tetrahydrofurans with an exocyclic methylene functionality can also be prepared from the appropriate alkynes, such as **32**, with $(TMS)_3SiH$ in refluxing benzene which afforded exclusive formation of the exomethylene in the *Z* conformation (Reaction 7.39) [50].

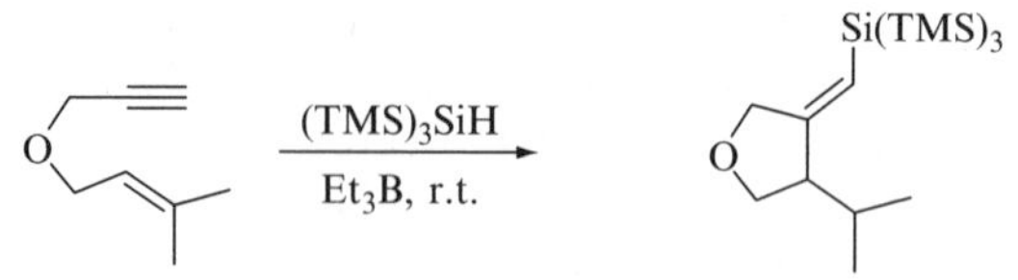

83%

(7.38)

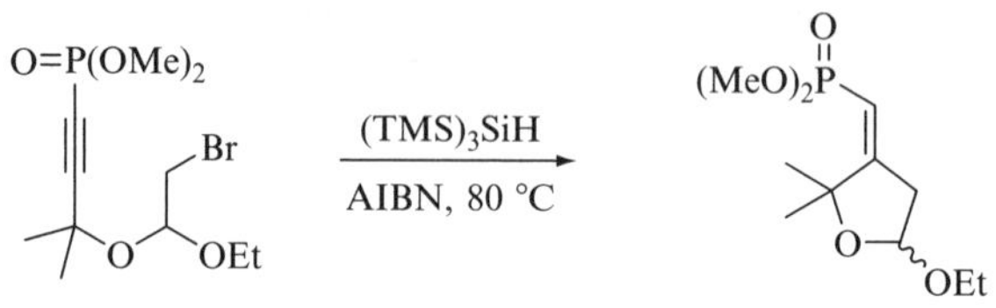

32 67%

(7.39)

A convenient preparation of dihydrobenzofurans has been achieved from the appropriately functionalized *ortho*-halophenol derivatives. Treatment of the aryl iodide (*S*)-**33** with $(TMS)_3SiH$ and Et_3B in the presence of air at room temperature, gave the aryl radical which cyclized in a 5-*exo-trig* mode and provided the bicyclic derivatives **34**/**35** as a 29:1 mixture of diastereoisomers in favour of **34** (Reaction 7.40) [51].

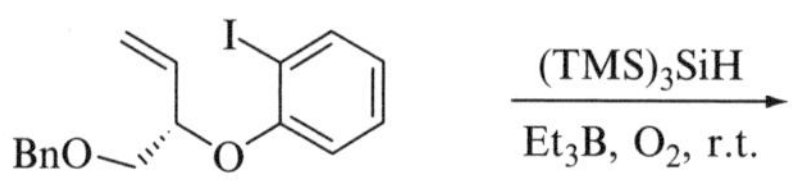

(*S*)-**33**

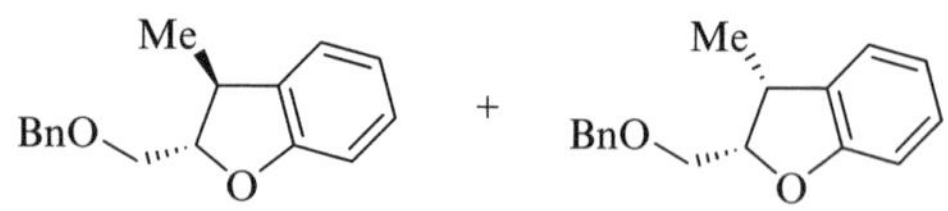

(7.40)

34 **35**

76%, **34**:**35** = 29:1

The cyclization strategy using silanes as mediators has been used for lactone formation by radical reaction of appropriately designed α, β-unsaturated or *ortho*-halobenzoate esters. Reaction (7.41) gives an example of the bromide **36** treated with $(TMS)_3SiH$ and AIBN in refluxing benzene, which affords the expected lactone in a good yield but with a insignificant diastereoselectivity (1.3:1) [52]. Two examples of aryl radical reactivity in suitable benzoate esters allowed for successful spirocyclization processes and the synthesis of a series of novel spirolactones. The keto spiro-γ-lactones (Reaction 7.42), spirodilactones, spirolactone-lactams and spirolactone-thiolactones (Reaction 7.43) have been prepared by this quite general route [53,54].

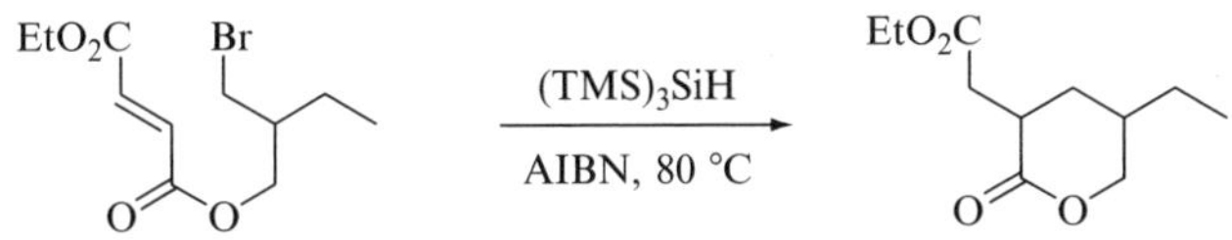

36 91%

(7.41)

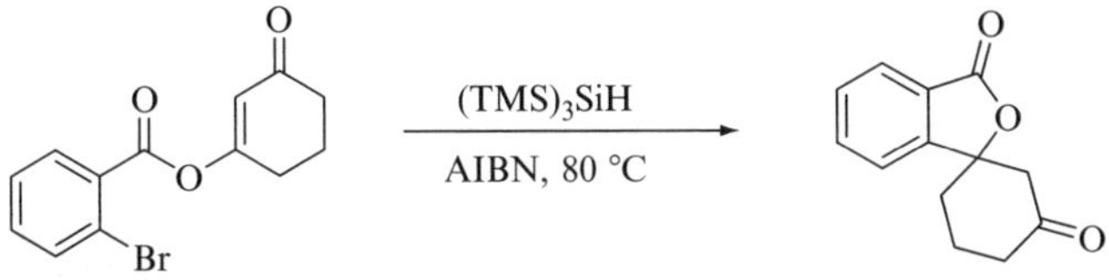

83%

(7.42)

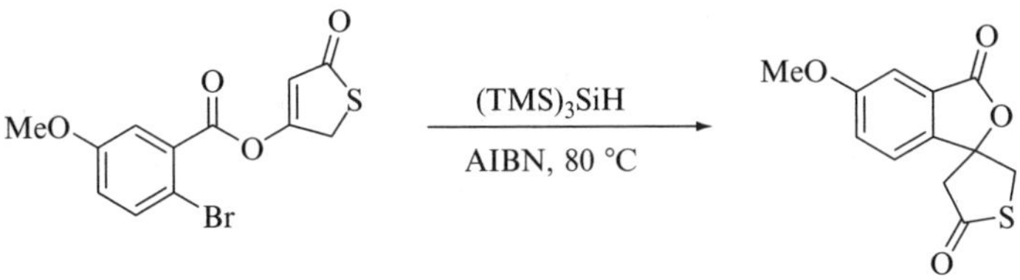

67%

(7.43)

7.3.3 CONSTRUCTION OF CYCLIC AMINES AND LACTAMES

The radical-based strategy has invaded the field of *N*-containing heterocycles. Cyclizations mediated by silyl radicals have been introduced as the key step in the synthesis of alkaloids and pharmacologically active compounds, with many advantages both in terms of selectivity and bio-compatibility. Some of the most significant and innovative examples are described in this section.

The construction of an indolizidine skeleton has been successfully obtained by radical cyclizations mediated by $(TMS)_3SiH$. Reaction (7.44) represented a key step in the total synthesis of (−)-slaframine. The two pairs of diastereomers were first separated and then hydrolysed to the corresponding alcohols in a 76% overall yield [55]. On the other hand the cyclization of the *N*-iodopropyl pyridinones in Reaction (7.45) occurs smoothly at room temperature using Et_3B/O_2 as initiator, to give the desired products with a trifluoromethyl group at the bridgehead position in a *syn/anti* ratio of 7:3 [56].

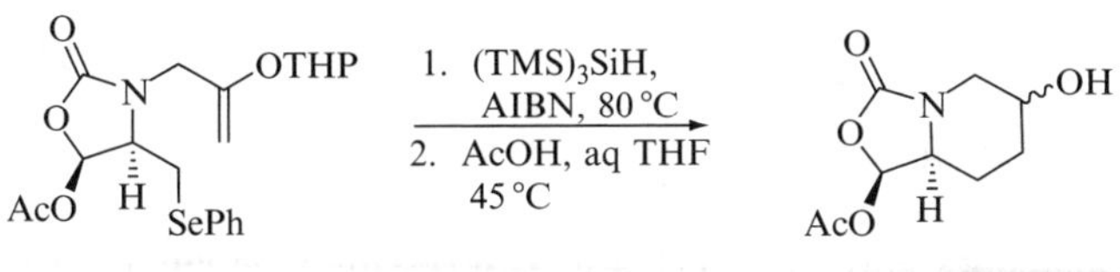

76%, α:β = 7:1

(7.44)

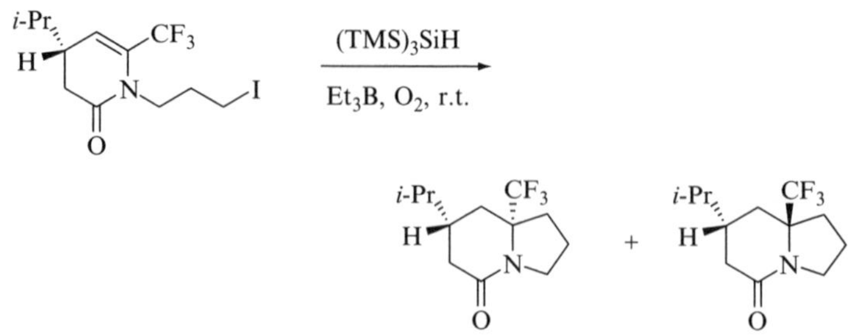

(7.45)

60%, *syn*:*anti* = 7:3

A synthetic strategy for the stereoselective construction of *trans*-2,5-disubstitutes cyclic amines has been reported [57]. The alkyl bromides **37** were prepared with an oxauracil moiety to function as the free radical acceptor. The reaction with $(TMS)_3SiH$ and Et_3B in the presence of air at room temperature provided the corresponding azabicycles **38** in high diastereomeric ratio (Reaction 7.46), as the consequence of the stereoselective intramolecular addition of alkyl radical to the oxauracil acceptor, driven by the rotamer preference. Similar results have been obtained by replacing the Me group with Ph in the chain. Further treatment of the azabicycles **38** afford the amino esters **39**, using *p*-methoxyphenol, or the dipeptides **40,** by reaction with $BnNH_2$ and then $Et_3N/PhC(O)Cl$.

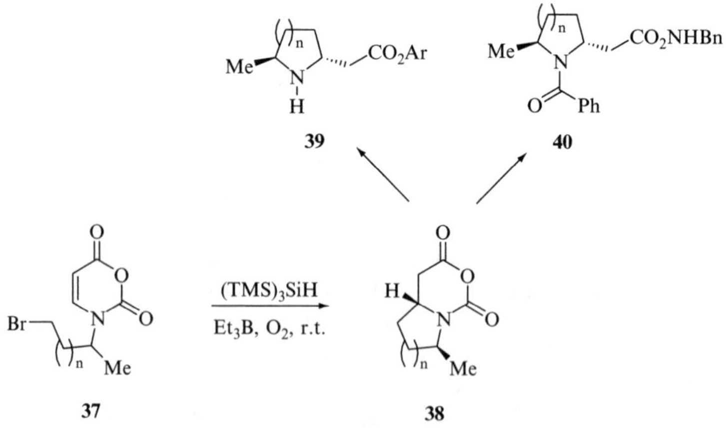

(7.46)

n = 1 97%, ds 37:1
n = 2 94%, ds 29:1
n = 3 58%, ds 24:1

The above-described synthetic strategy has also been adapted to the cyclization of acyl radicals [57]. Two examples are given in Reaction (7.47). The intramolecular addition of acyl radicals to the oxauracil moiety is also an efficient reaction for the construction of five-, six-, and seven-membered rings. By replacing the radical acceptor with oxathymine, an additional stereogenic centre at C5′ position is introduced.

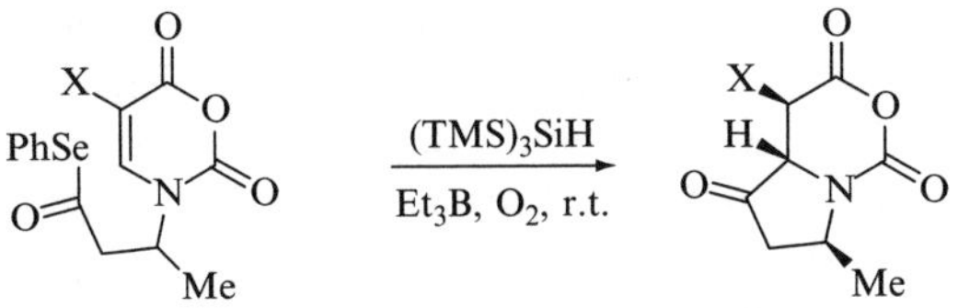

(7.47)

X = H, Me 92%, ds ≥19:1

The validity of the oxauracil strategy for a stereoselective construction of the azabicyclic core has been proven for the syntheses of biologically important polyguanidinium alkaloids, namely, batzelladine A and D [58]. Reaction (7.48) shows that treatment of **41** under the standard free-radical conditions furnished the desired azabicycle in a 77 % yield, in a high diastereoisomeric ratio. Here it can be seen the effect of H donation from the more accessible face of the molecule.

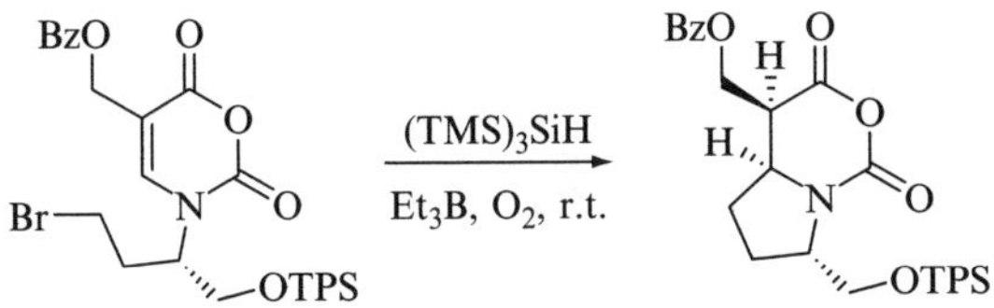

(7.48)

41 77%, ds 22:1

The aryl radical cyclization has been successfully used for the preparation of substituted dihydrobenzo[*b*]indoline derivatives [59]. An example is shown in Reaction (7.49). The diene **42** was preliminarily subjected to ring-closure metathesis using Grubbs catalyst and then treated with (TMS)$_3$SiH and Et$_3$B at −20 °C, in the presence of air, to provide the compound **43** with an excellent diastereoselectivity.

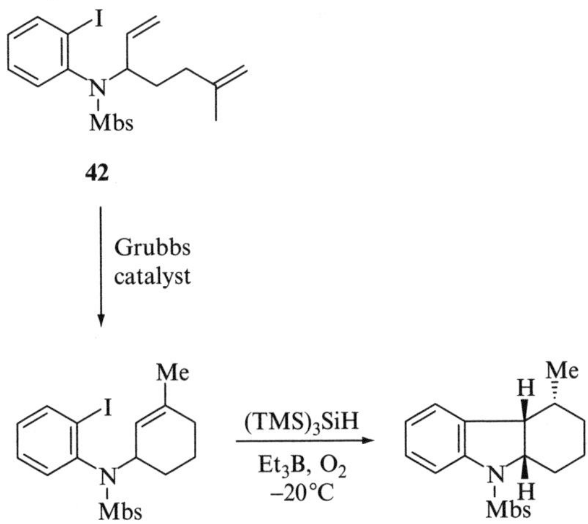

(7.49)

43, 85%, ds ≥19:1

Another strategy for the synthesis of *N*-containing heterocycles utilizes the azido group as the radical acceptor (Reaction 7.50). This is possible by using $(TMS)_3SiH$ which is relatively inert towards azides and can activate alkyl bromides or thionocarbonates as radical precursors [60,61]. An example is given in Reaction (7.51) where the amine product was tosylated before work-up. Alternatively, the carbon-centred radical derived by the addition of $(TMS)_3Si$• radical to the carbonyl group of **44** cyclized to the azido group to afford the bicyclic product in good yields, independently from the ring size (Reaction 7.52).

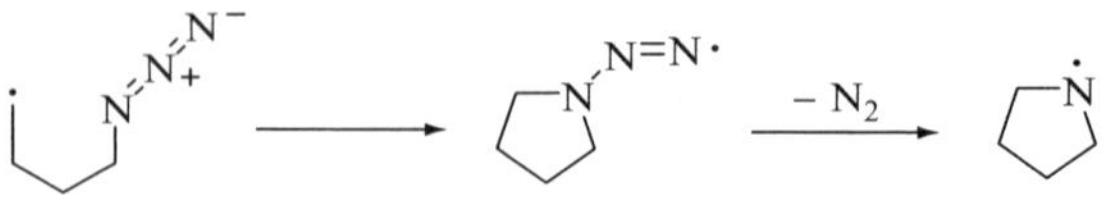

(7.50)

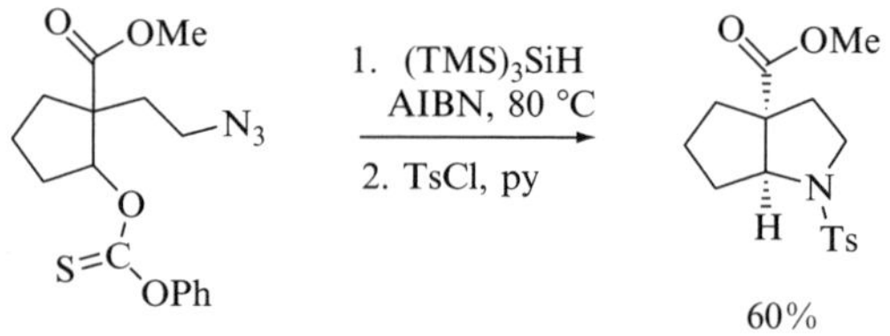

60%

(7.51)

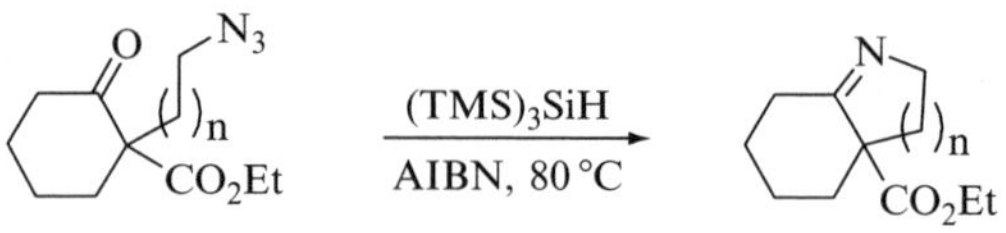

44, n = 1, 2

78%

(7.52)

The α-silyloxy alkyl radical generated by the addition of $(TMS)_3Si$• radical to the aldehyde moiety of **45** has been employed in radical cyclization of β-aminoacrylates (Reaction 7.53); the *trans*-hydroxy ester and the lactone in a 2.4:1 ratio were the two products [62].

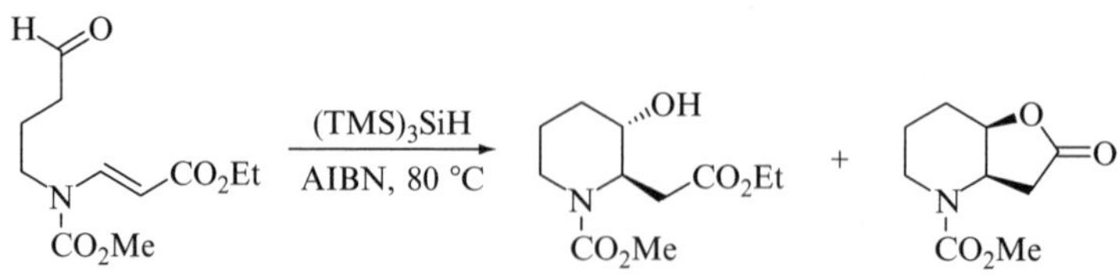

45

60%

25%

(7.53)

Analogously to the carbocycle and oxycycle synthesis, cyclic amines can be obtained by the hydrosilylation of a suitable enyne, such as **46** (Reaction 7.54), which gave the six-membered ring via a 6-*endo* cyclization of the vinyl radical onto the C═N bond [63]. In another example, the isothiocyanide functionality of compounds **47** or **48** reacts with silane under radical conditions and leads to the formation of cyclic pyroglutamate derivatives in excellent yields. Indeed, the products derived from the cyclization of α-silylthio imidoyl radical (Reaction 7.55) hydrolysed spontaneously during chromatography [64].

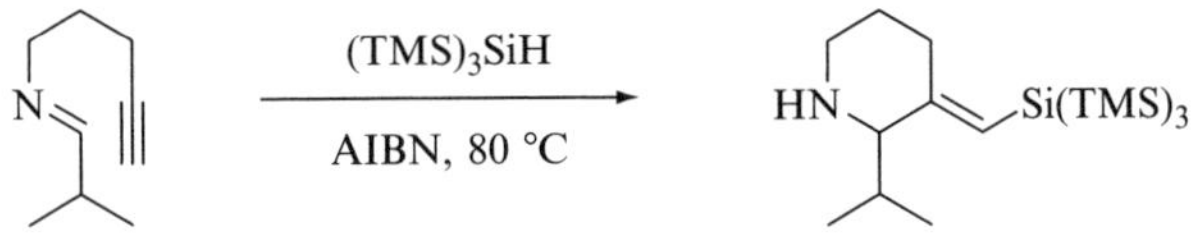

46

(7.54)

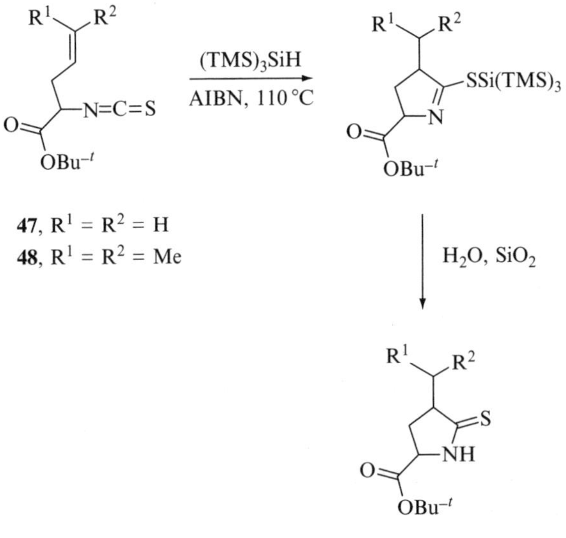

47, $R^1 = R^2 = H$
48, $R^1 = R^2 = Me$

93-96%

(7.55)

From all the data available so far, it is well estabilished that the geometries of the reactants play an important role in the regio- and stereochemical outcome of radical reactions since they are commonly involved in the early transition states. Previous attempts to affect rotamer populations during the reaction included, among others, control of temperature and addition of a Lewis acid. It was reported [65] that organotin halides, common by-products of radical reactions, act as Lewis acids and control the course of such reactions. An indicative example of this control is given in Scheme 7.6; the 5-*exo-trig* radical cyclization of *N*-enoyloxazolidinone **49**, mediated by $(TMS)_3SiH$ has been performed in the absence and in the presence of organotin halides. The results demonstrated that higher temperatures and weak Lewis acids were necessary, not only for inducing the conformational change from the stable *anti*-(*s*)-*Z* to the *syn*-(*s*)-*Z* required for the radical cyclization reaction, but also for obtaining high diastereoselectivity.

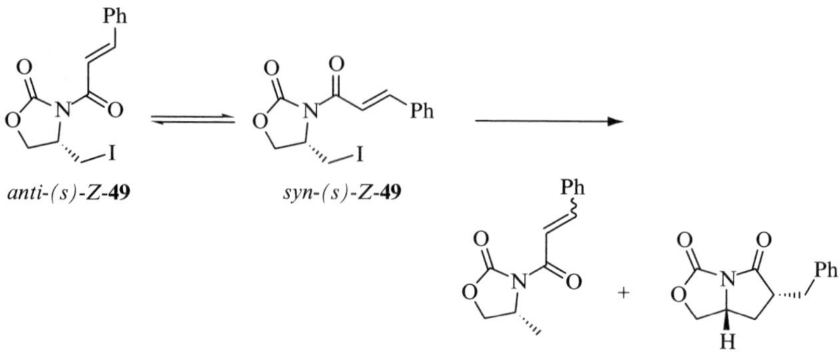

anti-(s)-Z-**49** *syn-(s)-Z*-**49**

	50 (*E*:*Z*)	**51** (de %)
$(TMS)_3SiH$, AIBN, 80 °C	78% (100:0)	13% (87)
$(TMS)_3SiH$, hν, r.t.	72% (71:29)	0%
$(TMS)_3SiH$, Bu_3SnCl, AIBN, 80 °C	0%	87% (>97)
$(TMS)_3SiH$, Bu_3SnCl, hν, r.t.	60% (60:40)	14% (86)

Scheme 7.6 Manipulation of rotamer population in *N*-enoyloxazolidinones

An example of the influence of structure and geometry on radical cyclization is shown by the stereoselectivity of the 5-*endo-trig* cyclization in compound **52** and 5-*exo-trig* cyclization in **53** using $(TMS)_3SiH$ as radical mediators [66]. In system **52** a mixture of *cis*-fused and *trans*-fused rings are obtained, whereas in system **53** the reaction proceeds in a stereoselective manner to give only the *cis*-fused product.

An extensive use of $(TMS)_3SiH$ can be found in the key steps of alkaloid syntheses. The synthesis of derivative **55**, as the key intermediate for the preparation of alkaloid (±)-pancracine [67], has been obtained from the reaction of **54** under normal conditions, in an 84 % yield as a single stereoisomer (Reaction 7.56).

52 X = SePh **53** X = Cl, SPh

PMB = *p*-$MeOC_6H_4CH_2$

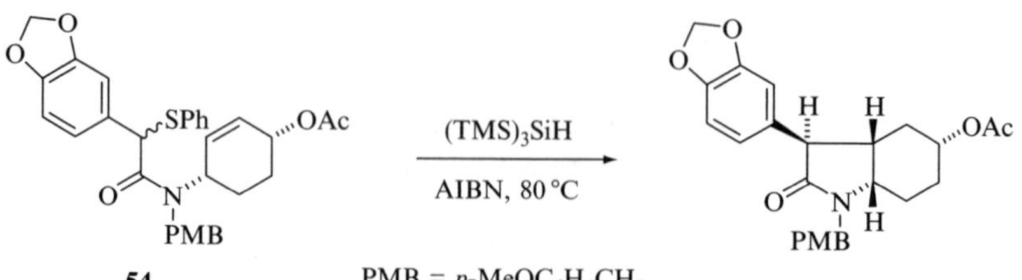

54 PMB = *p*-$MeOC_6H_4CH_2$ **55**, 84% (7.56)

The synthesis of tacamonine, an indole alkaloid of the *Iboga* type, was accomplished in both racemic and homochiral forms, by incorporating a classical 6-*exo-trig* radical cyclization in the key step of the synthesis (Reaction 7.57) [52]. The cyclization produced piperidinone in a 72% yield as a diastereomeric mixture.

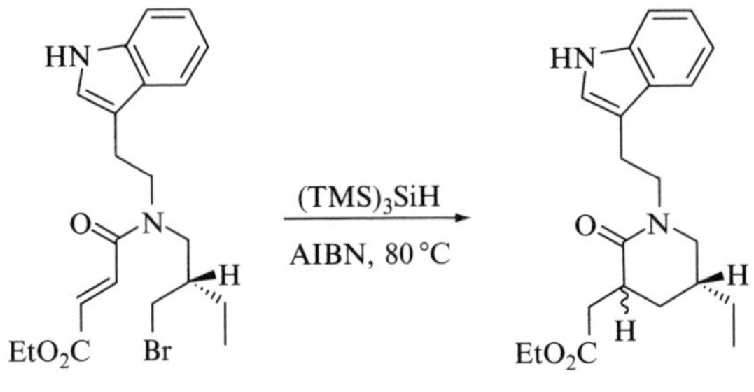

(7.57)

The reduction of the trichloroacetamide derivative **56** with $(TMS)_3SiH$ afforded **57** in a 60% yield (Reaction 7.58) [68]. The final product is the result of cyclization and further removal of two chlorine atoms. In fact, 3.5 equiv of silane and 1.1 equiv of AIBN are used.

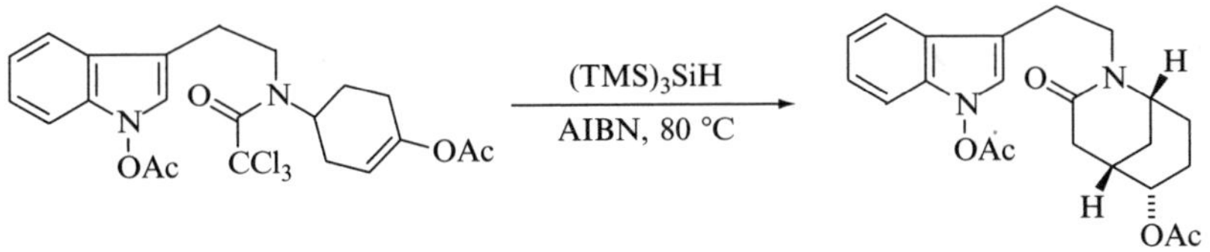

(7.58)

56 **57**, 60%

For the synthesis of tricyclic isoindolinones, the 5-*exo-trig* cyclization of aryl radicals has been applied [69]. Reaction (7.59) shows the example of a tricyclic system obtained in a 71% yield.

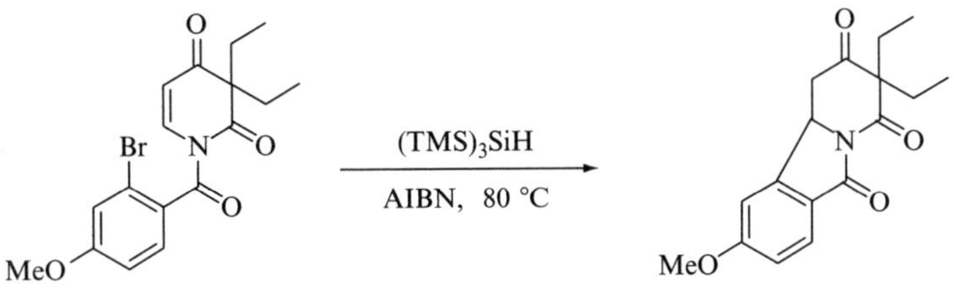

(7.59)

In model studies directed towards the synthesis of (±)-gelsemine, 5-*exo-trig* cyclization of an aryl radical, derived from the vinylogous uretane **58**, onto a methoxymethyl enolether resulted in partial fragmentation of the intermediate radical species with expulsion of a methoxymethyl radical and generation of the carbonyl group (Reaction 7.60) [70].

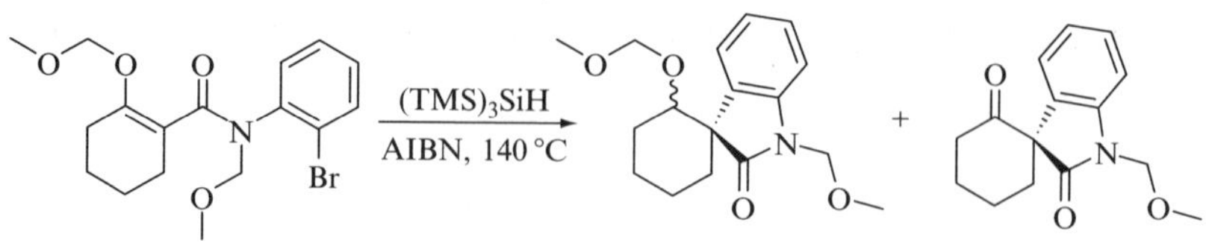

(7.60)

58 41% (7.2:1) 22%

Reaction (7.61) shows an example of macrocyclization which represents the key step in the total synthesis of the macrolide (−)-zearalenone [71].

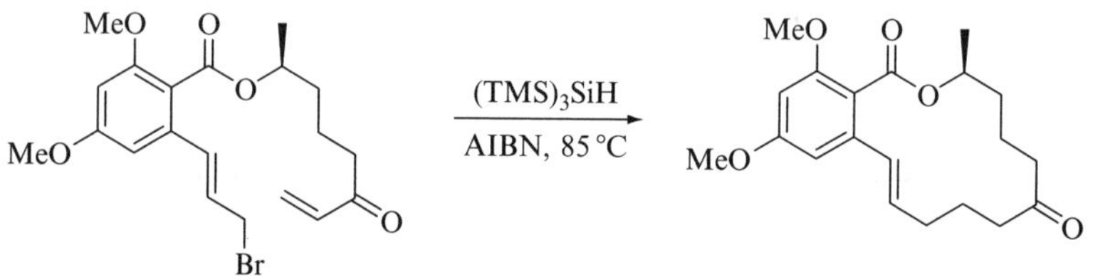

(7.61)

55%

7.4 FORMATION OF CARBON–HETEROATOM BONDS

Alkyl, vinyl and aryl radicals can give homolytic substitution at a heteroatom either inter- or intramolecularly thus forming C–heteroatom bonds.

A new entry for the synthesis of vinyl- and arylphosphonates has been achieved by reaction of vinyl or aryl halides with trialkyl phosphites in the presence of (TMS)$_3$SiH under standard radical conditions [72]. An example of this intermolecular C—P bond formation is given in Reaction (7.62). Interestingly, the reaction works well either under UV irradiation at room temperature or in refluxing toluene.

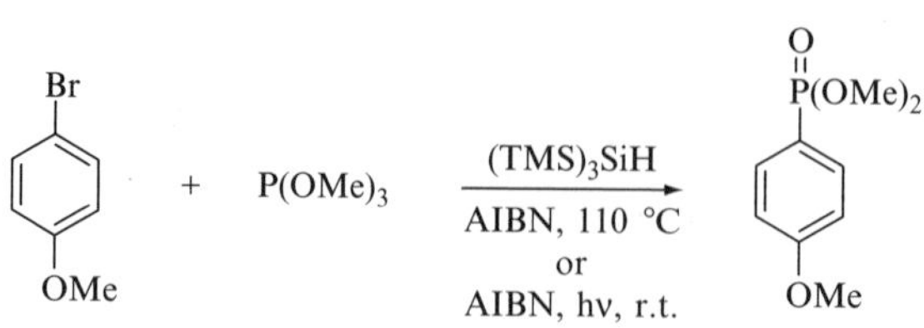

(7.62)

90-93%

Reaction (7.63) shows an example of C—S bond formation [73,74]. In fact, the aryl radical formed by iodine abstraction by (TMS)$_3$Si• radical rearranged by substitution to the sulfur atom, with expulsion of the acyl radical and concomitant formation of dihydrobenzothiophene (**60**). This procedure has been proposed as a versatile alternative to phenylseleno esters for the generation of acyl radicals [73]. Acyl radicals can in turn give addition to azides affording the cyclized products in satisfactory yields (Reaction 7.63) [74].

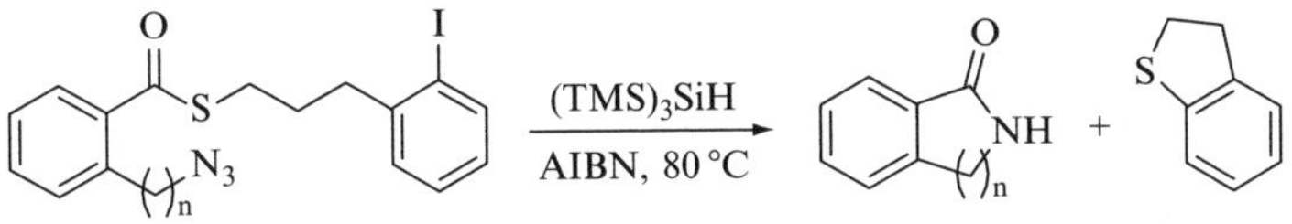

(7.63)

59 **60**

n = 1 81%
n = 2 75%

Similarly, C—Se bonds are formed by an internal homolytic substitution of aryl radicals at selenium, with the preparation of selenophenes and benzeneselenophenes [75]. Scheme 7.7 illustrates the reaction of aryl iodides **61** with $(TMS)_3SiH$, which afforded benzeneselenophenes in good yields. The presence of $(TMS)_3SiI$ after the reduction induced the final dehydration of the intermediate 3-hydroxyselenophenes, presumably through an intermediate silyl ether.

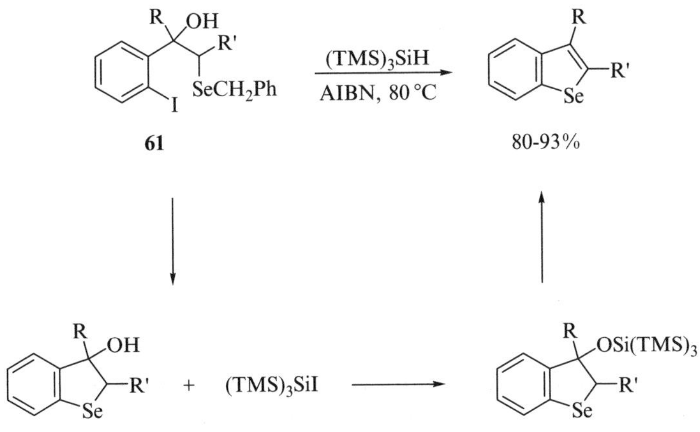

Scheme 7.7 Internal homolytic substitution at selenium

An example of C—Si bond formation concludes this overview of carbon–heteroatom bond formation. Reflux of bromide **62** in benzene and in the presence of small amounts of $(TMS)_3SiH$ and AIBN afforded the silabicycle **63** in 88% yield (Reaction 7.64) [76]. The key step for this transformation is the intramolecular homolytic substitution at the central silicon atom, which occurred with a rate constant of $2.4 \times 10^5\ s^{-1}$ at 80 °C (see also Section 6.4). The reaction has also been extended to the analogous vinyl bromide (Reaction 7.65) [49].

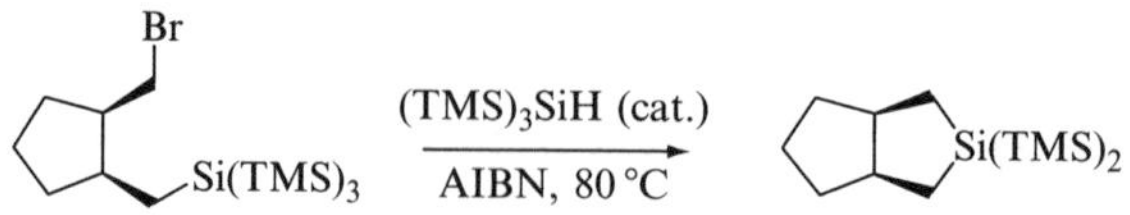

(7.64)

62 **63**, 88%

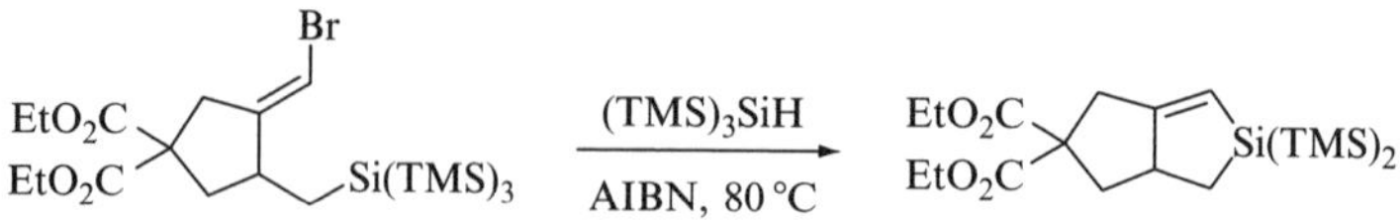

(7.65)

75%

7.5 OTHER USEFUL RADICAL REARRANGEMENTS

β-Silyl substituted carbon-centred radicals, which are produced when adding $R_3Si\bullet$ to unsaturated bonds can participate in consecutive reactions other than cyclization. A simple example is given in Reaction (7.66) where the adduct of silyl radical to β-pinene rearranged by opening the four-membered ring prior to H atom transfer [33,77].

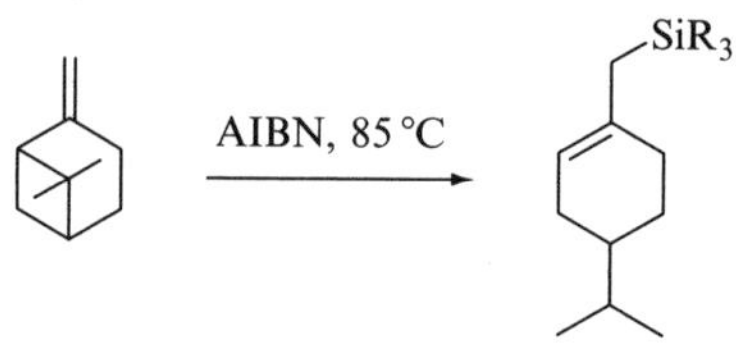

(7.66)

$(TMS)_3SiH$ 82%

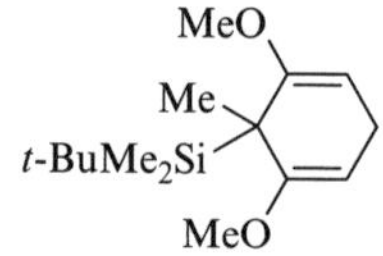

70%

Useful bicyclic ring systems are obtained by $(TMS)_3Si\bullet$ radical-mediated fragmentation of strained ketoalkene precursors. For example, the ketoalkene **64** reacted with 1.5 equiv of silane to give 95% of hydrindanone **65** (Reaction 7.67) [78]. $(TMS)_3Si\bullet$ radical adds first to the terminal alkene and the carbon-centred radical can relieve the strain by cleaving the adjacent C—C bond.

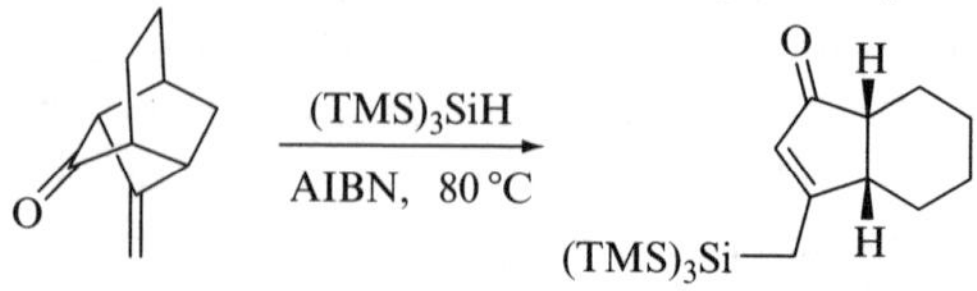

(7.67)

64 **65**, 95%

The addition of $(TMS)_3SiH$ in the 3-oxo-1,4-diene steroid-type derivative promoted C(9)—C(10) bond cleavage (Reaction 7.68) [79]. The removal of the silyl group was achieved by treating with dilute aqueous hydrochloric acid at room temperature and the final product was obtained in an 85% yield.

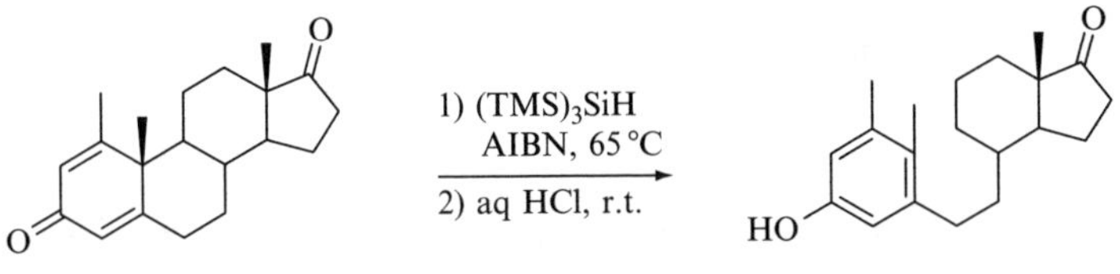

85%

(7.68)

The radical adduct obtained by the addition of $(TMS)_3Si\bullet$ to the keto azide **66** underwent an opening of the cyclopropyl ring, with formation of **67** as intermediate, prior to the intramolecular addition to the azido group [60]. After tosylation the final product was obtained in 61% yield.

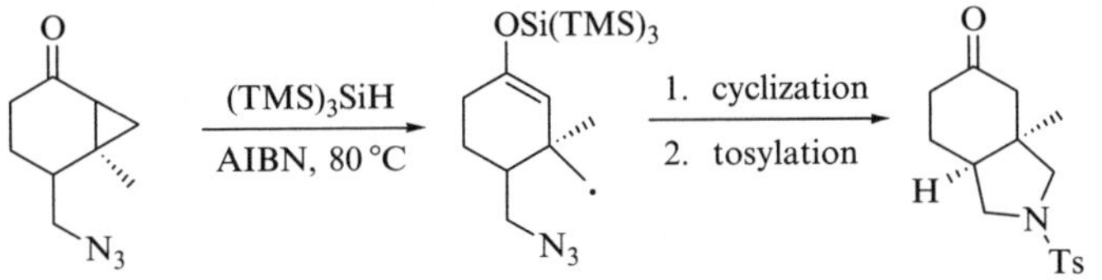

66 **67** 61%

(7.69)

A β-(acyloxy)alkyl radical rearrangement took place in the reduction of the bromonucleoside in Reaction (7.70), and allowed, through the generation of a C1′ radical species, for the stereoselective preparation of an α-ribonucleoside [80]. The use of $(TMS)_3SiH$ has been reported for a one-pot debromination–oxazine ring opening procedure, during the synthesis of narciclasine, an alkaloid screened for its antitumour activity [81]. The authors set their strategy in order to obtain oxazine **68** and further treated it under standard radical conditions (Reaction 7.71). The reduction of the vinylic bromide is accompanied by β-cleavage of the O—N bond prior to hydrogen abstraction to afford the unsaturated ketone **69** with a satisfactory yield.

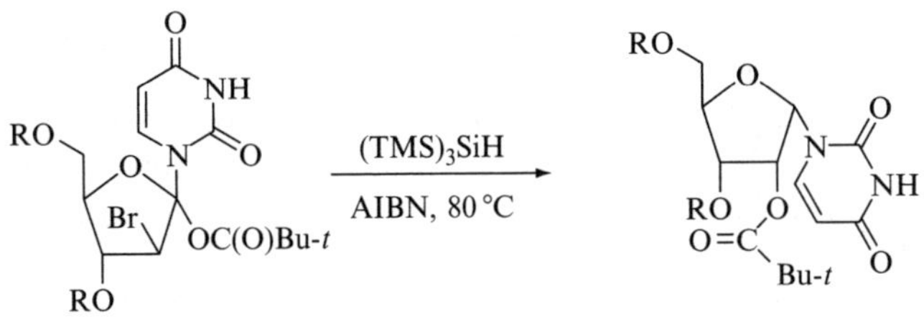

86%

(7.70)

R = t-$BuMe_2Si$

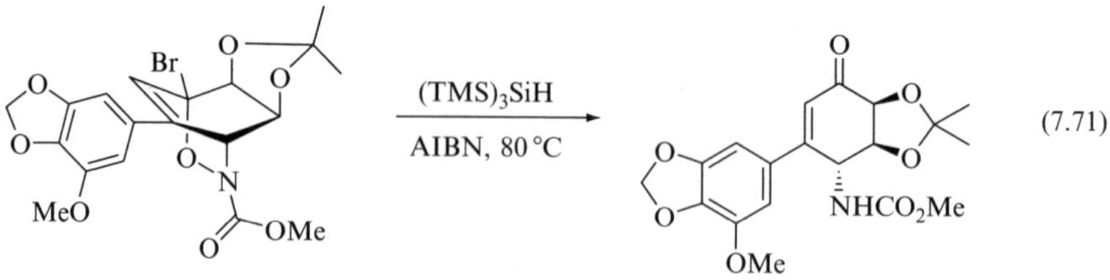

68 **69**, 80%

An interesting neophyl-type radical rearrangement process has been established for the synthesis of azabicycles, which are not readily accessible by other means. Barton–McCombie deoxygenation of xanthate **70** under slow addition of $(TMS)_3SiH$ and AIBN in refluxing toluene furnished the 2-azabenzonorbornane derivative in good yield (Reaction 7.72) [82].

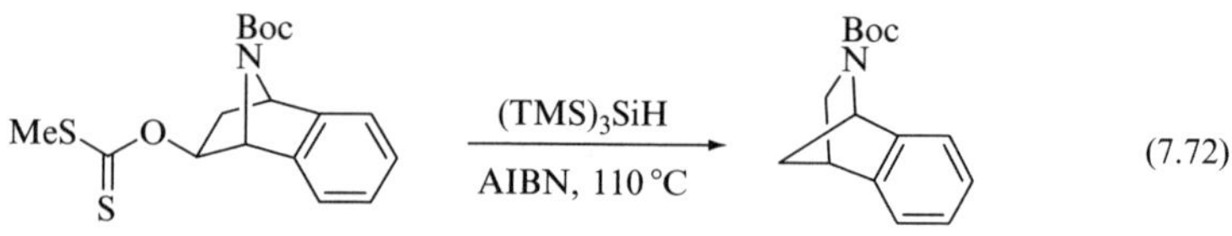

70 90%

α-Diazo ketones were found to react with $(TMS)_3SiH$ under free-radical conditions to give α-silyl ketones (Reaction 7.73) [83]. Mechanistic evidence has gathered for the attack of $(TMS)_3Si\bullet$ which takes place at carbon rather than nitrogen, with formation of the diazenyl radical adduct and decomposition to a α-silyl substituted radical and nitrogen.

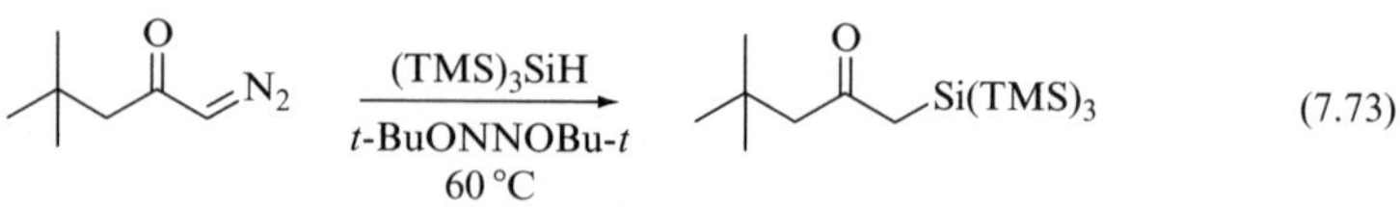

63%

7.6 ALLYLATIONS

Carbon–carbon bond formation using the radical allylation process is a very important method in radical-based synthetic applications. Several allylation methodologies exist, the most popular ones being those using allyltin derivatives. In analogy with allyltin derivatives, Scheme 7.8 shows the hypothetical propagation steps for using allylsilanes as the reagents for this purpose. The key step is the β-elimination of silyl radical, which should be the chain transfer agent. Only the $(TMS)_3Si\bullet$ radical is able to behave under standard experimental conditions.

Allyl tris(trimethylsilyl)silanes are obtained in high yields from the reactions of unsubstituted and 2-substituted allyl phenyl sulfides with $(TMS)_3SiH$

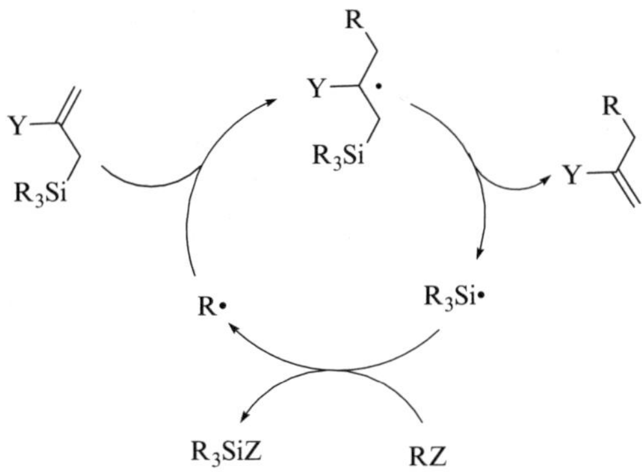

Scheme 7.8 Propagation steps for allylation reaction using allylsilanes

(Reaction 7.74) [84]. That is, $(TMS)_3Si$• radical added to the double bond of allyl sulfides, giving rise to a radical intermediate that undergoes β-scission with the ejection of the thiyl radical. Hydrogen abstraction from the silane completes the cycle of these chain reactions. 2-Functionalized allyl tris(trimethylsilyl)silanes (**71**) have been employed in the radical-based allylation reactions.

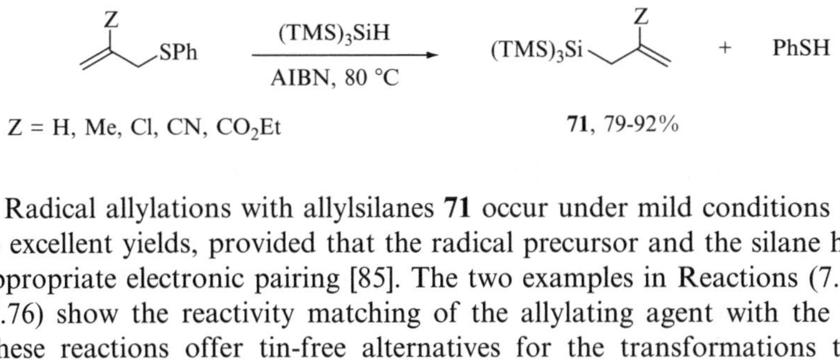

(7.74)

Radical allylations with allylsilanes **71** occur under mild conditions in good to excellent yields, provided that the radical precursor and the silane have the appropriate electronic pairing [85]. The two examples in Reactions (7.75) and (7.76) show the reactivity matching of the allylating agent with the radical. These reactions offer tin-free alternatives for the transformations that are currently carried out by allyl stannanes.

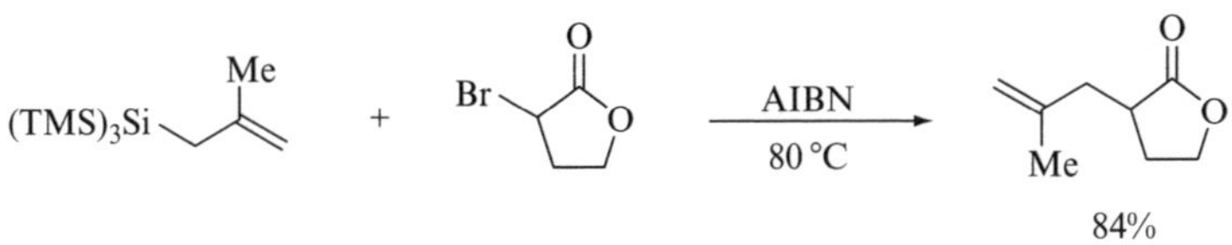

(7.75)

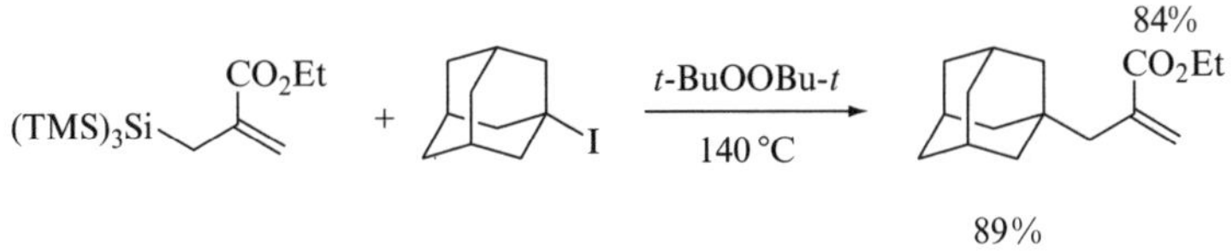

(7.76)

Allylsilylation procedures of carbon–carbon and carbon–oxygen double bonds under free-radical conditions have been established [86]. For example the reaction of **72** with a variety of monosubstituted alkenes (R = CN, CO_2 Me, Ph, n-C_8H_{17}, OBu, SPh) gives the corresponding allylsilylated products in moderate to very good yields, whereas with maleic anhydrides high stereoselectivity was observed. This methodology is also applicable to carbonyl derivatives (Scheme 7.9). Similar experiments have been performed by replacing alkenes with alkynes and the formation of bisallylic derivatives has been reported.

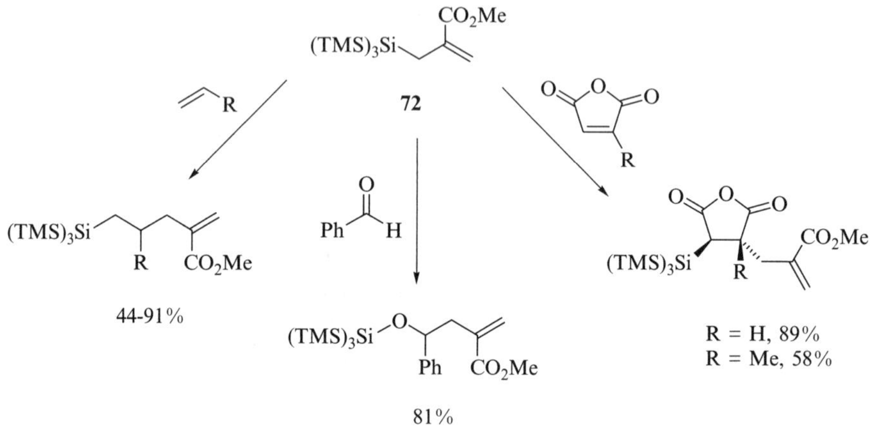

Scheme 7.9 Allysilylation of unsaturated bonds

7.7 APPLICATION TO TANDEM AND CASCADE RADICAL REACTIONS

The increasing popularity of radical reactions is certainly due to the so-called 'tandem or cascade reaction', i.e., the ability of forming and breaking several bonds in a one-pot procedure. Again, the concepts of disciplined processes and predictability are applied in these successful strategies, which show the elegance and power of the synthetic plan based on radical reactions.

Two examples of three-components coupling reaction are shown in Reactions (7.77) and (7.78) [27,87]. These radical chain reactions proceeded by the addition of an alkyl or vinyl radical onto carbon monoxide, generating an acyl radical intermediate, which, in turn, can further react with electron-deficient olefins to lead, after reduction, to a formal double alkylation of carbon monoxide. These three-components coupling reactions require the generation of four highly disciplined radical species, which have specific functions during the chain reaction.

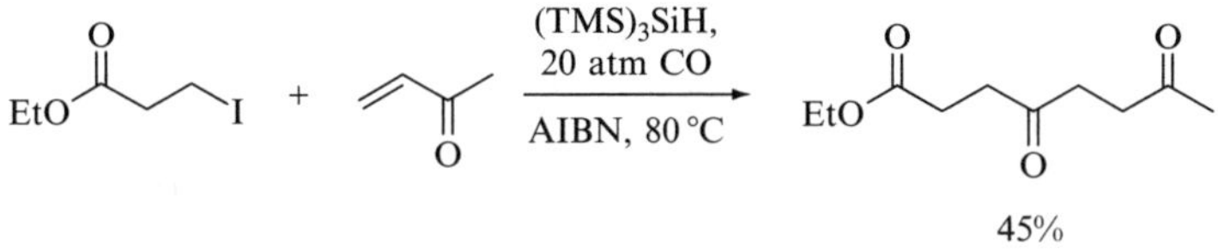

(7.77)

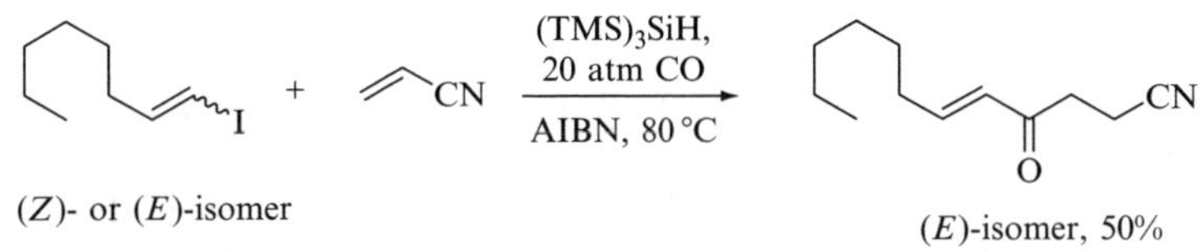

(7.78)

Supercritical carbon dioxide ($scCO_2$) is an emerging reaction medium for free-radical reactions for a number of reasons, such as the pressure-tunable cage effect, the absence of radical chain transfer to the medium [88], and unique properties such as high diffusivity and high miscibility of reactant gases [89]. Since $(TMS)_3SiH$ does not react with CO_2 and CO is highly soluble in $scCO_2$, it is thought that radical carbonylation in $scCO_2$ should be characterized by high tenability of product distribution [90]. Indeed, the extension of the three-components coupling reaction in this reaction medium gave excellent results as shown in Reaction (7.79).

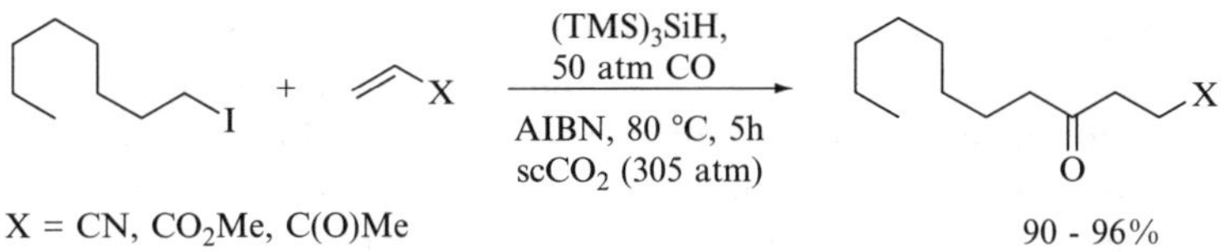

(7.79)

The addition of $(TMS)_3SiH$ to the substituted 1,5-dienes in the presence of CO (Reaction 7.80) is an intramolecular version of the above mentioned strategy and represents an example of (4 + 1) radical annulation process [91]. Indeed, the initial addition of silyl radicals to the least substituted terminus is followed by carbonylation and acyl radical cyclization prior to hydrogen abstraction. This methodology has been further developed for the preparation of cyclic ethers (Reaction 7.81). Indeed, the starting compound **73**, derived from the easy regioselective ring opening of appropriate epoxides with benzeneselenolate followed by hetero-Michael addition to ethyl propiolate, undergoes radical tandem carbonylation/cyclization reaction in the presence of $(TMS)_3SiH$ and CO (80 atm) [92].

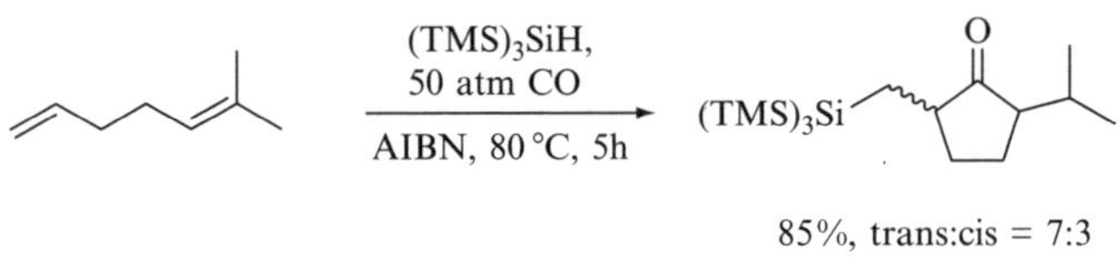

(7.80)

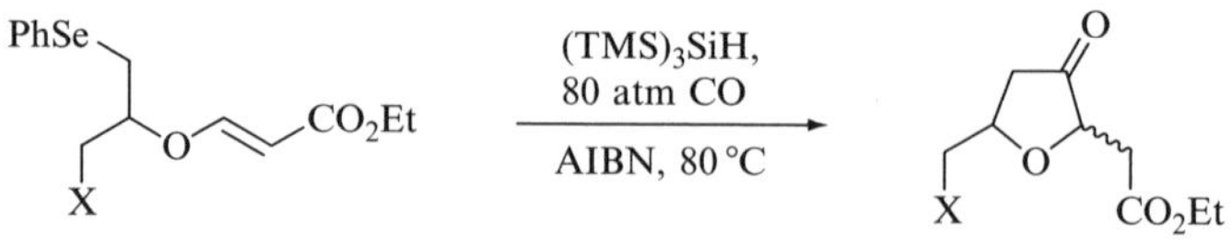

(7.81)

73

X = Ph　　86%, trans:cis = 9:1
X = OPh　　82%, trans:cis = 4:1

The field of alkyl radical macrocyclization reactions was further augmented with an $(n + 1)$ strategy, which incorporates a CO unit in the macrocycle [93]. Thus, in the presence of highly diluted (0.005–0.01 M) $(TMS)_3SiH$, ω-iodoacrylates underwent an efficient three-step radical chain reaction to generate 10- to 17-membered macrocycles in 28–78 % yields, respectively (Reaction 7.82).

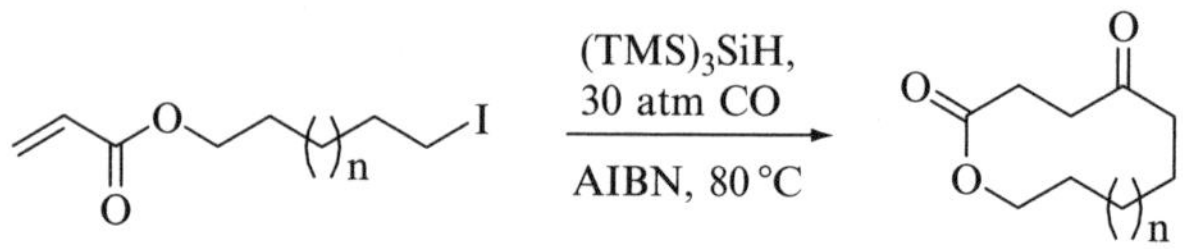

(7.82)

n = 1,2,4,6,7,8　　28-78%

Free-radical reaction of vinyl iodides having dienoate function in the presence of $(TMS)_3SiH$ and AIBN in refluxing benzene caused a tandem cyclization reaction to produce (4 + 1) and (4 + 2) annulated compounds [94]. Reaction (7.83) shows the transformation resulted from a tandem 5-*exo*, 6-*endo* cyclization to give the isoindole skeleton where the stereogenic centres were highly controlled, whereas Reaction (7.84) proceeded via a tandem 6-*exo*, 5-*exo* cyclization to furnish a (4 + 1) cycloadduct.

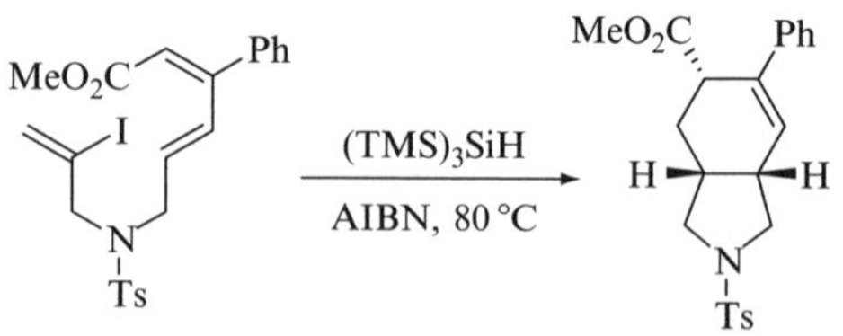

(7.83)

66%

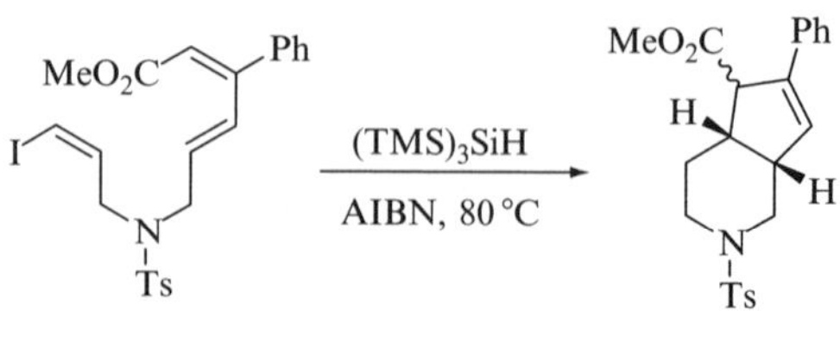

(7.84)

70%, α:β = 3:2

The plan for a tandem cyclization as the key step in the synthesis of morphine alkaloids has been successfully performed [95]. Three examples are given in Scheme 7.10. Specifically, the initially formed aryl radical, generated by bromine abstraction from compound **74**, underwent a tandem cyclization to construct the desired carbocyclic skeleton. Depending on the nature of the Z substituent to the double bond, the product radicals either abstracted hydrogen from silane (Z = CO_2Me or CN) or eliminated thiyl radical (Z = SPh).

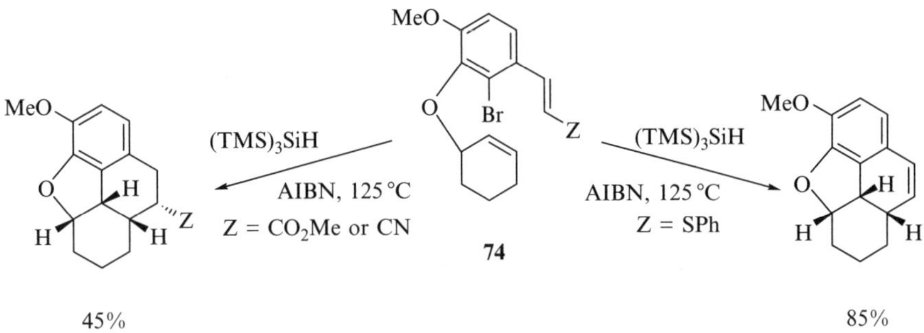

Scheme 7.10 Examples of tandem radical cyclizations

Novel bridged spirolactones have been synthesized via tandem radical cyclizations of enol ether radical. In Reaction (7.85) the first 5-*exo* spirocyclization is followed by a 6-*endo* cyclization to give the bridged derivative as a single diastereoisomer [96].

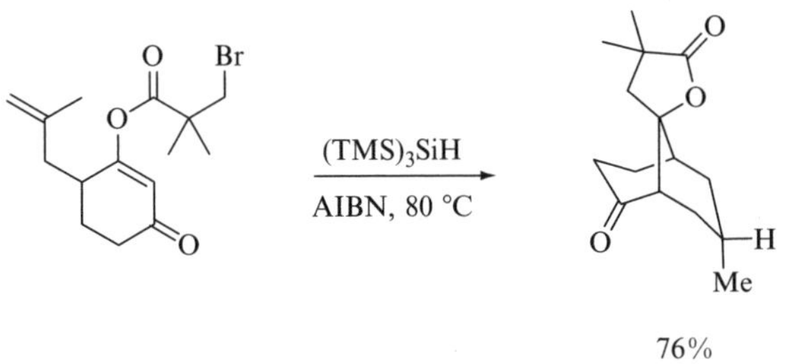

(7.85)

Tandem strategies have been successfully applied to synthesis of indole alkaloids, starting from aryl radical generation [97–99]. Iodoarylazides have been thoroughly used in protocols where the aryl radical formation is coupled with radical addition to alkenes and eventually a further cyclization onto an azide, with the elimination of nitrogen gas, as a driving force. The skeletons of (±)-horsfiline, (±)-aspidospermidine and (±)-vindoline have been achieved by this route. For example, cyclization of the iodoarylazide, such as **75**, mediated by $(TMS)_3SiH$ under standard experimental conditions, produced the N—$Si(TMS)_3$ protected alkaloid **76**, which after washing with dilute acid afforded the amine **77** in an overall 83% yield from **75** (Reaction 7.86). The formation of the labile N—Si(TMS)$_3$ bond was thought to arise from the reaction of the product amine **77** with the by-product (TMS)$_3$SiI. Similar results were obtained for some substituted analogues of **75** [98]. It is worth underlining that the field of alkaloid synthesis via tandem cyclizations has favoured the application of (TMS)$_3$SiH over other radical-based reagents, due to its very low toxicity and chemoselectivity.

(7.86)

75

76 R = Si(TMS)$_3$
77 R = H, 83%

A radical annulation strategy for the synthesis of heterocyclic fused quinolines has been proposed, starting from the readily available *N*-aryl thiocarbamates, thioamides and thioureas of general structure **78** (Reaction 7.87) [100]. (TMS)$_3$SiH (4 equiv) together with AIBN (1 equiv) were used under UV irradiation in benzene solutions, thus giving the cyclized products in good to excellent yields. The reaction mechansim consists of the addition of silyl radical to the thiocarbonyl function to give an α-thioalkylamino, which undergoes a tandem radical cyclization prior to rearomatization.

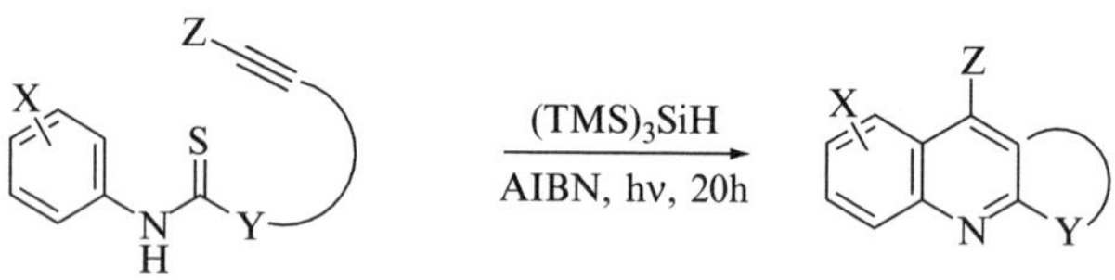

(7.87)

Y = CH$_2$, NH, O

78 44-88%

As model studies and radical reactivity control have improved, the so-called cascade (or domino) reactions have emerged as a very powerful method for natural product synthesis, since they offer a unique route to prepare complex backbones from appropriately designed but quite simple precursors. A few selected reactions will be presented here.

Treatment of iodo derivative **79** with (TMS)$_3$SiH starts a cascade of 6-*endo*, 6-*endo*, and 6-*exo* cyclizations to afford exclusively **80** in a 77% yield (Reaction

7.88) [101]. On the other hand, the analogous iodo derivative **81** with $(TMS)_3SiH$ at room temperature afforded a linear fused five-membered carbocycle in two isomeric forms and with an overall yield of 74% (Reaction 7.89) [102]. In this case, the cascade proceeds through 5-*exo*, 5-*exo*, and 5-*exo* modes of cyclization.

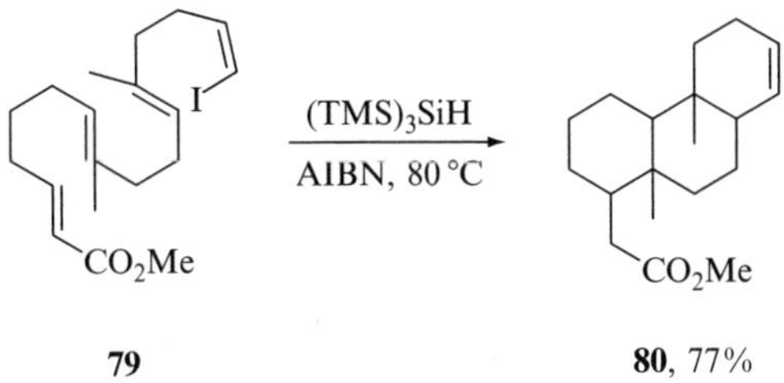

(7.88)

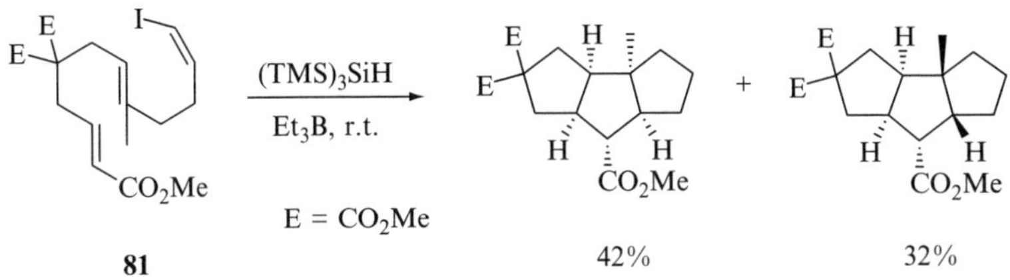

(7.89)

Radical cascade starting from bromomethyldimethylsilyl propargyl ethers has been utilized in a remarkable way [103–105]. Reduction of the bromide **82** in the presence of $(TMS)_3SiH$ and subsequent treatment with MeLi produced the functionalized cyclopentanone precursor **83** as a single diastereomer (Scheme 7.11). The formation of **83** has been explained by a series of reactions, indicated in Scheme 7.11, involving a 5-*exo* cyclization of the initial α-silyl alkyl radical (not shown) followed by a [1,5]-radical translocation of the generated σ-type vinyl radical onto the proximal acetal function and a final 5-*exo* cyclization process. Stereoselective hydrogen abstraction, dependent on the steric bulkiness of the hydrogen donor, followed by MeLi-induced opening of the Si—O bond afforded the final product. The introduction of different substituents on the skeleton, as in compound **84** resulted in a completely different reaction pattern (Reaction 7.90) [104,105]. In this case, the intermediate vinyl radical (cf. Scheme 7.11) underwent a [1,5]-hydrogen abstraction from the non-activated C—H bond of the proximal isopropyl group. Furthermore, the resulting primary alkyl radical underwent a unique, stereoselective 5-*endo* cyclization onto the adjacent double bond to generate a tertiary radical, which is a precursor of the highly substituted cyclopentanols **85** and **86**.

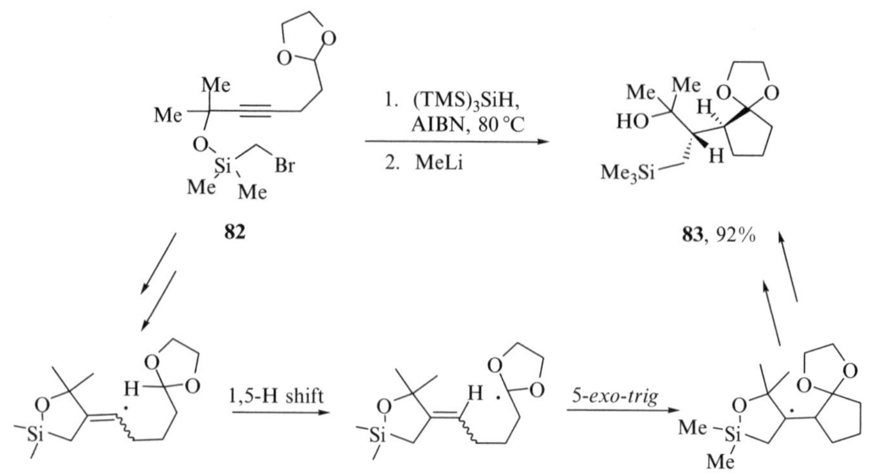

Scheme 7.11 Some details of a radical cascade reaction

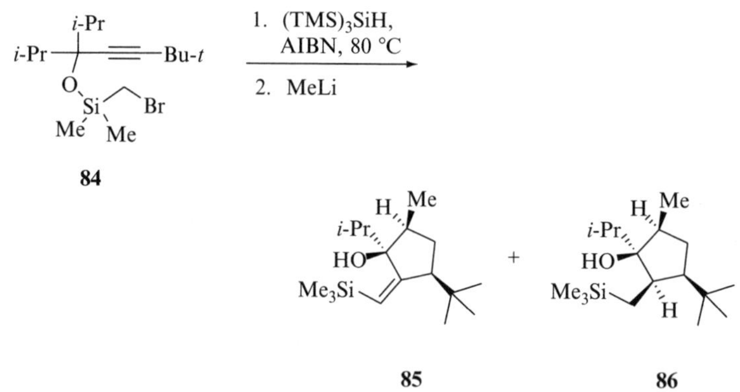

(7.90)

95%, **85**:**86** = 90:10

A highly efficient route to azacoumarins has been obtained by treatment of compounds **87** with $(TMS)_3SiH$ under standard experimental conditions (Reaction 7.91) [69]. The mechanism based on an unusual double intramolecular 1,5-*ipso* and 1,6-*ipso* substitution with the final ejection of MeO• radical.

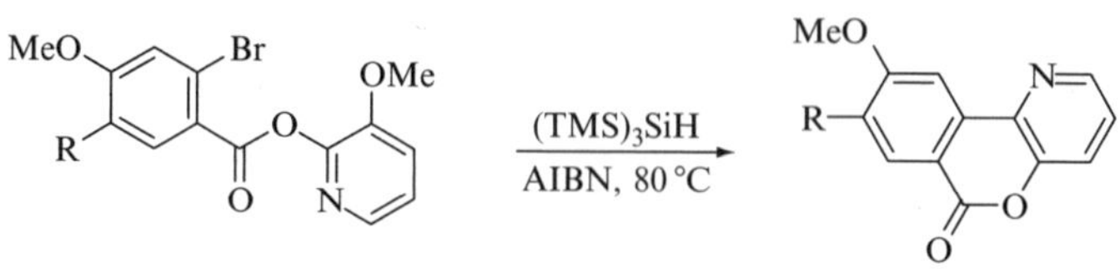

(7.91)

87

R = H	62%
R = OMe	75%

The number of consecutive processes in some radical cascade processes can be really astonishing [106,107]. Two representative examples are described below [106]. Treatment of the acetylene derivative **88** with $(TMS)_3SiH$ leads, in one pot, to the bicyclic compound **89** in a 70% yield (Reaction 7.92). The proposed mechanism involves $(TMS)_3Si•$ radical addition to the triple bond to form a vinyl radical followed by a remarkable cascade of radical 6-*exo* cyclization, fragmentation, transannulation, ring expansion, and termination via ejection of the $(TMS)_3Si•$ radical to afford the bicyclic product. When a solution of vinyl bromide **90** was irradiated with a sunlamp in the presence of $(TMS)_3SiH$ it gave the corresponding triquinane oxime **91** as a 1:1 mixture of α- and β-methyl diastereomers resulting from a cascade radical 6-*exo* cyclization, aminyl radical fragmentation, 5-*exo* radical transannulation, 5-*exo* cyclization sequence (Reaction 7.93).

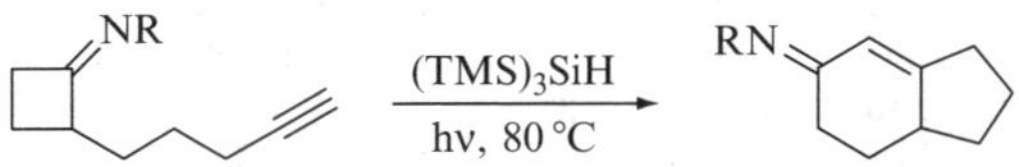

(7.92)

88, R = OCH_2Ph **89**, 70%

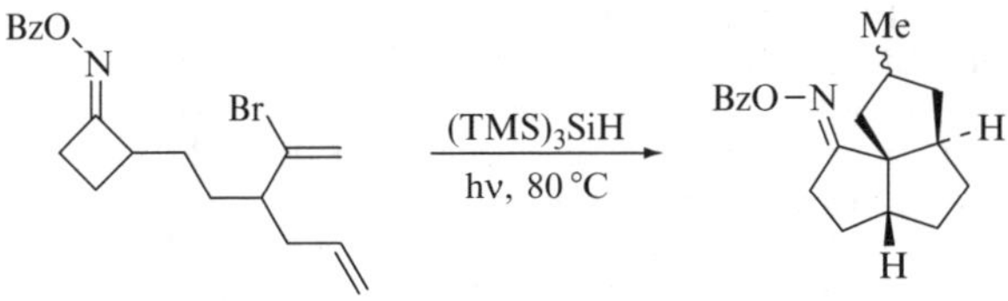

(7.93)

90 **91**, 38%

7.8 REFERENCES

1. Renaud, P., and Sibi, M.P. (Eds), *Radicals in Organic Synthesis*, Wiley-VCH, Weinheim, 2001.
2. Curran, D.P., Porter, N.A., and Giese, B., *Stereochemistry of Radical Reactions: Concepts, Guidelines, and Synthetic Applications*, VCH, Weinheim, 1996.
3. Renaud, P., and Gerster, M., *Angew. Chem. Int. Ed. Engl.*, 1998, **37**, 2562.
4. Sibi, M.P., and Porter, N.A., *Acc. Chem. Res.*, 1999, **32**, 163.
5. Barton, D.H.R., *Half a Century of Free Radical Chemistry*, Cambridge University Press, Cambridge, 1993.
6. Giese, B., Kopping, B., and Chatgilialoglu, C., *Tetrahedron Lett.*, 1989, **30**, 681.
7. Chatgilialoglu, C., Giese, B., and Kopping, B., *Tetrahedron Lett.*, 1990, **31**, 6013.
8. Schneider, H., Fiander, H., Harrison, K.A., Watson, M., Burton G.W., and Arya, P., *Bioorg. Med. Chem. Lett.*, 1996, **6**, 637.
9. Urabe, H., Kobayashi, K., and Sato, F., *J. Chem. Soc. Chem. Commun.*, 1995, 1043.
10. Junker, H.-D., Phung, N., and Fessner, W.-D., *Tetrahedron Lett.*, 1999, **40**, 7063. Junker, H.-D., and Fessner, W.-D., *Tetrahedron Lett.*, 1998, **39**, 269.

11. Togo, H., Sugi, M., and Toyama, K., *C.R. Acad. Sci. Paris, Ser. IIc: Chem.*, 2001, 539.
12. Togo, H., Matsubayashi, S., Yamazaki, O., and Yokoyama, M., *J. Org. Chem.*, 2000, **65**, 2816.
13. Dang, H.-S., Elsegood, M.R.J., Kim, K.-M., and Roberts, B.P., *J. Chem. Soc., Perkin Trans. 1,* 1999, 2061.
14. Curran, D.P., Xu, J., and Lazzarini, E., *J. Am. Chem. Soc.*, 1995, **117**, 6603.
15. Curran, D.P., Xu, J., and Lazzarini, E., *J. Chem. Soc., Perkin Trans. 1,* 1995, 3049.
16. Ryu, I., Hasegawa, M., Kurihara, A., Ogawa, A., Tsunoi, S., and Sonoda, N., *Synlett*, 1993, 143.
17. Bennasar, M.-L., Roca, T., Griera, R., Bassa, M., and Bosch, J., *J. Org. Chem.*, 2002, **67**, 6268.
18. Chatgilialoglu, C., Alberti, A., Ballestri, M., Macciantelli, D., and Curran, D.P., *Tetrahedron Lett.*, 1996, **35**, 6391.
19. Martin, J. Jaramillo G., L.M., and Wang, P.G., *Tetrahedron Lett.*, 1998, **39**, 5927.
20. Minisci, F., Fontana, F., Pianese, G., and Yan, Y.M., *J. Org. Chem.*, 1993, **58**, 4207.
21. Togo, H., Hayashi, K., and Yokoyama, M., *Bull. Chem. Soc. Jpn.*, 1994, **67**, 2522.
22. Beckwith, A.L.J., and Schiesser, C.H., *Tetrahedron*, 1985, **41**, 3925.
23. Spellmeyer, D.C., and Houk, K.N., *J. Org. Chem.*, 1987, **52**, 959.
24. Bailey, W.F., and Longstaff, S.C., *Org. Lett.*, 2001, **3**, 2217.
25. Usui, S., and Paquette, L.A., *Tetrahedron Lett.*, 1999, **40**, 3495.
26. Paquette, L.A., and Usui, S., *Tetrahedron Lett.*, 1999, **40**, 3499.
27. Chatgilialoglu, C., Crich, D., Komatsu, M., and Ryu, I., *Chem. Rev.*, 1999, **99**, 1991.
28. Lowinger, T.B., and Weiler, L., *J. Org. Chem.*, 1992, **57**, 6099.
29. Stalinski, K., and Curran, D.P., *J. Org. Chem.*, 2002, **67**, 2982.
30. Batey, R.A., and MacKay, D.B., *Tetrahedron Lett.*, 1998, **39**, 7267.
31. Salazar, K.L., and Nicholas, K.M., *Tetrahedron*, 2000, **56**, 2211.
32. Kraus, G.A., and Liras, S., *Tetrahedron Lett.*, 1990, **31**, 5265.
33. Amrein, S., Timmermann, A., and Studer, A., *Org. Lett.*, 2001, **3**, 2357.
34. Kulicke, K.J., and Giese, B., *Synlett*, 1990, 91.
35. Evans, P.A., and Manangan, T., *J. Org. Chem.*, 2000, **65**, 4523.
36. Zahouily, M., Journet, M., and Malacria, M., *Synlett*, 1994, 366.
37. Lee, E., and Choi, S.J., *Org. Lett.*, 1999, **1**, 1127.
38. Arya, P., and Wayner, D.D.M., *Tetrahedron Lett.*, 1991, **32**, 6265.
39. Yamazaki, O., Yamaguchi, K., Yokoyama, M., and Togo, H., *J. Org. Chem.*, 2000, **65**, 5440.
40. Cossy, J., and Sallé, L., *Tetrahedron Lett.*, 1995, **40**, 7235.
41. Evans, P.A., and Roseman, J.D., *J. Org. Chem.*, 1996, **61**, 2252.
42. Evans, P.A., Murthy, V.S., Roseman, J.D., and Rheingold, A.L., *Angew. Chem. Int. Ed.*, 1999, **38**, 3175.
43. Evans, P.A., and Roseman, J.D., *Tetrahedron Lett.*, 1997, **38**, 5249.
44. Evans, P.A., Roseman, J.D., and Garber, L.T., *J. Org. Chem.*, 1996, **61**, 4880.
45. Pandey, G., Rao, K.S.S.P., Palit, D.K., and Mittal, J.P., *J. Org. Chem.*, 1998, **63**, 3968.
46. Pandey, G., Rao, K.S.S.P., and Rao, K.V.N., *J. Org. Chem.*, 2000, **65**, 4309.
47. Sasaki, K., Kondo, Y., and Maruoka, K., *Angew. Chem. Int. Ed.*, 2001, **40**, 411.
48. Ooi, T., Hokke, Y., and Maruoka, K., *Angew. Chem. Int. Ed. Engl.*, 1997, **36**, 1181.
49. Miura, K., Oshima, K., and Utimoto, K., *Bull. Chem. Soc. Jpn.*, 1993, **66**, 2348.
50. Jiao, X.-Y., and Bentrude, W.G., *J. Am. Chem. Soc.*, 1999, **121**, 6088.
51. Evans, P.A., and Leahy, D.K., *J. Am. Chem. Soc.*, 2000, **122**, 5012.
52. Ihara, M., Setsu, F., Shohda, M., Taniguchi, N., Tokunaga, Y., and Fukumoto, K., *J. Org. Chem.*, 1994, **59**, 5317.

53. Zhang, W., and Pugh, G., *Tetrahedron Lett.*, 1999, **40**, 7595.
54. Zhang, W., *Tetrahedron Lett.*, 2000, **41**, 2523.
55. Knapp, S., and Gibson, F.S., *J. Org. Chem.*, 1992, **57**, 4802.
56. Okano, T., Fumoto, M., Kusukawa, T., and Fujita, M., *Org. Lett.*, 2002, **4**, 1571.
57. Evans, P.A., Manangan, T., and Rheingold, A.L., *J. Am. Chem. Soc.*, 2000, **122**, 11009.
58. Evans, P.A., and Manangan, T., *Tetrahedron Lett.*, 2001, **42**, 6637.
59. Evans, P.A., Robinson, J.E., and Moffett, K.K., *Org. Lett.*, 2001, **3**, 3269.
60. Kim, S., Joe, G.H., and Do, J.Y., *J. Am. Chem. Soc.*, 1994, **116**, 5521.
61. Kim, S., *Pure Appl. Chem.*, 1996, **68**, 623.
62. Lee, E., Kang, T.S., Joo, B.J., Tae, J.S., Li, K.S., and Chung, C.K., *Tetrahedron Lett.*, 1995, **36**, 417.
63. Ryu, I., Ogura, S., Minakata, S., and Komatsu, M., *Tetrahedron Lett.*, 1999, **40**, 1515.
64. Bachi, M.D., Balanov, A., Bar-Ner, N., Bosch, E., Denenmark, D., and Mizhiritskii, M., *Pure Appl. Chem.*, 1993, **65**, 595.
65. Sibi, M.P., and Ji, J., *J. Am. Chem. Soc.*, 1996, **118**, 3063.
66. Ikeda, M., Hamada, M., Yamashita, T., Matsui, K., Sato, T., and Ishibashi, H., *J. Chem. Soc., Perkin Trans. 1*, 1999, 1949.
67. Ikeda, M., Hamada, M., Yamashita, T., Ikegami, F., Sato, T., and Ishibashi, H., *Synlett*, 1998, 1246.
68. Quirante, J., Escolano, C., Diaba, F., and Bonjoch, J., *J. Chem. Soc., Perkin Trans. 1*, 1999, 1157.
69. Zhang, W., and Pugh, G., *Tetrahedron*, 2003, **59**, 3009.
70. Hart, D.J., and Kuzmich, D., *J. Chin. Chem. Soc.*, 1995, **42**, 873.
71. Hitchcock, S.A., and Pattenden, G., *J. Chem. Soc., Perkin Trans. 1*, 1992, 1323.
72. Jiao, X.-Y., and Bentrude, W.G., *J. Org. Chem.*, 2003, **68**, 3303.
73. Crich, D., and Yao, Q., *J. Org. Chem.*, 1996, **61**, 3566.
74. Benati, L., Leardini, R., Minozzi, M., Nanni, D., Spagnolo, P., Strazzari, S., and Zanardi, G., *Org. Lett.*, 2002, **4**, 3079.
75. Lyons, J.E., Schiesser, C.H., and Sutej, K., *J. Org. Chem.*, 1993, **58**, 5632.
76. Kulicke, K.J., Chatgilialoglu, C., Kopping, B., and Giese, B., *Helv. Chim. Acta*, 1992, **75**, 935.
77. Kopping, B., Chatgilialoglu, C., Zehnder, M., and Giese, B., *J. Org. Chem.*, 1992, **57**, 3994.
78. Dufour, C., Iwasa, S., Fabré, A., and Rawal, V.H., *Tetrahedron Lett.*, 1996, **37**, 7867.
79. Künzer, H., Sauer, G., and Wieckert, R., *Tetrahedron Lett.*, 1991, **32**, 7247.
80. Gimisis, T., Ialongo, G., Zamboni, M., and Chatgilialoglu, C., *Tetrahedron Lett.*, 1995, **36**, 6781.
81. Hudlicky, T., Rinner, U., Gonzalez, D., Akgun, H., Schilling, S., Siengalewicz, P., Martinot, T.A., and Pettit, G.R., *J. Org. Chem.*, 2002, **67**, 8726.
82. Hodgson, D.M., Bebbington, M.W.P., and Willis, P., *Org. Lett.*, 2002, **4**, 4353.
83. Dang, H.-S., and Roberts, B.P., *J. Chem. Soc., Perkin Trans. 1*, 1996, 769.
84. Chatgilialoglu, C., Ballestri, M., Vecchi, D., and Curran, D.P., *Tetrahedron Lett.*, 1996, **37**, 6383.
85. Chatgilialoglu, C., Ferreri, C., Ballestri, M., and Curran, D.P., *Tetrahedron Lett.*, 1996, **37**, 6387.
86. Miura, K., Saito, H., Nakagawa, T., Hondo, T., Tateiwa, J., Sonoda M., and Hosomi, A., *J. Org. Chem.*, 1998, **63**, 5740.
87. Ryu, I., and Sonoda, N., *Angew. Chem. Int. Ed. Engl.*, 1996, **35**, 1050.
88. Tanko, J.M., and Blackert, J.F., *Science*, 1994, **263**, 203.

89. Jessop, P.G., Hsiao, Y., and Noyori, R., *Chem. Rev.*, 1999, **99**, 475.
90. Kishimoto, Y., and Ikariya, T., *J. Org. Chem.*, 2000, **65**, 7656.
91. Ryu, I., Nagahara, K., Kurihara, A., Komatsu M., and Sonoda, N., *J. Organomet.Chem.*, 1997, **548**, 105.
92. Berlin, S., Ericsson, C., and Engman, L., *Org. Lett.*, 2002, **4**, 3.
93. Nagahara, K., Ryu, I., Yamazaki, H., Kambe, N., Komatsu, M., Sonoda, N., and Baba, A., *Tetrahedron*, 1997, **53**, 14615.
94. Takasu, K., Kuroyanagi, J., and Ihara, M., *Org. Lett.*, 2000, **2**, 3579.
95. Parker, K.A. and Fokas, D., *J. Org. Chem.*, 1994, **59**, 3927.
96. Zhang, W., and Pugh, G., *Tetrahedron Lett.*, 2001, **42**, 5617.
97. Kizil, M., Patro, B., Callaghan, O., Murphy, J.A., Hursthouse, M.B., and Hibbs, D., *J. Org. Chem.*, 1999, **64**, 7856.
98. Zhou, S., Bommezijn, S., and Murphy, J.A., *Org. Lett.*, 2002, **4**, 443.
99. Lizos, D.E., and Murphy, J.A., *Org. Biomol. Chem.*, 2003, **1**, 117.
100. Du, W., and Curran, D.P., *Org. Lett.*, 2003, **5**, 1765.
101. Takasu, K., Kuroyanagi, J., Katsumata, A., and Ihara, M., *Tetrahedron Lett.*, 1999, **40**, 6277.
102. Takasu, K., Maiti, S., Katsumata, A., and Ihara, M., *Tetrahedron Lett.*, 2001, **42**, 2157.
103. Bogen, S., Journet, M., and Malacria, M., *Synlett*, 1994, 958.
104. Bogen, S., and Malacria, M., *J. Am. Chem. Soc.*, 1996, **118**, 3992.
105. Bogen, S., Gulea, M., Fensterbank, L., and Malacria, M., *J. Org. Chem.*, 1999, **64**, 4920.
106. Hollingworth, G.J., Pattenden, G., and Schulz, D.J., *Aust. J. Chem.*, 1995, **48**, 381.
107. Pattenden, G., and Wiedenau, P., *Tetrahedron Lett.*, 1997, **38**, 3647.

8 Silyl Radicals in Polymers and Materials

8.1 POLYSILANES

Polysilanes (or polysilylenes) consist of a silicon-catenated backbone with two substituents on each silicon atom (Structure **1**). The groups R and R′ attached to the silicon chain can be of a large variety. Polysilanes with alkyl and/or aryl substituents have been the most thoroughly investigated [1–3], whereas polysilanes having at least a heteroatom substitution such as H, Cl, OR, NR_2 have received much less attention [4]. The number of silicon atoms is usually from several hundreds to several thousands.

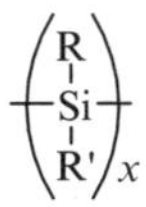

1

Polysilanes have been shown to possess a number of interesting chemical and physical properties attributed to the extensive delocalization of σ-electrons along the silicon backbone [1–3]. Indeed, the delocalized Si—Si σ-electrons in polysilanes give rise to σ–σ* electronic excitations in the near-UV region, whose energies depend strongly on the substituents, degree of polymerization and chain conformation. For example, high molecular weight poly(dialkylsilane)s and poly(alkylarylsilane)s generally show a λ_{max} ranging from 305–325 and 335–355 nm, respectively. Because the conformation of the polysilane chain may change with environmental factors many of these materials are chromotropic. The effects of changes in temperature, leading to thermochromism, have been the most extensively studied both in solution and as solids. However,

Organosilanes in Radical Chemistry C. Chatgilialoglu
© 2004 John Wiley & Sons, Ltd ISBN: 0-471-49870-X

solvatochromism, ionochromism, piezochromism and electrochromism have also been investigated to some extent.

Polysilanes can be regarded as one-dimensional analogues to elemental silicon, on which nearly all of modern electronics is based. They have enormous potential for technological uses [1–3]. Nonlinear optical and semiconductive properties, such as high hole mobility, photoconductivity, and electrical conductivity, have been investigated in some detail. However, their most important commercial use, at present, is as precursors to silicon carbide ceramics, an application which takes no advantage of their electronic properties.

Among the various synthetic procedures for polysilanes is the Harrod-type dehydrogenative coupling of $RSiH_3$ in the presence of Group 4 metallocenes (Reaction 8.1) [5,6]. One of the characteristics of the product obtained by this procedure is the presence of Si—H moieties, hence the name poly(hydrosilane)s. Since the bond dissociation enthalpy of Si—H is relatively weak when silyl groups are attached at the silicon atom (see Chapter 2), poly(hydrosilane)s are expected to exhibit rich radical-based chemistry. In the following sections, we have collected and discussed the available data in this area.

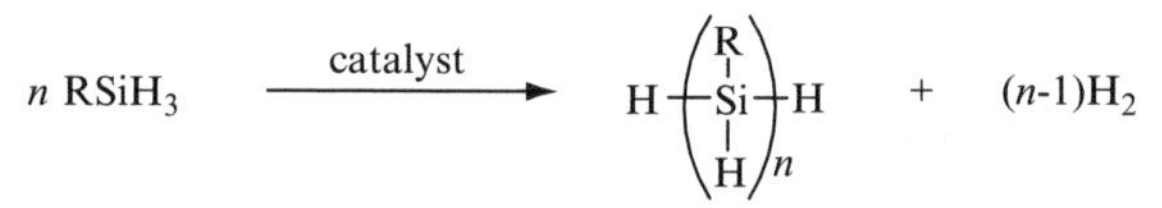

(8.1)

8.1.1 POLY(HYDROSILANE)S AND RELATED SILYL RADICALS

Both the catalyst and the silane monomer involved in dehydrocoupling polymerization have been studied in some detail. Two approaches have been used to obtain active catalysts: presynthesized complexes (generally of low stability but with definite properties) and *in situ* generation (limited choice of substituents but more convenient from practical and economic points of view). Zirconocene alkyl complexes are generally the most effective and efficient catalysts [6]. The nature of the substituents plays an important role in the induction period as well as in chain elongation [6,7]. For example, the induction period obtained for Cp_2ZrMe_2 is eliminated by using $Cp_2Zr(Me)Si(SiMe_3)_3$ and a significant chain elongation is observed by using a $CpCp^*Zr$ fragment ($Cp = \eta^5\text{-}C_5H_5$ and $Cp^* = \eta^5\text{-}C_5Me_5$). Most of the work to date has been limited to poly (phenylhydrosilane) (**2**), although a variety of arylsilanes [8] and alkylsilanes [9] have also been successfully polymerized. Despite of the modest molecular weights obtained by these reactions, catalytic routes offer the potential for preparing polymers of defined tacticity [10,11].

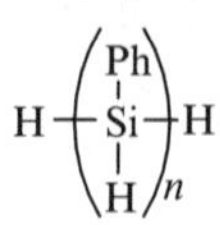

2

Figure 8.1 shows the GPC analysis of polysilane **2** and its radical-based degradation [12]. Line A shows a typical bimodal distribution ($M_W = 2320$, $M_W/M_n = 1.67$ against a polystyrene standard), the low molecular weight portions being mainly cyclic. Lines B and C show the outcome of the reaction between thermal generated *t*-BuO• radical with **2** at 50 °C/5 h and 140 °C/1 h, respectively. The M_W of the starting materials decreased from 2320 to 840 and 750, respectively, whereas the polydispersity index (M_w/M_n) decreased from 1.67 to 1.28 in both experiments. In the absence of radical initiators, any decomposition of polysilane **2** was found upon heating. Radical inhibitors such as quinones, considerably slow down the polysilane degradation, thus indicating that silyl radicals are involved as intermediates.

It was proposed that an initially generated silyl radical **3**, by reaction of *t*-BuO• radical and polysilane **2**, attacks another silicon atom in the same backbone to give a cyclic polysilane that contains an acyclic chain and another silyl radical (Scheme 8.1) [12]. The last silyl radical can either cyclize or abstract a hydrogen atom from another macromolecule, thus propagating the chain degradation. The reaction in Scheme 8.1 is an example of intramolecular homolytic substitution (S_Hi), a class of reactions discussed in Chapter 6.

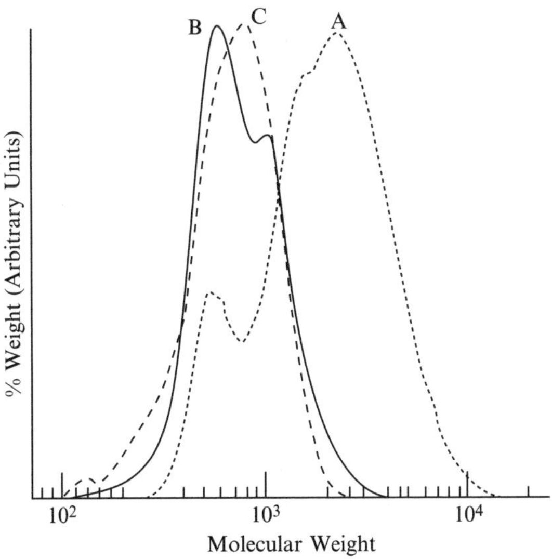

Figure 8.1 Gel permeation chromatograms related to the radical initiated degradation of poly(phenylhydrosilane) (**2**). (A) Initial sample with $M_w = 2320$ and $M_w/M_n = 1.67$. (B) Sample obtained after 5 h heating at 50 °C of a benzene solution containing 0.09 M of **2** in the presence of *t*-BuONNOBu-*t* (10 mol %); $M_w = 840$ and $M_w/M_n = 1.28$. (C) Sample obtained after 1 h heating at 140 °C of a *tert*-butylbenzene solution containing 0.09 M of **2** in the presence of *t*-BuOOBu-*t* (10 mol %); $M_w = 750$ and $M_w/M_n = 1.29$.

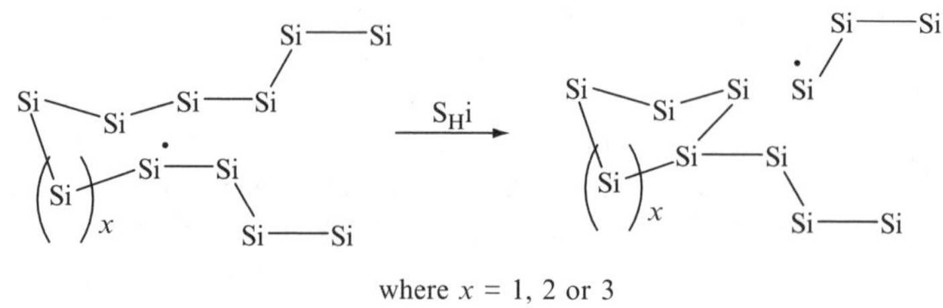

3

Scheme 8.1 Radical-based degradation of polysilanes

Direct evidence that a hydrogen abstraction from poly(hydrosilane)s affords the corresponding silyl radical was obtained by EPR spectroscopy [13]. The photochemical behaviour of poly(*n*-hexylhydrosilane) (**4**) in the presence and absence of *t*-BuOOBu-*t* is shown in Figures 8.2a and 8.2b, respectively. When irradiation was cut off in the experiment with peroxide, the remaining signal is that of Figure 8.2b, which persisted for several hours. The spectrum in Figure 8.2c, obtained by subtracting spectrum b from spectrum a, is characterized by the coupling of the unpaired electron with two hydrogens (7.10 G) and

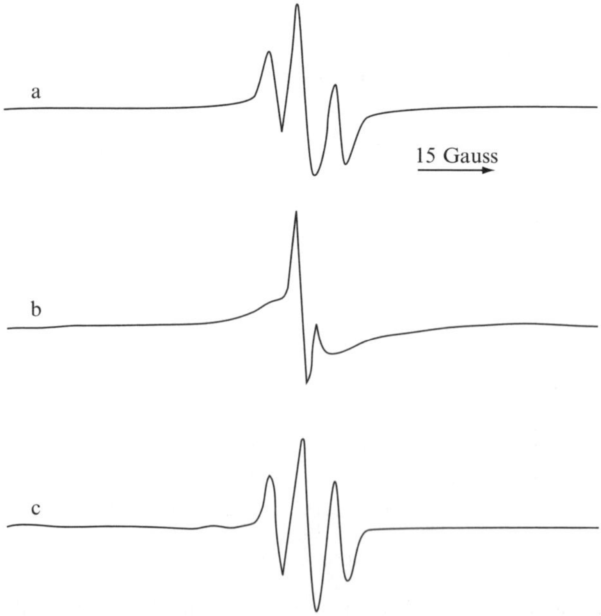

Figure 8.2 (a) EPR spectrum obtained by photolyzing a solution of poly(*n*-hexylhydrosilane) and *t*-BuOOBu-*t* in cyclohexane; (b) Spectrum ($g = 2.0057$) observed when irradiation was discontinued; (c) Spectrum ($g = 2.0047$) obtained by subtracting spectrum b from spectrum a. Reprinted with permission from Reference [13]. Copyright 1997 Elsevier.

a *g* factor of 2.0047, which is assigned to radical **5** obtained by Reaction (8.2). The observed alternating line width effect with decreasing temperature is indicative of a restricted rotational motion around the C—Si bond between the radical centre and the *n*-hexyl substituent.

(8.2)

4 Hx = *n*-hexyl **5**

Figure 8.2b corresponds to the superimposition of signals from at least two persistent paramagnetic species. The nature of these species is still unclear even if the *g* factors are consistent only with silyl radicals carrying three silyl groups as in **6**. Mechanistic schemes for their formation could be suggested based on the addition of transient silyl radicals to the transient disilenes generated photolytically [14].

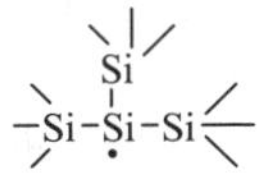

6

8.2 OXIDATION STUDIES ON SILYL-SUBSTITUTED SILICON HYDRIDES

8.2.1 POLY(HYDROSILANE)S

Poly(hydrosilane)s are stable compounds and can be manipulated in the air only for a short period since they are oxygen sensitive. In order to study the oxidation products, a xylene solution of poly(phenylhydrosilane) ($M_W = 2340$, $M_W/M_n = 1.72$) was refluxed (140 °C) for 12 h in a system exposed to the air [15]. Only minor changes were observed by GPC analysis whereas FTIR showed characteristic absorptions due to siloxane-type structures on the polymer backbone. A detailed NMR analysis, based on ^{1}H NMR, ^{29}Si INEPT and ^{1}H–^{29}Si HMQC spectroscopies, indicated that the oxidized material contains the units **7**–**10** shown in Scheme 8.2. In particular, units **7**, **8** and **9** + **10** were present in relative percentages of 27 %, 54 % and 19 %, respectively, which mean that more than 70 % of the catenated silicons were altered. It has also been reported that silyl hydroperoxides and peroxides are not found as products in the autoxidation of poly(phenylhydrosilane) [16].

Two different kinetic approaches have been used to obtain mechanistic information [15,16]. Kinetic studies carried out by EPR spectroscopy using fusinite as a paramagnetic probe of the oxygen concentration allowed the oxidizability of poly(hydrosilane)s to be obtained [15]. Oxidizability values of 1.8×10^{-2} and

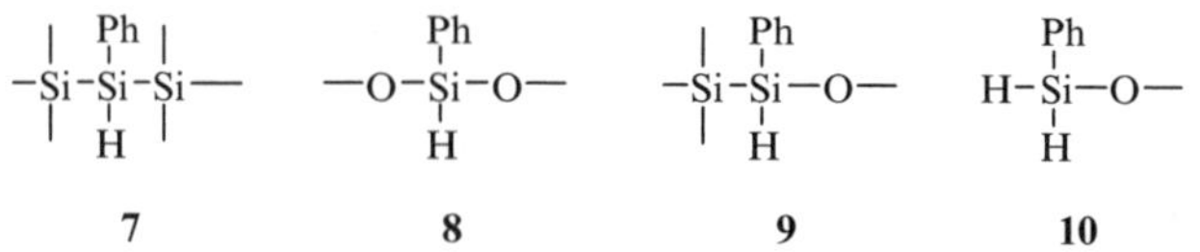

Scheme 8.2 Type of units assigned in oxidized poly(phenylhydrosilane) by ^{1}H detected ^{1}H–^{29}Si HMQC spectroscopy

$1.2 \times 10^{-2}\,M^{-1/2}\,s^{-1/2}$ (referring to each SiH moiety) were found for poly(phenylhydrosilane) and poly(*n*-hexylhydrosilane), respectively, when a benzene solution of polysilane was oxidized under air at 50 °C using AIBN as the radical initiator. Moreover, the kinetics of poly(phenylhydrosilane) oxidation in benzene using AIBN as the radical initiator was investigated in the temperature range of 58–77 °C by following oxygen uptake [16]. The reaction rate showed the following dependences: a first-order on polysilane, a zero-order on oxygen and a half-order on radical initiator. The chain length of oxidation is found to be relatively long ($\nu > 10$). An oxidizability value of $10.9 \times 10^{-2}\,M^{-1/2}\,s^{-1/2}$ was measured at 58 °C for the polysilane by this technique, which corresponds to $0.6 \times 10^{-2}\,M^{-1/2}\,s^{-1/2}$ for each (PhSiH) group.

Based on these investigations the radical chain reactions for the oxidation of poly(phenylhydrosilane) can be described by the reaction sequence shown in Scheme 8.3 (see also the following section). The initially formed ROO• radical is able to abstract hydrogen from the polysilane. Indeed, a rate constant of $24.7\,M^{-1}\,s^{-1}$ (referring to each SiH moiety) was measured for the reaction of cumylperoxyl radical with poly(phenylhydrosilane) [17]. In the propagation steps, the reaction of silyl radical with oxygen should be *fast* and the resulting silylperoxyl radical is expected to rearrange to disilyloxysilyl radical also *fast*, whereas the *slow* step is expected to be the hydrogen abstraction from the starting polysilane by the disilyloxysilyl radical. Using the usual steady-state approximation, the overall rate of oxidation is given by Equation (8.3), from which the oxidizability, $k_p/(2k_t)^{1/2}$ can be obtained.

$$-\frac{d[O_2]}{dt} = R_i + \frac{k_p}{(2k_t)^{1/2}}[R_3SiH](R_i)^{1/2} \tag{8.3}$$

The values of oxidizability obtained for poly(hydrosilane)s are comparable with the $1.9 \times 10^{-2}\,M^{-1/2}\,s^{-1/2}$ of methyl linoleate, an example of easily oxidizable substrate.

8.2.2 $(Me_3Si)_3SiH$ AND $(Me_3Si)_2Si(H)Me$ AS MODEL COMPOUNDS

$(Me_3Si)_3SiH$ as a pure material or in solution reacts spontaneously and slowly at ambient temperature with air or molecular oxygen, to form a major product that contains a siloxane chain (Reaction 8.4) [18]. The percentages of

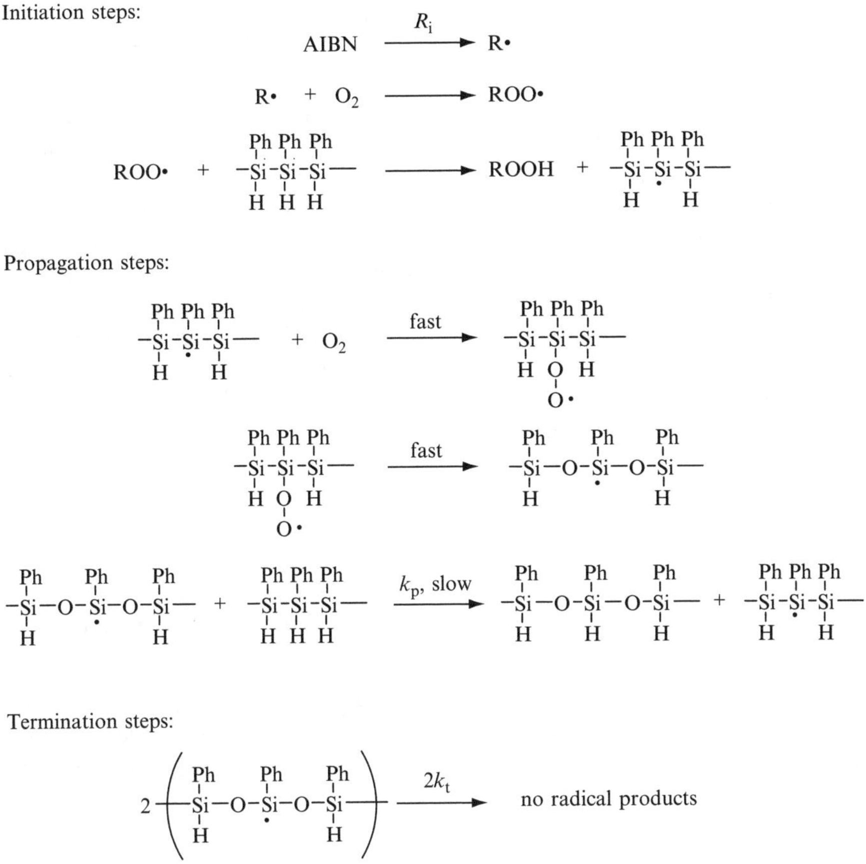

Scheme 8.3 The reaction mechanism for the radical chain oxidation of poly(phenylhydrosilane)

conversion and yield depend upon the experimental conditions. For example, 50 % conversion and yield > 95 % have been obtained after 1 h of air bubbled into 1 ml of pure silane. The reaction is a radical chain process since it is accelerated by radical initiation and retarded by common radical inhibitors.

(8.4)

The homologous $(Me_3Si)_2Si(H)Me$ does not react spontaneously with air or molecular oxygen at room temperature. However, a reaction takes place at 80 °C when air or molecular oxygen is bubbled into the pure material or its solutions to form a major product that contains the siloxane chain (Reaction 8.5) [15]. In general, yields of siloxane are about 50 %. Also in this case the reaction is accelerated by radical initiation and retarded by common radical inhibitors.

(8.5)

Tetrakis(trimethylsilyl)silane, $(Me_3Si)_4Si$, cannot be oxidized under the same conditions [18]. This excludes any direct insertion of an oxygen molecule on the Si—Si bond followed by a molecular rearrangement.

Oxygen-labelling experiments were also performed in order to distinguish whether the two oxygen atoms in the final product arise from two different oxygen molecules or from the same oxygen molecule. Both $(Me_3Si)_3SiH$ and $(Me_3Si)_2Si(H)Me$ were treated with a mixture of $^{16}O_2$ and $^{18}O_2$ (Reaction 8.6) and the crude products were analysed by mass spectrometry to determine the isotopic distribution. The fact that similar label distribution was observed in the products as in the reactants for both cases, indicates an *intramolecular* mechanism, i.e., two oxygen atoms in the final product arise from the same oxygen molecule [15,18].

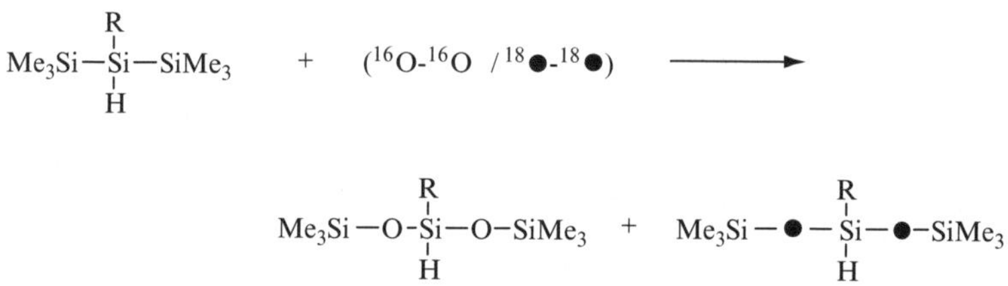
(8.6)

The reaction sequence shown in Scheme 8.4 is in accord with all the experimental observations and involves at least two consecutive unimolecular steps [15,18]. Silyl radical **11** adds to molecular oxygen to form the peroxyl radical **12**.

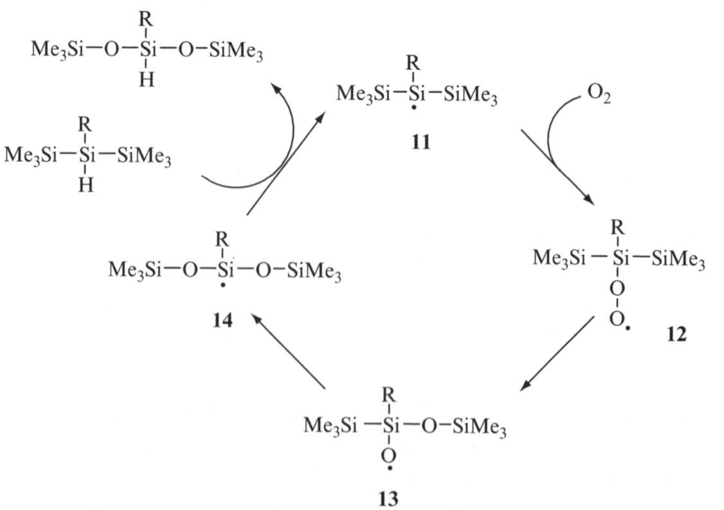

Scheme 8.4 Proposed propagation steps of the autoxidation of silicon hydrides containing at least two silyl substituents (R = Me or $SiMe_3$)

This silylperoxyl radical undergoes an unusual rearrangement to **13** followed by a 1,2-shift of the Me_3Si group to give **14**. Hydrogen abstraction from the silane by radical **14** gives the desired product and another silyl radical **11**, thus completing the cycle of this chain reaction.

The reaction mechanism of $(Me_3Si)_3SiH$ oxidation has been studied in some detail [19]. The absolute rate constant for the spontaneous and strongly endothermic (ca 130 kJ/mol) reaction of $(Me_3Si)_3SiH$ with molecular oxygen (Reaction 8.7) has been determined to be $(0.48 \pm 0.18) \times 10^{-4}\,M^{-1}\,s^{-1}$ at 70 °C by using an inhibitor-based (hydroquinone) method.

$$(Me_3Si)_3SiH + O_2 \longrightarrow (Me_3Si)_3Si\bullet + HOO\bullet \qquad (8.7)$$

A rate constant of $8.3 \times 10^3\,s^{-1}$ at 70 °C for the rearrangement of radical **12** is obtained by kinetic analysis of the autoxidation of $(Me_3Si)_3SiH$ and based on the assumption that the termination rate constant for radical **12** is $2k_t = 2.4 \times 10^8\,M^{-1}\,s^{-1}$ at 70 °C, i.e., the corresponding value for $Ph_3SiOO\bullet$ radical. It is worth mentioning that the hydrogen abstraction from $(Me_3Si)_3SiH$ of cumylperoxyl radical (Reaction 8.8) occurs with a rate constant of $66.3\,M^{-1}\,s^{-1}$ at 70 °C, which means that in pure silane the rearrangement of silylperoxyl radical will be ca 30 times faster. Obviously, this is the reason for the absence of silyl hydroperoxide in the product studies.

$$ROO\bullet + (Me_3Si)_3SiH \longrightarrow ROOH + (Me_3Si)_3Si\bullet \qquad (8.8)$$

The rearrangement **12**→**13** is not straightforward and two alternatives routes have been suggested. One of them involves a 1,3 transfer from silicon to oxygen to give the silyl radical **15** followed by S_Hi reaction on the peroxide moiety [15]. The other indicated by PM3 calculations is a dioxirane-like three-membered intermediate **16** having a pentacoordinated central silicon followed by a 1,2 silyl migration to afford radical **13** [19].

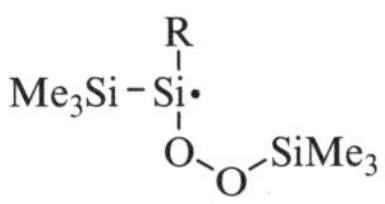

15

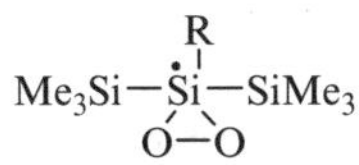

16

The 1,2 shift of Me_3Si group from silicon to oxygen (**13** → **14**) is expected to be a very fast process ($>10^7\,s^{-1}$) as demonstrated in analogous reactions (see Chapter 6). Interestingly, the hydrogen abstraction from $(Me_3Si)_3SiH$ by the disilyloxysilyl radical **14** is expected to be the *slow* step (i.e., k_p in the oxidizability value $k_p/(2k_t)^{1/2}$) although it cannot be very slow. In fact, the silyl radical **14** rather prefers to abstract a hydrogen atom from $(Me_3Si)_3SiH$ than to add to another oxygen molecule. Product studies of $(Me_3Si)_3SiH$ oxidation under high dilution showed the formation of siloxane **19**, which is probably derived through the intermediacy of peroxyl radical **18** (Scheme 8.5) [12]. Therefore, the rate constant for addition of radical **17** to molecular oxygen should be much slower than a diffusion-controlled rate constant.

EPR investigation on the oxidation of $(Me_3Si)_2Si(H)Me$ in benzene at 50 °C showed some interesting features too [15]. First of all, the measured oxidizability value $k_p/(2k_t)^{1/2} = 1.8 \times 10^{-2}\,M^{-1/2}\,s^{-1/2}$ for this silane is similar to that of poly(hydrosilane)s when referring to each Si—H moiety. Assuming a

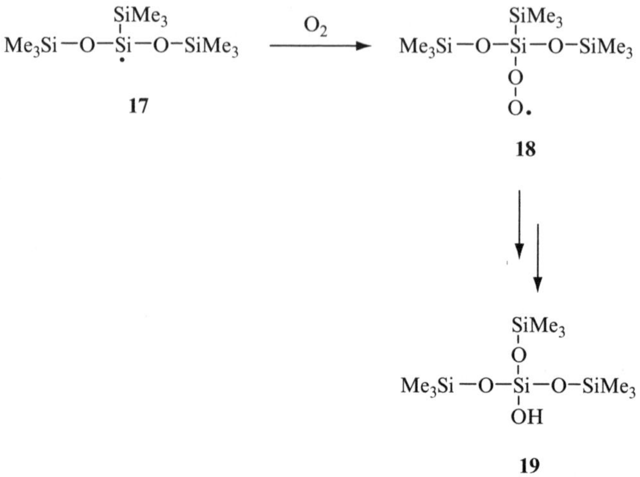

Scheme 8.5 Reaction sequence for the complete oxidation of $(Me_3Si)_3SiH$

$2k_t = 1 \times 10^9\,M^{-1}\,s^{-1}$ for radical **14** (R = Me), the rate constants (k_p) for the hydrogen abstraction from $(Me_3Si)_2Si(H)Me$ by radical **14** is $6 \times 10^2\,M^{-1}\,s^{-1}$. Oxygen consumption of $(Me_3Si)_2Si(H)Me$ initiated by photogenerated *t*-BuO• radicals in the presence of 14 mM of α-tocopherol, indicated that not all the silylperoxyl radicals **12** are trapped. Since the rate constant for the reaction of peroxyl radical with α-tocopherol (under similar experimental conditions) is $4.1 \times 10^6\,M^{-1}\,s^{-1}$ [20], we calculated a rate constant of ca $10^3\,s^{-1}$ for the unimolecular reaction of silylperoxyl radical **12**, where R = Me.

8.3 FUNCTIONALIZATION OF POLY(HYDROSILANE)S

8.3.1 HALOGENATION

It was reported that the reaction of poly(phenylhydrosilane) ($M_w = 3004$, $M_w/M_n = 1.79$) with CCl_4 under room light selectively affords poly(phenylchlorosilane) along with $CHCl_3$ and CH_2Cl_2 as by-products (Reaction 8.9) [21]. The replacement of Si—H moieties by Si—Cl is found to be >84 % or >95 % for reaction periods of 28 h and 5 days, respectively. Vapour pressure osmometry revealed that some degradation of polysilane backbone also accompanies chlorination [22]. A similar reaction in which CCl_4 was replaced by CBr_4 afforded the corresponding poly(phenylbromosilane) [21].

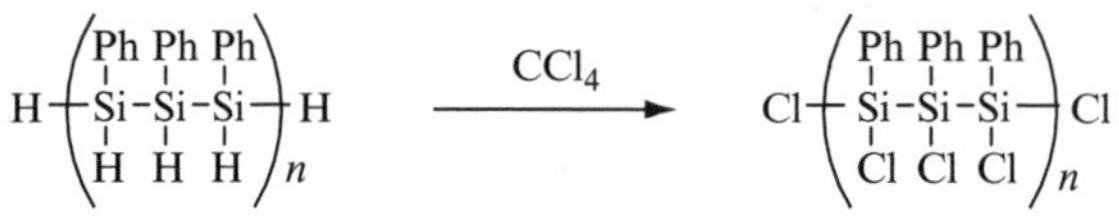

(8.9)

Poly(phenylhydrosilane) ($M_w = 2440$, $M_w/M_n = 1.89$) and poly(n-hexylhydrosilane) ($M_w = 1310$, $M_w/M_n = 1.08$) have been used as radical-based reducing agents for organic halides (RX, where X = Cl, Br, I) and they rival the effectiveness of the $(Me_3Si)_3SiH$ (see Chapter 4) [13]. These reactions are accelerated by radical initiators and retarded by substituted phenols, quinones or nitroxides indicating a free-radical process, in which the two propagation steps are the hydrogen abstraction from the polysilane by an alkyl radical (Reaction 8.10) and the halogen abstraction from alkyl halide by the silyl radical on the polysilane chain. A rate constant (k_H referring to each SiH moiety) between 5×10^4 and $6 \times 10^5\,M^{-1}\,s^{-1}$ is found for Reaction (8.10), where R = Ph, using the radical clock methodology [13].

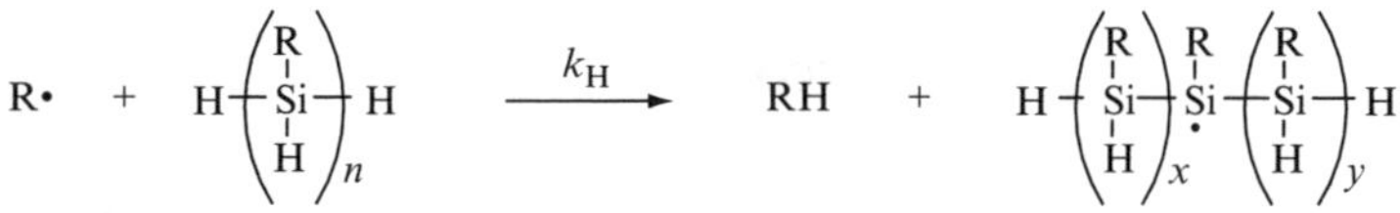

(8.10)

Poly(phenylchlorosilane) can be considered as a versatile synthon for the preparation of a variety of functionalized polysilanes. Indeed, its reactions with MeOH or MeMgBr afforded polysilane containing Si—OMe or Si—Me moieties, respectively, whereas its reaction with $LiAlH_4$ regenerated the starting poly(phenylhydrosilane) [21,22].

8.3.2 ADDITION OF UNSATURATED COMPOUNDS

The addition of a variety of monosubstituted olefins on poly(phenylhydrosilane) (**2**) ($M_w = 1960$, $M_w/M_n = 1.66$) was carried out in refluxing toluene or 2,5-dimethyl-THF by using AIBN as the radical initiator [23,24]. These reactions allowed the synthesis of functional polysilanes **20** derived from the addition of silyl radical to the less crowded end of the double bond and with a high degree of substitution (Reaction 8.11). Indeed, the replacement of Si—H moieties ranged from 84% to 93%. Vapour pressure osmometry revealed only a slight loss in molecular weight following hydrosilylation. The incorporation of polar side groups renders some of these polysilanes also soluble in environmentally friendly solvents (water and alcohols). No examples of hydrosilylation of poly(hydrosilane)s with electron-poor olefins are reported. Probably such reactions suffer of the competitive polymerization of olefins.

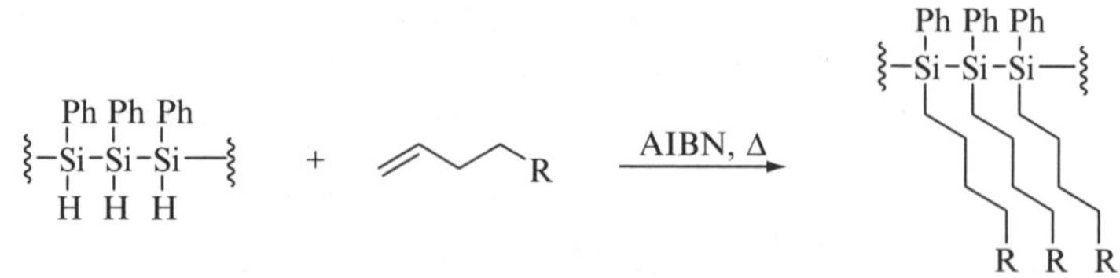

R = Et, CO_2Me, CO_2H, CH_2OH, CH_2NMe_2

(8.11)

2 **20**

The addition of *gem*-disubstituted olefins, $CH_2{=}CXY$, on polysilane **2** also worked well [23,24]. For example, the addition of 2-methoxypropene and methylenecyclohexane afforded the expected adducts with 73% and 77% degrees of substitution, although a higher loss of molecular weight with respect to the hydrosilylation of monosubstituted olefins is observed. Copolymer **21**, containing both mono- and disubstituted olefins, was made from **2** in a single reaction by adding 50 mol% vinyl acetic acid and an excess of 2-methoxypropene to the THF–polymer solution [24].

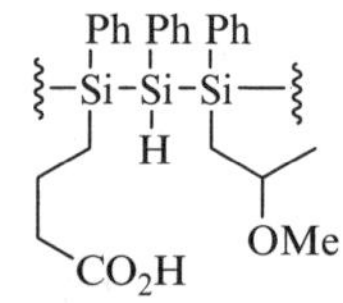

21

Information about the individual steps of these radical chain hydrosilylations is very scarse. Silyl radicals obtained from both phenyl and *n*-hexyl substituted poly(hydrosilane)s add to the less crowded end of C=C bond of 1,1-diphenylethylene and 9-methyleneanthranone to give silyl adducts whose EPR spectra are similar to that found for the $Ph_3Si^\bullet$ adduct [13]. Competition experiments between $CH_2{=}CHR$ and $BrCH_2CH_2R$ (where $R = C_8H_{17}$) for the reaction with silyl radicals from **2** showed that the bromide is only 2.9 times more reactive than the alkene [13], whereas for $Et_3Si^\bullet$ and $(Me_3Si)_3Si^\bullet$ radicals, the bromide is at least 100 times more reactive than the alkene (cf. Chapters 4 and 5). In the hydrogen abstraction step it is reasonable to think in terms of intramolecular reactions also because of the high degree of observed substitution. Although no experimental evidence for the radical translocation have yet been found, 1,4- and 1,5-hydrogen shifts are expected to be the best candidates due also to relatively long Si—Si bonds (Scheme 8.6).

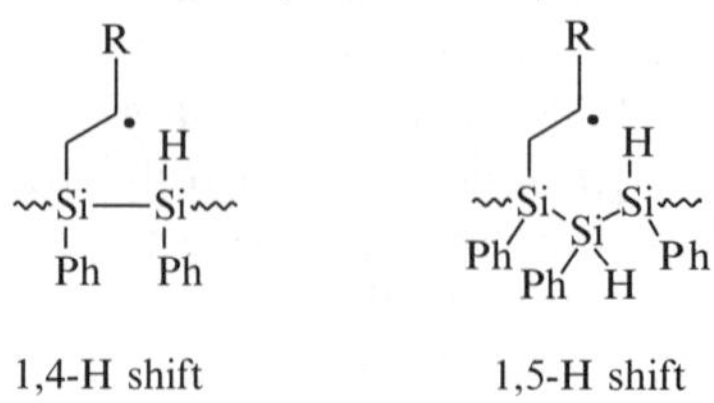

Scheme 8.6 Possible geometries for hydrogen shift

Analogously to terminal alkenes, the reaction of **2** with valeraldehyde and cyclohexanone under radical-based conditions allowed for the preparation of the corresponding functional polysilanes **22** (Reaction 8.12). The efficiency of Si—H bond replacement was 80–85% [21].

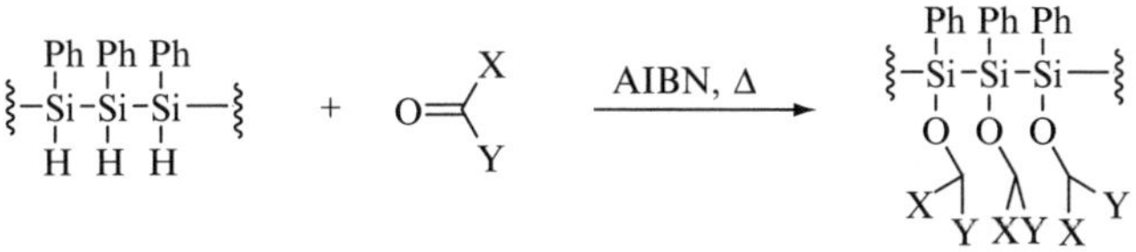

(8.12)

2 **22**

Silyl radicals obtained by reaction of **2** or **4** with thermally generated *t*-BuO• radicals add readily to a variety of carbonyl compounds (aromatic ketones and quinones) to give the corresponding adducts for which EPR spectra have been recorded [13]. It is worth mentioning that adducts with 9,10-phenanthroquinone (**23**) are characterized by the magnetic equivalence of the two aromatic rings. This means that a fluxional motion connected with silyl group migration between the two oxygens occurs faster than similar ones obtained for Et_3Si•, Ph_3Si• and $(Me_3Si)_3Si$• radicals (see Section 5.3.1).

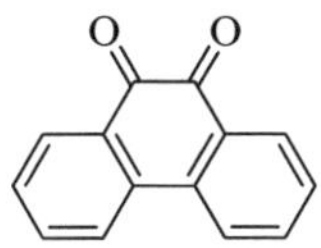

23

The importance of poly(hydrosilane)s in radical chemistry is manifested in their applications as processing stabilizers for organic polymeric materials subject to oxidative degradation [25]. The degradation of polyolefins during processing takes place by a widely accepted free-radical mechanism. The ability of poly(hydrosilane)s, and in particular poly(phenylhydrosilane), to stabilize polypropylene during multiple extrusion is a relevant property. It is believed to be due to the combined action of the effectiveness of hydrogen donation of these polysilanes together with their capabilities to scavenge the traces of oxygen present during the extrusion process. On the other hand, the introduction of sterically hindered amine groups through radical hydrosilylation reactions allowed the preparation of polysilanes used as stabilizers of polyolefins during their life [26]. The hydrosilylation was carried out with simple molecules such as olefin **24** or ketone **25** as well as with larger olefins such as **26** and the degree of Si—H substitution was up to 90% depending on the experimental conditions. In this context, the free-radical chemistry associated with poly(hydrosilane) domains has been used in conjuction with the well known activity of HALS (hindered amines light stabilizer) as a synergistic effect.

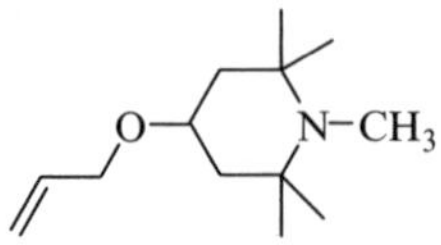

24

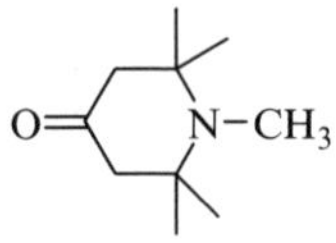

25

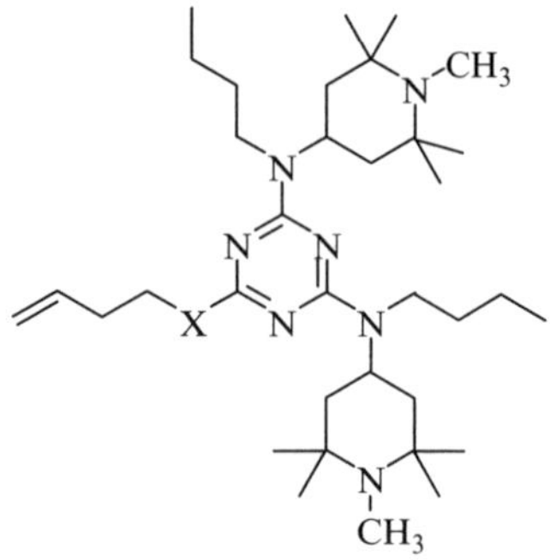

26, X = O, NH

8.3.3 OTHER USEFUL RADICAL REACTIONS

Silyl radicals (*Si*) obtained by reaction of **2** or **4** with thermally generated *t*-BuO• radicals add readily to pyrazine and nitroalkanes to form the corresponding adducts **27** and **28** for which EPR spectra have been recorded [13].

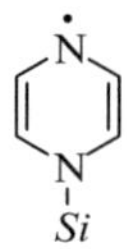

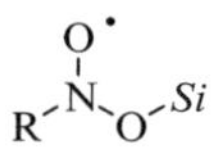

27 **28**

Polysilanes **2** and **4** reacted with nitroxides (TEMPO) under free-radical conditions and in the absence of molecular oxygen to give the corresponding amine in good yields (Reaction 8.13) [15]. The propagation steps for these radical chain processes are thought to be similar to those reported for the analogous reduction using $(Me_3Si)_3SiH$ (see Section 4.3.4).

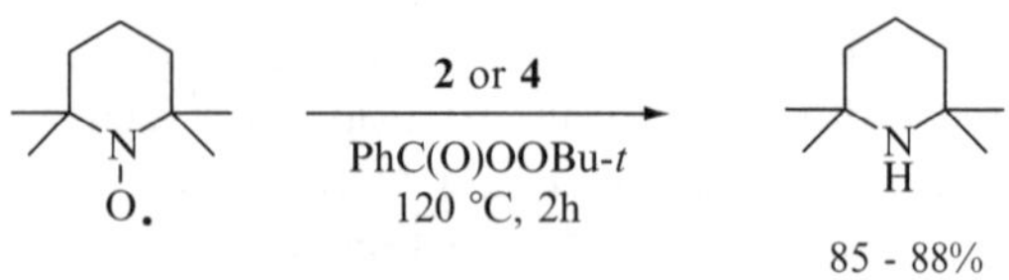

(8.13)

8.4 SILYLATED FULLERENES

The chemical modification of fullerenes has received considerable attention in the last decade in order to achieve new applications to material sciences [27]. Fullerene-bonded polysilane derivatives might be expected to show high conductivity since C_{60}-doped polysilane is found to be a good photoconductor [28]. Therefore, a variety of silylated derivatives have been obtained to date, although the available methods are limited to the photoinduced addition of various silanes to C_{60}.

EPR studies showed that $R_3Si\bullet$ radicals (R = Me, Et, *i*-Pr and *t*-Bu), obtained by hydrogen abstraction from the parent silanes by photogenerated *t*-BuO• radicals, add to C_{60} to form the corresponding adducts [29]. The spectra of Me_3Si- and *t*-Bu_3Si-adducts have hyperfine structure due to 9 and 27 equivalent protons, respectively, at 27 °C suggesting free rotation of Si—C bonds on the EPR time scale. On the other hand, the middle members of the series, Et_3Si- and *i*-Pr_3Si-adduct radicals, showed a unexpected hyperfine manifold which has been accommodated with free rotation about the Si—C_{60} and frozen rotation about the Si—R bonds on the EPR time scale.

Photolysis of some disilanes having phenyl substituents or mixed phenyl/*tert*-butyl substituents in the presence of C_{60} with a low-pressure mercury lamp (254 nm) resulted exclusively in the formation of 1,16-adducts **29** in 54–62% yield (Reaction 8.14) [30–32]. Evidence that the initial step is the homolytic cleavage of Si—Si bond, followed by silyl radical addition to fullerene has been obtained. Similar 1,16-adducts were obtained by mixing the silyllithium derivative $(t\text{-Bu})_2PhSiLi$ or $(Me_3Si)_3SiLi$ with C_{60} at –78 °C in 69–72% yield [33]. Although the mechanism of this reaction is unknown, the formation of a 1,16-adduct suggested an involvement of radicals or radical ions [34]. The redox properties of 1,16-adduct **29** when R = Ph and R′ = *t*-Bu consist of three reversible reductions (−1.19, −1.59 and −2.18 V) and two irreversible oxidations (+0.33 and +1.20 V) in 1,2-dichlorobenzene [32]. It is worth underlining the remarkably low oxidation potential of this adduct (+ 0.33 V) compared to C_{60} (+1.21 V).

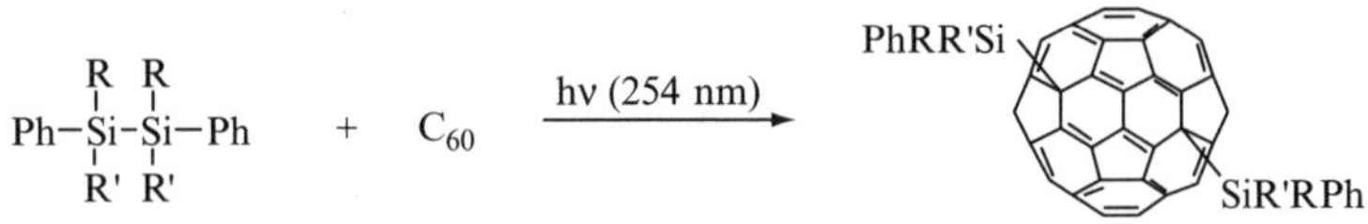

(8.14)

R = R' = Ph
R = Ph, R' = *t*-Bu
R = R' = *t*-Bu
R = R' = $SiMe_3$

29

For R = R′ = Me_3Si, the 1,16-adduct **29** is formed in a 43% yield (Reaction 8.14) together with the adduct **30** as a by-product in a 24% yield (Reaction 8.15). This unusual 1,2-adduct is observed in higher yields as the only product by replacing one silyl group by phenyl, or both by isopropyl groups [30,31].

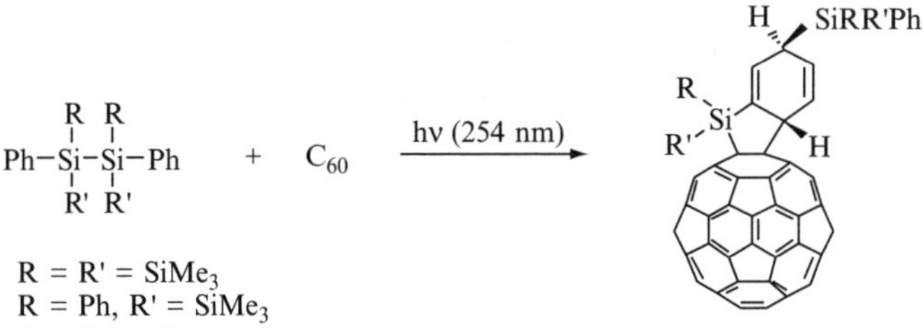

(8.15)

R = R' = $SiMe_3$
R = Ph, R' = $SiMe_3$
R = R' = *i*-Pr

30

Interestingly, 1,2-adducts **33, 34** and **35** have also been obtained in a 69% yield and a ratio of 1:10:9, when $PhSi(SiMe_3)_3$ is photolysed in the presence of C_{60}, which clearly indicates the intermediacy of silyl-adduct radical **31** followed by an intramolecular addition to the phenyl ring (**32**) prior to combining with $Me_3Si\bullet$ radicals to provide these unusual adducts (Scheme 8.7) [30,31].

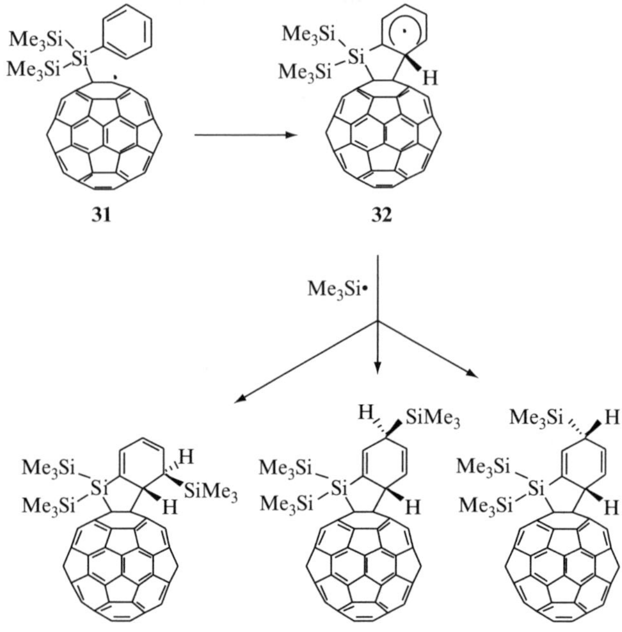

Scheme 8.7 Reaction paths for the formation of 1,2-adducts **33, 34** and **35** obtained from the photolysis of $PhSi(SiMe_3)_3$ in the presence of C_{60}

The photoreaction of polysilanes with C_{60} has also been investigated [35]. Reaction (8.16) shows an example in which the irradiation in benzene with a low-pressure mercury-arc lamp afforded a product that contains 14 wt% of C_{60} into the polysilane chain. The incorporation of C_{60} into the polysilane backbone has not been observed upon irradiation with $\lambda > 300$ nm, when the cleavage of Si—Si bond does not take place. The adduct obtained from Reaction (8.16) has a lower oxidation potential than C_{60}(+0.77 vs +1.21 V) and a lower reduction potential than polysilane (−1.24 vs −2 V).

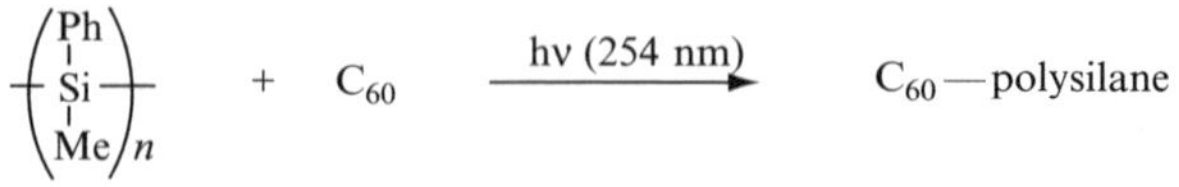

(8.16)

Laser flash photolysis experiments showed that the C_{60} triplet state ($^3C^*_{60}$), obtained by 532 nm excitation, reacts with poly(methylphenylsilane) by an electron transfer mechanism (Reactions 8.17 and 8.18) [36]. The electron transfer has been revealed by following the decay of ${}^3C_{60}^*$ or, alternatively, the rise of the radical anion of C_{60} and the radical cation of polysilane, which both absorb in the near-IR region. The k_{ET} for the latter reaction is $2.0 \times 10^8\ M^{-1}\ s^{-1}$ in benzene–acetonitrile (2:1) and depends on the polarity of the medium.

$$C_{60} \xrightarrow{h\nu\ (532\ nm)} {}^3C_{60}^* \qquad (8.17)$$

$$ {}^3C_{60}^* + \left(\begin{array}{c} Ph \\ | \\ Si \\ | \\ Me \end{array} \right)_n \xrightarrow{k_{ET}} {}^3C_{60}^{\bullet -} + \left(\begin{array}{c} Ph \\ | \\ Si \\ | \\ Me \end{array} \right)_n^{\bullet +} \qquad (8.18)$$

The photochemical addition of some cyclic oligosilanes to C_{60} has also been found interesting. Scheme 8.8 shows some examples of such a transformation. Irradiation ($\lambda > 300$ nm) of a toluene solution of disilirane **36** with C_{60} afforded the fullerene derivative **37** in a 82% yield [37]. The reaction mechanism is still unknown. When toluene is replaced by benzonitrile the bis-silylated product of the solvent was obtained in good yields. In these experiments a photoinduced electron transfer between **36** and C_{60} is demonstrated, indicating the role of C_{60} as sensitizer [38]. The photoinduced reactions of disilirane **36** with higher fullerenes such as C_{70}, $C_{76}(D_2)$, $C_{78}(C_{2v})$and $C_{84}(D_2)$ have also been reported [39,40].

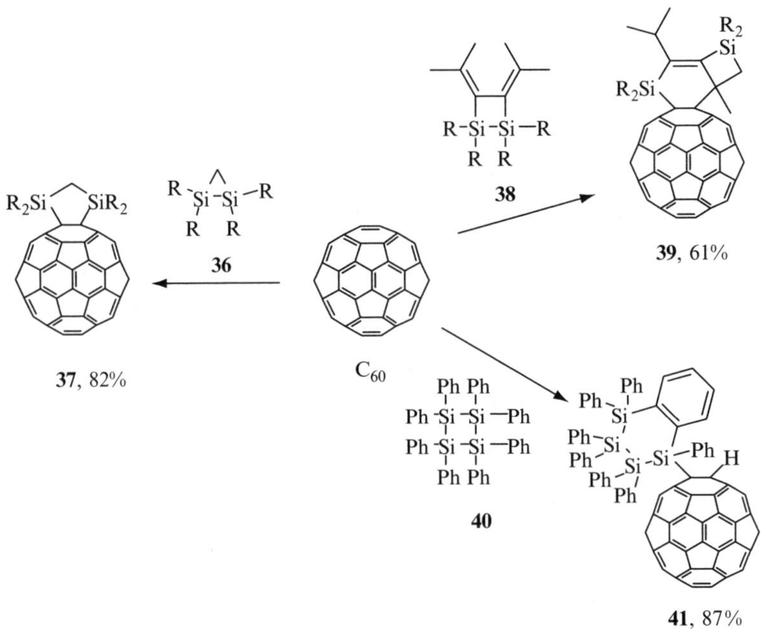

Scheme 8.8 Photochemical ($\lambda > 300$ nm) addition of cyclic oligosilanes to C_{60}

Irradiation ($\lambda > 300$ nm) of a toluene solution of disilacyclobutane **38** with C_{60} afforded the fullerene derivative **39** in a 61 % yield, resulting from an unexpected rearrangement of the disilacyclobutane unit [41]. Also this reaction mechanism is still unknown.

On the other hand, cyclotetrasilane **40** reacted with C_{60} under identical experimental conditions, affording the 1:1 adduct **41** in an 87 % yield [42]. It is worth underlining that the rearrangement of the cyclotetrasilane unit derived from the addition of a silyl radical moiety to one of the terminal phenyl rings. Photoreaction of C_{60} with higher homologous cyclic oligosilanes such as $(Ph_2Si)_5$, $(Et_2Si)_5$ and $(Me_2Si)_6$ has also been investigated using 254 nm excitation [35]. In these cases, 1,4-addition products having the formula $C_{60}(SiR_2)_4$ have been obtained.

Laser flash photolysis experiments showed that the ${}^3C_{60}^*$ (Reaction 8.17) reacts with cyclic oligosilanes **42, 43** and **44** in benzonitrile by an electron transfer mechanism [43]. The rate constant (k_{ET}) for the three-membered cyclic compound **42** is found to be $7.0 \times 10^8\ M^{-1}\ s^{-1}$, whereas for the other two compounds it was more than two orders of magnitute lower, i.e., $(1\text{–}2) \times 10^6\ M^{-1}\ s^{-1}$.

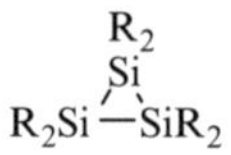

42, R = mesityl

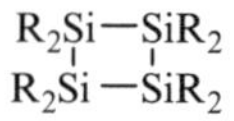

43, R = *i*-Pr

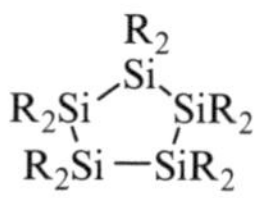

44, R = Ph

The chemical reactivity of silylated fullerenes has hardly been explored. The fluorescence quantum yield of 1,16-adduct **29** (R = Ph, R′ = *t*-Bu) was found slightly larger than that of C_{60}, whereas its triplet excited state and quenching rates were very similar to those of C_{60} [44]. From the silyl radical chemistry point of view, it is worth underlining that the reaction of silylated fullerene **41** with bromine in carbon disulfide under daylight irradiation, afforded C_{60} in a 90 % yield, together with the unexpected 9,10-disilaanthracene derivative **47**, whose structure was determined by X-ray crystallography, in a 43 % yield (Scheme 8.9) [45]. Although the reaction mechanism is unknown, the formation of intermediate silyl radical **45** was suggested, which rearranges by intramolecular addition to a pendant phenyl moiety to ultimately provide **46** (cf. Section 6.2). Further oxidation of the central Si—Si bond is expected to give the observed product.

8.5 RADICAL CHEMISTRY ON SILICON SURFACES

Silicon is the most technologically important material utilized today owing to its unique role in the fabrication of semiconductor devices and microprocessor chips. The understanding and control of silicon surfaces is of great importance in the production of silicon-based electronic devices, since the fraction of atoms

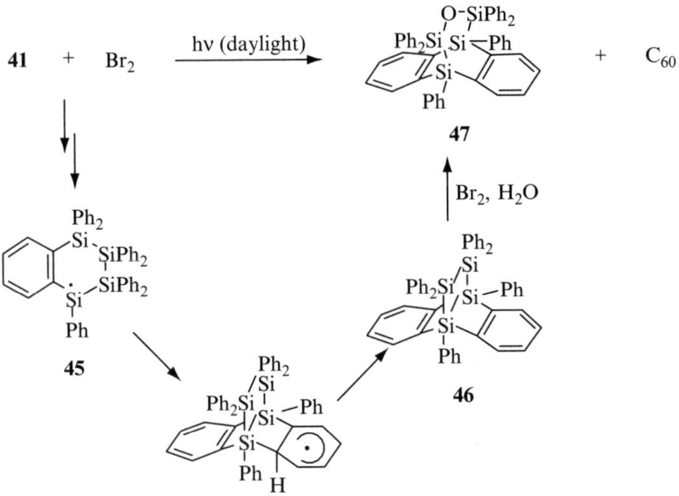

Scheme 8.9 Proposed reaction mechanism for the formation of 9,10-disilaanthracene derivative **47**

residing on or near the surface became significant. Although the success of silicon is mainly due to the presence of robust native oxide on silicon surfaces, much attention is being directed towards the synthesis of an organic monolayer, which can be modified upon demand for specific requirements. In this section, we deal with the chemistry of hydrogen-terminated silicon surfaces and, in particular, with radical reactions that have been found to be the most convenient methods for organic modification of silicon surface [46,47]. Indeed, the chemistry can be understood in many cases by analogy with radical reactions of organosilicon hydrides described in other chapters of this book.

Structural properties of hydrogen-terminated silicon surfaces are of critical importance for their chemical behaviour. Flat (single crystal) and porous silicon surfaces are available. The Si(111) and Si(100) orientation of a single crystal are shown in Figure 8.3 together with porous silicon as examples of silicon hydride-terminated surfaces [47]. The porous silicon is terminated with SiH, SiH_2 and SiH_3 moieties in a variety of different local orientations and environments. These materials are reasonably stable and can be prepared and manipulated in air for tens of minutes as well as in a number of organic solvents. However, on prolonged exposure to air, single crystal silicon becomes coated with a thin, native oxide that can be removed chemically from Si(111) using 40 % aqueous NH_4F or from Si(100) and porous silicon using dilute aqueous HF. Under ultrahigh vacuum conditions it is also possible to produce uniform monohydride H—Si(100) surfaces (see below). It is worth underlining that the monohydride terminal surface of H—Si(111) resembles $(Me_3Si)_3SiH$. For example, the Si(111) surface after fluoride ion treatment exhibits a sharp peak at 2084 cm^{-1} with p-polarized infrared light due to Si—H stretching absorption

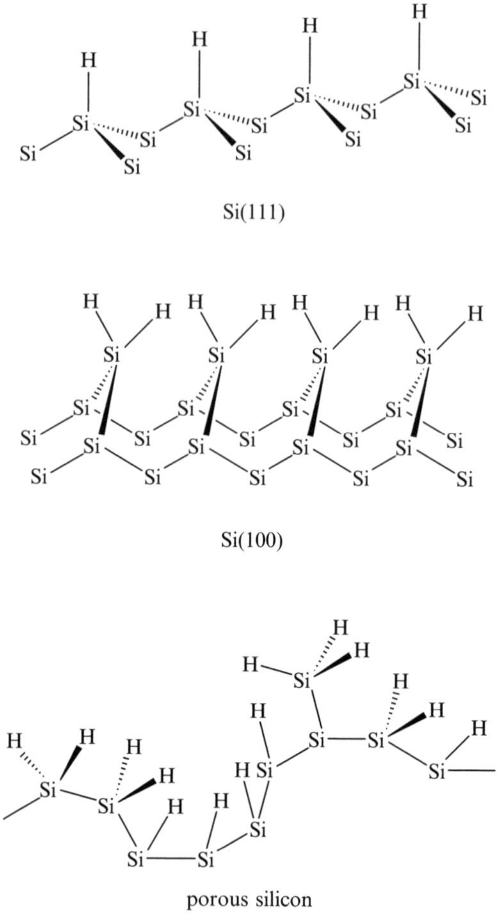

Figure 8.3 Hydrogen-terminated Si(111), Si(100) and porous silicon surfaces.

[46], which is comparable with the $\nu_{Si—H}$ of 2052 and 2075 cm^{-1} for $(Me_3Si)_3Si—H$ and $(Me_3Si)_2MeSi—H$, respectively. Therefore, it is not surprising that several of $(Me_3Si)_3SiH$ reactions have been adopted and applied to surfaces, as described here.

This section will cover aspects of monohydride terminal surface reactions that were carried out under free-radical conditions. The description will be circumscribed to the reactions with molecular oxygen and monounsaturated compounds. Mechanistic information for these reactions is scarce mainly due to the complexity of the system, and mechanistic schemes are often proposed in analogy with radical chemistry of organosilane molecules. H—Si(111) has a band gap of about 1.1 eV while the HOMO–LUMO gap in $(Me_3Si)_3SiH$ is within 8–11 eV and, therefore, has very important consequences for the reactions with nucleophilic and electrophilic species where frontier orbital interactions determine the reactivity and selectivity of the reaction [46]. On the other hand, in radical chemistry consequences should be attenuated.

8.5.1 OXIDATION OF HYDROGEN-TERMINATED SILICON SURFACES

The reaction of hydrogen-terminated Si(111), H—Si(111), with molecular oxygen under photochemical conditions has been investigated in some detail [48]. Figure 8.4 highlights the photoinduced reactivity of H—Si(111) in air followed by FTIR. Spectrum (a) shows that the exposure of material to air in the dark does not affect its stability, whereas after 30 min of photolysis with a mercury lamp, the area of the original peak decreases to 14% (spectrum d). X-ray photoelectron spectroscopy (XPS) of the irradiated surface shows a strong O(1s) signal and a shifted Si(2p) signal indicating the formation of SiO_2. Ellipsometry studies reveal the presence of an 8 Å film. When the surface is exposed to 350 nm light for the same total energy, the loss of Si—H bonds is decreased appreciably (spectrum c). With 450 nm light no detectable loss of Si—H is observed (spectrum b).

It is worth mentioning that the photooxidation of porous silicon behaves differently [49]. Indeed, FTIR spectra show that there is a tremendous increase in $\nu_{Si—O}$, without a correspondingly large loss of $\nu_{Si—H}$ peak intensity. The decrease of the $\nu_{Si—H}$ band is offset by an increase in the $\nu_{OSi—H}$ band, resulting in no net loss of hydride species on the surface during the course of the photooxidation reaction. These data apparently suggest that oxidation does not result in the removal of H atoms, implying that Si—Si bonds are attacked directly.

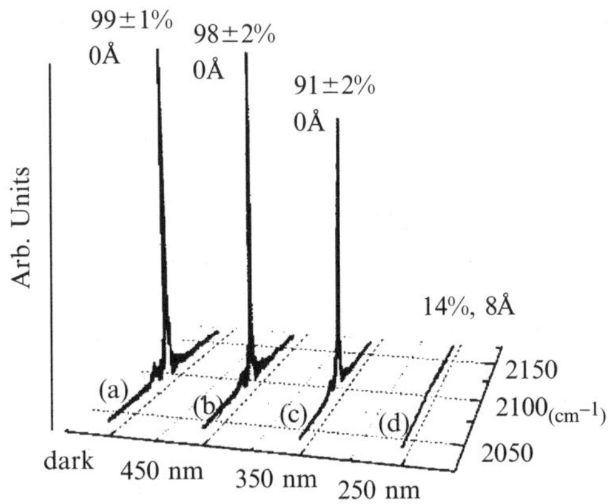

Figure 8.4 ATR-FTIR spectra of H—Si(111) after exposure to air (a) in the dark, (b) with 450 nm (broadband) illumination, (c) with 350 nm (broadband) illumination, and (d) with Hg lamp illumination. Reprinted with permission from Reference [48]. Copyright 2000 American Chemical Society.

The mechanism for the photooxidation of silicon surfaces is not well understood [48,49]. An early mechanistic proposal in which O_2 absorbs to silicon surface atoms and dissociates before it inserts into Si—Si bonds seems to have been abandoned. More recently, it was suggested that surface silyl radicals (**48**), formed by UV irradiation of H—Si(111), react with oxygen to form a peroxyl radical (**49**) that can abstract a neighbouring hydrogen to produce a new surface dangling bond (**50**) (Scheme 8.10) [48]. How the migration of oxygen into the Si—Si backbone of the lattice occurs and how the regeneration of surface silyl radical or Si—H bond arises is not yet understood. If we keep the analogy with the oxidation of $(Me_3Si)_3Si$—H and $(Me_3Si)_2MeSi$—H and poly(hydrosilane)s described earlier in this chapter, the mechanism reported in Scheme 8.10 can be proposed. The peroxyl radical **49** rearranges to silyloxyl radical **51** by the oxygen insertion step, which is found to be ca $10^4\ s^{-1}$ for the analogous reaction in $(Me_3Si)_3Si$—H. The alternative neighbouring hydrogen abstraction to give **50** is expected to be much slower, for analogy with cumylperoxyl radical hydrogen abstraction from $(Me_3Si)_3Si$—H occurring with a rate constant of $66\ M^{-1}\ s^{-1}$ at 73 °C. In turn radical **51** is expected to undergo a fast 1,2-silyl shift to give silyl radical **52**, which can add to oxygen to give peroxyl radical **53** followed by oxygen insertion to the remaining γ(Si—Si) bond. The resulting silyloxyl radical **54** is ready to undergo 1,5 hydrogen shift to give another surface silyl radical (**55**). Radical **51** can also abstract neighbouring hydrogen via a six-membered transition state from the side that already inserted an oxygen atom to give another surface silyl radical (**56**). The latter could add to oxygen to give **57** and continue the chain, eventually until complete oxidation of the surface.

Both 1,5 hydrogen and 1,2 silyl shifts from radical **51** are expected to be very fast. The bimolecular rate constant for hydrogen abstraction from $(Me_3Si)_3SiH$ by alkoxyl radical is $1 \times 10^8\ M^{-1}s^{-1}$ which suggests that the intramolecular version could be even two orders of magnitude faster. On the other hand, a rate constant $> 10^8\ s^{-1}$ is estimated for the 1,2 silyl shift in the oxidation of $(Me_3Si)_3SiH$. Therefore, it is likely that the preferred path will strongly depend on entropic factors determined by the rigidity of the surface. It should also be noticed that the silyl radical **52** has two silyloxyl substituents and a 1,*x* H shift from an $(Si)_3Si$—H moiety is expected to be strongly exothermic. When favourable transition states are formed such 1,*x* H shifts should be quite fast to give **59** (the silyl radical moiety is not shown). Therefore, the reaction mechanism for the surface oxidation given in Scheme 8.10 could lead to Si—OH or Si—H as terminal groups depending on the type of silicon surface.

The oxidation of hydrogen-terminated silicon surfaces by molecular oxygen also occurs without irradiation. Scanning tunnelling microscopy (STM) investigation shows that the exposure of H—Si(111) to O_2 induces surface modification that is assigned to the insertion of oxygen atoms into the Si—Si backbone [50]. However, the exposure of hydrogen-terminated silicon surfaces either to dry molecular oxygen or to deoxygenated water gives low oxide growth rates, whereas the combination of water and oxygen results in a significantly faster

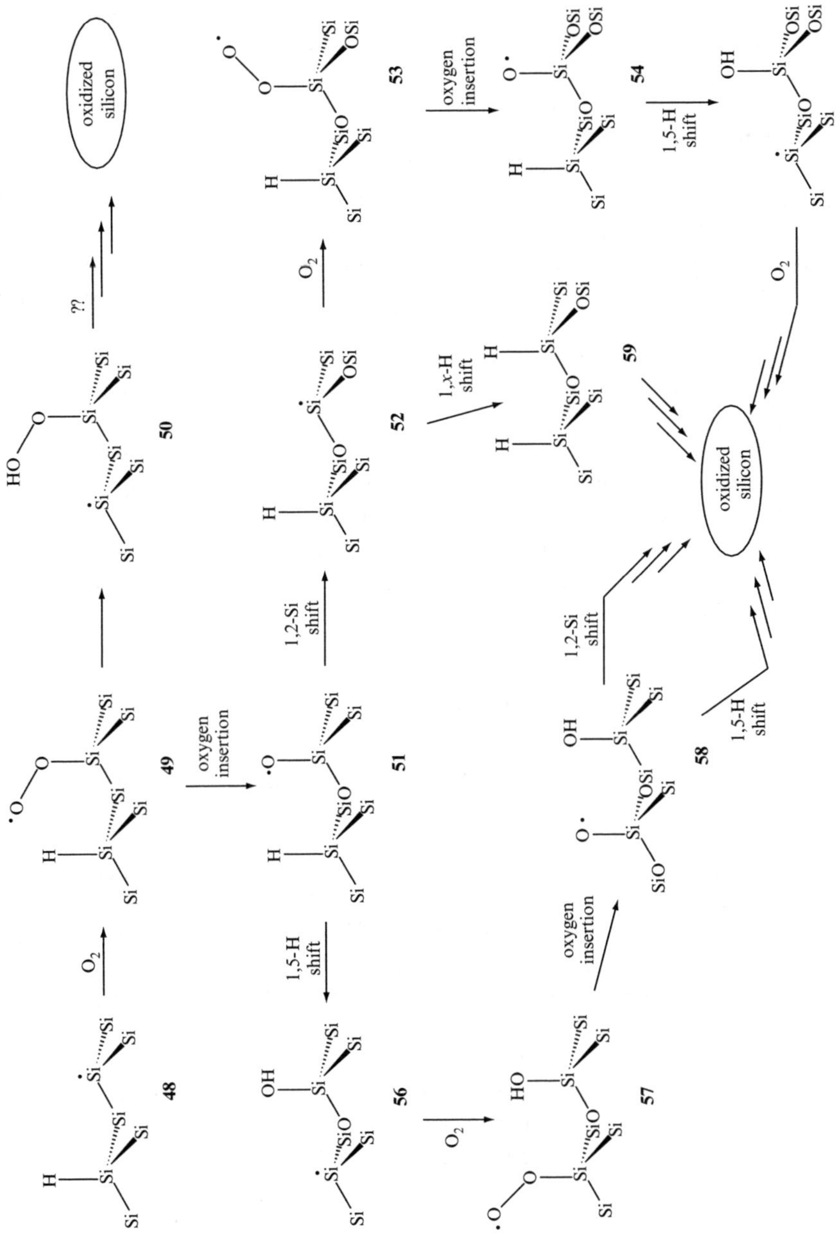

Scheme 8.10 Proposed mechanism for the oxidation of silicon surfaces

oxide growth rate. Indeed, infrared studies have shown that the half-life of the Si—H stretch in air of an H—Si(111) surface is humidity dependent. It was suggested that the oxidation of Si—H to Si—OH by H_2O is the rate-limiting step in the native oxide formation [51]. On the other hand, dissolved oxygen in 40% aqueous ammonium fluoride solution initiates the formation of etch pits in the terraces of the otherwise ideal H—Si(111) surface. Removal of oxygen from the fluoride solution by sparging with argon substantially reduces the initiation of etch pits. It was suggested that oxygen molecules are reduced to superoxide radicals ($O_2^{-\bullet}$) at the negative open-circuit potential of the silicon surface and that $O_2^{-\bullet}$ abstracts hydrogen atoms from the H—Si(111) terraces to form silicon radicals [52].

Whatever the initial step of formation of surface silyl radicals, the mechanism for the oxidation of silicon surfaces by O_2 is expected to be similar to the proposed Scheme 8.10. This proposal is also in agreement with the various spectroscopic measurements that provided evidence for a peroxyl radical species on the surface of silicon [53] during thermal oxidation (see also references cited in [50]). The reaction being a surface radical chain oxidation, it is obvious that temperature, efficiency of radical initiation, surface precursor and oxygen concentration will play important roles in the acceleration of the surface oxidation and outcome of oxidation.

8.5.2 HALOGENATION OF H—Si(111)

The analogy drawn between radical reactions of organosilanes and those of H—Si(111) led to the exploration of halogenation reactions. Chlorine-terminated Si(111) surfaces, Cl—Si(111), were prepared by treating H—Si(111) with PCl_5 in chlorobenzene under free-radical conditions using either thermal decomposition of dibenzoyl peroxide [54] or UV illumination under reflux [55]. Cl—Si(111) was also obtained by reaction of Cl_2 gas with H—Si(111) activated either by heating at 80 ° C for 10 min or by irradiating with a 300 W tungsten lamp for 2 min at room temperature [56]. Analogously, H—Si(111) surfaces were brominated by reaction with *N*-bromosuccinimide or $BrCCl_3$ to give Br—Si(111) using radical initiating conditions [57]. Halogen-terminated Si(111), X—Si(111), were characterized by several spectroscopic methods (XPS; AES; ATR-FTIR, HREELS, and NEXAFS) and have been further used in reactions with organolithium reagents (RLi) to obtain monolayer films R—Si(111) [54,55,57].

8.5.3 ADDITION OF UNSATURATED COMPOUNDS ON H—Si(111)

Wet chemical procedures for forming covalently bonded monolayer films on H—Si(111) are well developed. Among various methods, the radical approach is the most common one.

H—Si(111) undergoes radical-activated reactions with a variety of terminal olefins to afford densely packed monolayers [48,57,58]. Reactions were carried out in neat deoxygenated alkenes using thermal decomposition of diacyl peroxides as the radical initiation. In the absence of radical initiation, it was found that the hydrosilylation process occurs either by heating at temperatures >150 °C or by UV irradiation, probably through a homolytic cleavage of the Si—H bond. ATR-FTIR and XPS spectroscopies as well as scanned-energy photoelectron diffraction (PED) provide evidence for a covalent bond between carbon and silicon. The modified surface could withstand exposure to boiling water, boiling $CHCl_3$ and sonication in CH_2Cl_2, which also suggests chemisorption and not physisorption of the monolayer to the silicon. Approximately 50% of the H—Si(111) groups reacted with 1-alkene and in agreement with the outcome of molecular modelling [48,59]. The alkyl chains are tilted ca 30° with respect to surface normal and are mainly in the all-*trans* conformation [60]. It is also worth mentioning that similar monolayers on H—Si(111) surface were obtained by Lewis acid-catalysed hydrosilylation of alkenes or by direct reaction with alkylmagnesium bromide [61].

The reaction is formally a hydrosilylation process analogous to the homogeneous reactions described in Chapter 5. Scheme 8.11 shows the proposed H—Si(111) surface-propagated radical chain mechanism [48]. The initially formed surface silyl radical reacts with alkene to form a secondary alkyl radical that abstracts hydrogen from a vicinal Si—H bond and creates another surface silyl radical. The best candidate for the radical translocation from the carbon atom of the alkyl chain to a silicon surface is the 1,5 hydrogen shift shown in Scheme 8.11. Hydrogen abstraction from the neat alkene, in particular from the

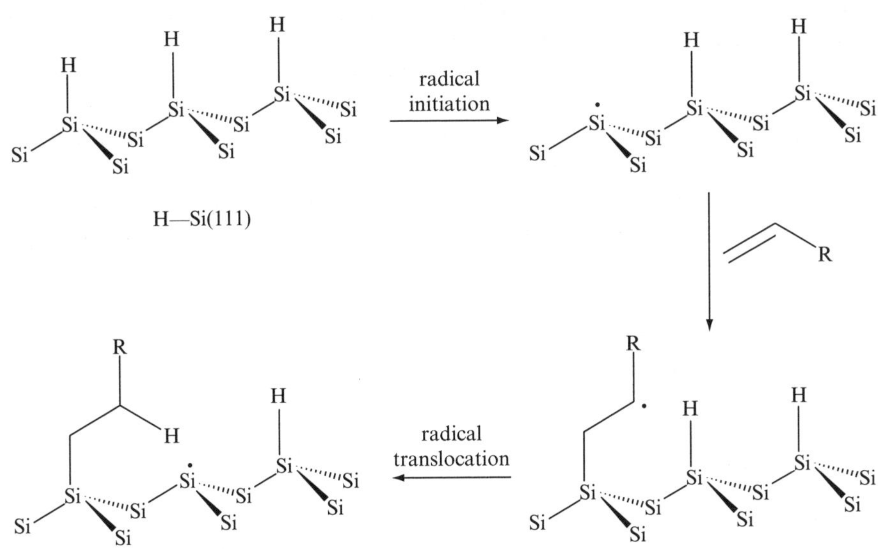

Scheme 8.11 Hydrosilylation of an alkene by hydrogen terminated Si(111)

allylic position, may be competitive. However, lack of significant polymerization when styrene is used as 1-alkene, indicates the efficiency of the 1,5 hydrogen shift with respect to carbon–carbon bond formation [48]. Electrons from a scanning tunnelling microscope in ultrahigh vacuum have been used to create surface isolated silyl radical on Si(111) and their exposure to styrene leads to the formation of compact islands containing multiple styrene adsorbates bonded to the surface through individual C—Si bonds [62]. These observations support the radical chain reaction mechanism of Scheme 8.11 and indicate that the chain reaction does not propagate in a single direction on H—Si(111). The fact that UV irradiation of O_2-saturated terminal olefins and H—Si(111) leads to the formation of oxidized silicon and a partial hydrocarbon monolayer [48] indicates competitive reactions of oxygen and olefin with surface silyl radical.

Terminal acetylenes, such as 1-octyne or phenylacetylene, also form monolayers on H—Si(111) when initiated by diacyl peroxide decomposition or UV irradiation [48,58]. Evidence that a vinyl group is attached to the Si surface is obtained from the ATR-FTIR and XPS spectra. Also in this case, the proposed mechanism is a surface propagation chain reaction in which a vinyl radical, formed by the addition of alkyne to a surface silyl radical, abstracts a hydrogen atom from the adjacent site of silicon backbone. The efficiency of this hydrogen transfer is expected to depend also on the shape of the intermediate carbon-centred radical. Figure 8.5 shows the shapes of radical-adducts derived from 1-octyne (**60**) and phenylacetylene (**61**) being a σ-type and π-type radical, respectively (see also Section 5.2.1). The decrease in packing density observed on going from 1-octyne to phenylacetylene [48] could also be due to different radical chain efficiencies and, in particular, of the hydrogen transfer step.

The functionalization of H—Si(111) surfaces has been extended to the reaction with aldehydes. Indeed, H—Si(111) reacts thermally (16 h at 85 °C) with decanal to form the corresponding Si—OCH_2R monolayer that has been characterized by ATR-FTIR, XPS and atomic force microscopy (AFM) [63]. The reaction is thought to proceed either by a radical chain mechanism via adventitious radical initiation or by nucleophilic addition/hydride transfer. On the other hand, the reaction of H—Si(111) with octadecanal activated by irradiation with a 150 W mercury vapour lamp (21 h at 20–50 °C) afforded a

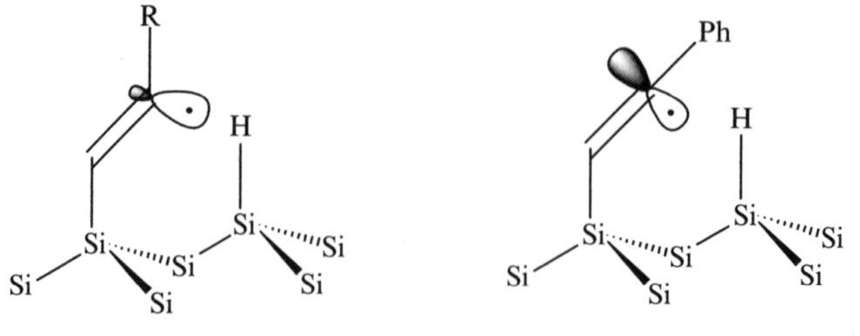

60 **61**

Figure 8.5 σ-type (**60**) and π-type (**61**) vinyl radicals.

well-ordered monolayer with surprisingly high coverage (97 %) [64]. For comparison, the analogous reaction with 1-octadecene afforded a 55 % coverage.

The dependence of monolayer formation on the wavelength of the irradiation light (400–350 nm) and the reaction time has been studied in some details [64]. Interestingly, the optimum wavelength for both C═O and C═C additions was found to be about 385 nm, whereas later it was reported that the reaction efficiency increases monotonically with decreasing wavelength up to 254 nm (see Figure 8.4). However, the ability to photoinitiate the reaction of unsaturated compounds with H—Si(111) opens the possibility of direct photopatterning of organics on the surface. Figure 8.6 shows schematically the light-promoted hydrosilylation with 1-alkene in the presence of a mask that can pattern the surface. The first photolithographic accomplishment to write >100 μm features, was obtained by exposing to 385 nm light through a structured mask octadecanal with an H—S(111) surface [64]. It was further demonstrated that the unexposed areas remained completely reactive.

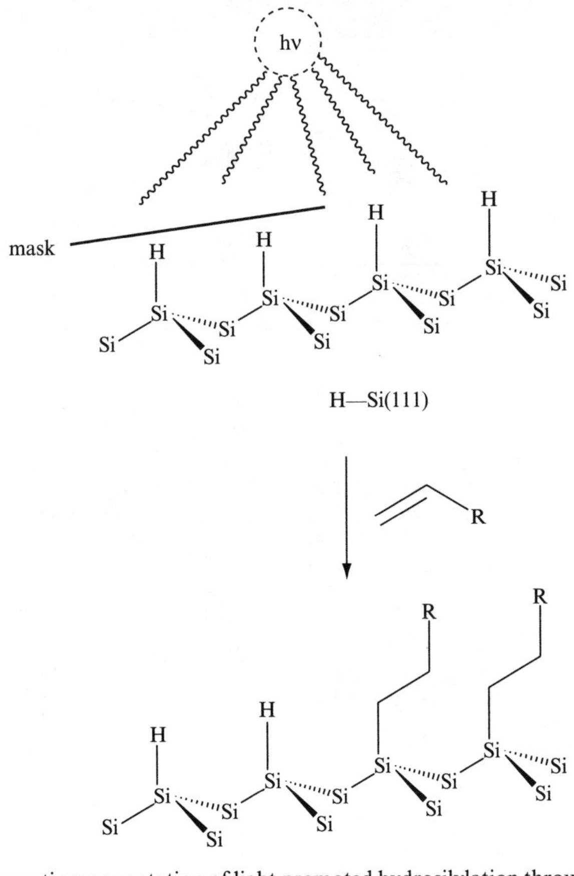

Figure 8.6 Schematic representation of light promoted hydrosilylation through masking procedure.

The fabrication of Si(111) surfaces patterned with organic monolayers and oxide on the 10 μm scale has also been reported [65]. An H—Si(111) surface was irradiated at 254 nm through a gold minigrid mask (Figure 8.7a) in air for 30 min to create a pattern of silicon oxide on an otherwise hydrogen-terminated Si(111) surface. Immersion of this sample in deoxygenated 1-decene and irradiation (300 nm) for 3 h induces hydrosilylation in the remaining hydrogen-terminated surface and results in an oxide/alkyl pattern. AFM and CCD optical images of these patterned surfaces are shown in Figure 8.7. Figure 8.7b reveals a well-ordered pattern after exposure to water vapour with the oxide (squares) being hydrophilic and 1-decene-modified surface (lines) being hydrophobic. Figure 8.7c shows an organic surface that is topographically higher than the oxide, whereas in the frictional force mode image (Figure 8.7d) the contrast is improved by highlighting the silicon oxide squares.

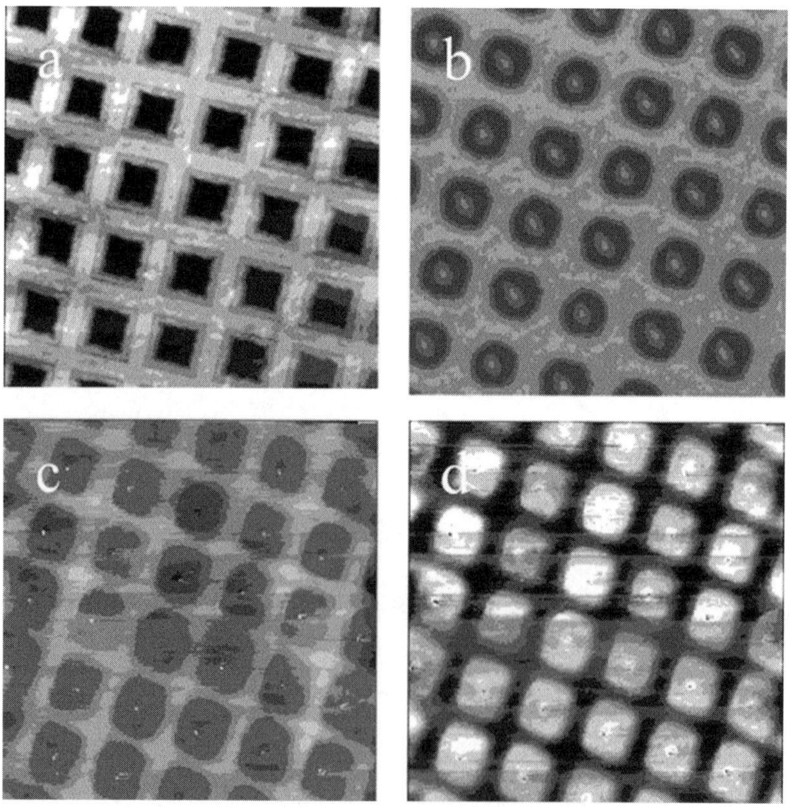

Figure 8.7 140 μm × 140 μm images of (a) AFM tapping mode image of the gold minigrid mask, (b) CCD optical image of the patterned Si(111) surface after exposure to water vapor, (c) AFM height image of the patterned surface, and (d) AFM image taken in frictional force mode. Reprinted with permission from Reference [65]. Copyright 2001 American Chemical Society.

8.5.4 ADDITION OF ALKENES ON Si(100) SURFACES

The method of heating at 200 °C without radical initiator has been extended to the monolayer formation on a Si(100) surface (see Figure 8.3) [66]. The reaction works also well with ω-functionalized alkenes and allows further manipulation of the monolayers, such as the removal of protecting groups of acids or alcohols. A major limitation of the above mentioned hydrosilylation procedures, is the large excess of alkene required since the freshly prepared fluoride ion-etched silicon hydride surface was placed in pure liquid alkene. It was reported that it is possible to work in solution with dilute alkene but the choice of solvent has an influence on the molecular ordering of the alkyl monolayers [67]. Mesitylene was the only solvent in the study that provided well ordered monolayers and it was suggested that it is too large to fit in pinholes of the forming film and thus, it cannot interfere with the monolayer formation process.

It is worth underlining two fundamental aspects of organosilicon hydrides in solution that may be also important to surface radical chain reactions: (i) The dissociation enthalpy of H—Si bond in the surface should be strongly dependent on the nature of the surface. Based on the available thermochemical data of organosilicon hydrides reported in Chapter 2, the H—Si(111) is expected to be about 10 kJ/mol weaker than the Si—H bond in dihydride-terminated Si(100) surfaces. Consequently, the rate constant for hydrogen abstraction should vary substantially depending on the nature of the silicon surface (cf. Chapter 3). (ii) The hydrogen donor abilities of solvent could also play an important role like that observed in reduction reactions (cf. Chapter 4). For example, it would be not very surprising to discover that mesitylene is also a mediator in the hydrosilylation of dihydride-terminated Si(100) surface with 1-alkenes. Therefore, attention should be paid to extrapolating a mechanistic scheme from one to another silicon surface or in comparing two similar reaction products derived from different silicon material.

Under ultrahigh vacuum conditions and exposure to hydrogen atoms it is possible to produce the so-called Si(100) 2×1 dimer surface, in which the Si—H surface bonds decrease from two per silicon to only one (Scheme 8.12) [46,48]. It was shown that the addition of styrene to H—Si(100) surface could be initiated from isolated surfaces silyl radicals created using the tip of the scanning tunnelling microscope [68]. Sequence of STM images showed that by increasing the exposure to styrene, growth lines up to 13 nm long (corresponding to 34 adsorption sites) were observed. However, growth of the lines was observed to stop at pre-existing defects of the surface. It was proposed that the reaction proceeds through a surface radical chain reaction as shown in Scheme 8.12. The mechanism consists of the addition of surface silyl radical to styrene followed by a 1,5 hydrogen shift to generate another surface silyl radical that continues the chain. The interesting consequence of the anisotropic nature of the H—Si(100) surface is the preferential growth along one edge of the silicon dimer row, whereas in the analogous H—Si(111) surface experiment compact

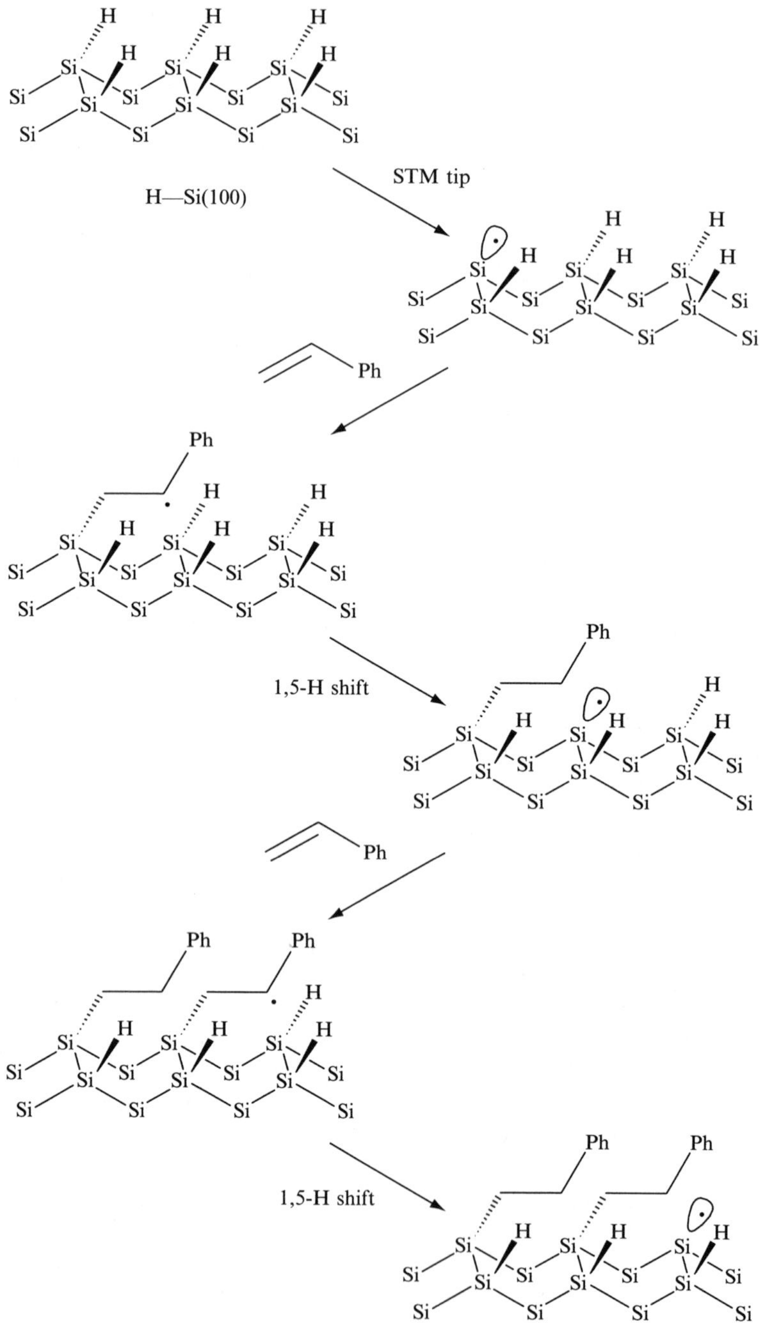

Scheme 8.12 Proposed mechanism of the growth of styrene line along the edge of a dimer of a H—Si(100) surface

islands were observed (see above). It is also worth mentioning that, although there is clear preference for lines to grow along the edge of a dimer row, occasionally double lines are observed suggesting that hydrogen atom abstraction across a dimer row is also possible (separation of silicon atoms within a dimer is 2.34 Å). Abstraction across a single dimer has not been observed and is likely to be a prohibitively strained process (separation of silicon atoms on either edge is 5.34 Å) [46].

8.5.5 SOME EXAMPLES OF TAILORED EXPERIMENTS ON MONOLAYERS

Organic monolayers bonded to silicon in order to be of significant utility often require chemical functionality other than the methyl groups that terminate the most easily prepared monolayers of alkyl chains. The use of α,ω-derivatized long-chain hydrocarbons has been a successful methodology, although not all functional groups are compatible for this purpose. For example, the reaction of Si(111) with 11-bromo-1-undecene has been used to prepare the anchor —$(CH_2)_{11}Br$ on the surface. Then, end-grafted polysilanes have been synthesized by reaction of polysilane having a silyl anion at the terminus with the bromoalkyl group [69]. This end-graft technique, gave the opportunity for direct AFM observation of polysilane single-molecule structures and allowed the physical properties of isolated polysilane chains to be measured. Another example is the reaction of Si(111) with ethyl undecylenate to afford the anchor —$(CH_2)_{10}C(O)OEt$ on the surface. Then, the ester groups were converted into di(2-thienyl)carbinol moieties by reaction with 2-thienyllithium and subsequently photoanodically oxidized in the presence of thiophene to yield a strongly adherent and smooth conducting film [70]. It is also worth mentioning that monolayers prepared from 1-octadecene on Si(111) were found to undergo chlorination exclusively on the terminal methyl group, when exposed to Cl_2 with illumination at 350 nm [71]. This unusual high selectivity of radical chlorination demonstrates a strong steric effect due to the densely packed array of alkyl chains that favors attack at the chain ends. Other free radical reactions are expected to be convenient ways to functionalize methyl-terminated alkyl monolayers.

8.6 REFERENCES

1. West, R., *J. Organomet. Chem.*, 1986, **300**, 327.
2. Miller, R.D., and Michl, J., *Chem. Rev.*, 1989, **89**, 1359.
3. West, R. In *The Chemistry of Organic Silicon Compounds*, Volume 3, Z. Rappoport and Y. Apeloig (Eds), Wiley, Chichester, 2001, pp. 541–563.
4. Koe, J.R., and Fujiki, M., *Silicon Chem.*, 2002, **1**, 77.
5. Harrod, J.F., *Prog. Catal.*, 1992, 147.

6. Tilley, T.D., *Acc. Chem. Res.*, 1993, **26**, 22.
7. Yamashita, H., and Tanaka, M., *Bull. Chem. Soc. Jpn.*, 1995, **68**, 403.
8. Banovetz, J.P., Suzuki, H., and Waymouth, R.M., *Organometallics*, 1993, **12**, 4700.
9. Woo, H-G., Kim, S-Y., Han, M-K., Cho, E.J., and Jung, I.N., *Organometallics*, 1995, **14**, 2415.
10. Banovetz, J.P., Stein, K.M., and Waymouth, R.M., *Organometallics*, 1991, **10**, 3430.
11. Grimmond, B.J., Rath, N.P., and Corey, J.Y., *Organometallics*, 2000, **19**, 2975.
12. Chatgilialoglu, C. Unpublished results.
13. Chatgilialoglu, C., Ferreri, C., Vecchi, D., Lucarini, M., and Pedulli, G.F., *J. Organomet. Chem.*, 1997, **545/546**, 475.
14. McKinley, A.J., Karatsu, T., Wallraff, G.M., Thompson, D.P., Miller, R.D., and Michl, J., *J. Am. Chem. Soc.*, 1991, **113**, 2003.
15. Chatgilialoglu, C., Guerrini, A., Lucarini, M., Pedulli, G.F., Carrozza, P., Da Roit, G., Borzatta, V., and Lucchini, V., *Organometallics*, 1998, **17**, 2169.
16. Zaborovskiy, A.B., Timokhin, V.I., and Chatgilialoglu, C., *Polym. Sci., Series A*, 2003, **45**, 612.
17. Chatgilialoglu, C., Timokhin, V.I., Zaborovskiy, A.B., Lutsyk, D.A., and Prystansky, R.E., *J. Chem. Soc., Perkin Trans. 2*, 2000, 577.
18. Chatgilialoglu, C., Guarini, A., Guerrini, A., and Seconi, G., *J. Org. Chem.*, 1992, **57**, 2207.
19. Zaborovskiy, A.B., Lutsyk, D.A., Prystansky, R.E., Kopylets, V.I., Timokhin, V.I., and Chatgilialoglu, C., Unpublished results.
20. Pratt, D.A., DiLabio, G.A., Brigati, G., Pedulli, G.F., and Valgimigli, L., *J. Am. Chem. Soc.*, 2001, **123**, 4625.
21. Banovetz, J.P., Hsiao, Y-L., and Waymouth, R.M., *J. Am. Chem. Soc.*, 1993, **115**, 2540.
22. Hsiao, Y-L., Banovetz, J.P., and Waymouth, R.M., *ACS Symp. Ser.*, 1994, **572**, 55.
23. Hsiao, Y-L., and Waymouth, R.M., *J. Am. Chem. Soc.*, 1994, **116**, 9779.
24. Beach, J.V., Loy, D.A., Hsiao, Y-L., and Waymouth, R.M., *Polym. Prepr., (Am. Chem. Soc., Div. Polym. Chem.)*, 1995, **72**, 205.
25. For the use of poly(hydrosilane)s as process stabilizers, see: Carrozza, P., Borzatta, V., Chatgilialoglu, C. PCT Int. Appl. WO97/02322, 1997; U.S. patent 6005036, 1999; EP patent 836635, 2001.
26. For the use of polysilane stabilizers containing sterically hindered amine groups, see: Carrozza, P., Borzatta, V., Chatgilialoglu, C., and Da Roit, G. PCT Int. Appl. WO00/64965, 2000; US patent 6538055, 2003; EP patent 1185577, 2003.
27. Hirsch, A., *Fullerenes and Related Structures*, Springer, Berlin, 1998.
28. Wang, Y., West, R., Yuan, and C.-H., *J. Am. Chem. Soc.*, 1993, **115**, 3844.
29. Keizer, P.N., Morton, J.R., Preston, K.F., and Krusic, P.J., *J. Chem. Soc., Perkin Trans. 2*, 1993, 1041.
30. Kusukawa T., and Ando, W., *Organometallics*, 1997, **16**, 4027.
31. Kusukawa K., and Ando, W., *J. Organomet. Chem.*, 1998, **559**, 11.
32. Akasaka, T., Suzuki, T., Maeda, Y., Ara, M., Wakahara, T., Kobayashi, K., Nagase, S., Kako, M., Nakadaira, Y., Fujitsuka M., and Ito, O., *J. Org. Chem.*, 1999, **64**, 566.
33. Kusukawa K., and Ando, W., *Angew. Chem. Int. Ed. Engl.*, 1996, **35**, 1315.
34. Ando, W., and Kusukawa K. In *The Chemistry of Organic Silicon Compounds,* Volume 2, Z. Rappoport and Y. Apeloig (Eds), Wiley, Chichester, 1998, pp. 1929–1960.
35. Wakahara, T., Kondo, T., Okamura, M., Akasaka, T., Hamada, Y., Suzuki, T., Kako, M., and Nakadaira, Y., *J. Organomet. Chem.*, 2000, **611**, 78.
36. Watanabe, A., and Ito, O., *J. Phys. Chem.*, 1994, **98**, 7736.

37. Akasaka, T., Ando, W., Kobayashi, K., and Nagase, S., *J. Am. Chem. Soc.*, 1993, **115**, 10366.
38. Maeda, Y., Sato, R., Wakahara, T., Okamura, M., Akasaka, T., Fujitsuka, M., Ito, O., Kobayashi, K., Nagase, S., Kako, M., Nakadaira, Y., and Horn, E., *J. Organomet. Chem.*, 2000, **611**, 414.
39. Han, A., Wakahara, T., Maeda, Y., Niino, Y., Akasaka, T., Yamamoto, K., Kako, M., Nakadaira, Y., Kobayashi, K., and Nagase, S., *Chem. Lett.*, 2001, 974.
40. Wakahara, T., Han, A., Niino, Y., Maeda, Y., Akasaka, T., Suzuki, T., Yamamoto, K., Kako, M., Nakadaira, Y., Kobayashi, K., and Nagase, S., *J. Mater. Chem.*, 2002, **12**, 2061.
41. Kusukawa, T., Kabe, Y., Erata, T., Nestler, B., and Ando, W., *Organometallics*, 1994, **13**, 4186.
42. Kusukawa, T., Kabe, Y., and Ando, W., *Organometallics*, 1995, **14**, 2142.
43. Sasaki, Y., Konishi, T., Fujitsuka, M., Ito, O., Maeda, Y., Wakahara, T., Akasaka, T., Kako, M., and Nakadaira, Y., *J. Organomet. Chem.*, 2000, **599**, 216.
44. Fujitsuka, M., Ito, O., Maeda, Y., Kako, M., Wakahara, T., and Akasaka, T., *Phys. Chem. Chem. Phys.*, 1999, **1**, 3527.
45. Kusukawa, T., Ohkubo, K., and Ando, W., *Organometallics*, 1997, **16**, 2746.
46. Wayner, D.D.M., and Wolkow, R.A., *J. Chem. Soc., Perkin Trans. 2*, 2002, 23.
47. Buriak, J.M., *Chem. Rev.*, 2002, **102**, 1271.
48. Cicero, R.L., Linford, M.R., and Chidsey, C.E.D., *Langmuir*, 2000, **16**, 5688.
49. Harper, J., and Sailor, M.J., *Langmuir*, 1997, **16**, 5688.
50. Hamers, R.J., and Wang, Y., *Chem. Rev.*, 1996, **96**, 1261.
51. Miura, T., Niwano, M., Shoji, D., and Miyamoto, N., *J. Appl. Phys.*, 1996, **79**, 4373.
52. Wade, C.P., and Chidsey, C.E.D., *Appl. Phys. Lett.*, 1997, **71**, 1679.
53. Vezenov, D.V., Gobulev, V.B., and Melnikov, V.P., *Macromolecules*, 2003, **36**, 1819.
54. Bansal, A., Li, X., Lauermann, I., Lewis, N.S., Yi, S.I., and Weinberg, W.H., *J. Am. Chem. Soc.*, 1996, **118**, 7225.
55. Okubo, T., Tsuchiya, H., Sadakata, M., Yasuda, T., and Tanaka, K., *Appl. Surf. Sci.*, 2001, **171**, 252.
56. Zhu, X.-Y., Boiadjiev, V., Mulder, J.A., Hsung, R.P., and Major, R.C., *Langmuir*, 2000, **16**, 6766.
57. He, J., Patitsas, S.N., Preston, K.F., Wolkow, R.A., and Wayner, D.D.M., *Chem. Phys. Lett.*, 1998, **286**, 508.
58. Linford, M.R., Fenter, P., Eisenberger, P.M., and Chidsey, C.E.D., *J. Am. Chem. Soc.*, 1995, **117**, 3145.
59. Terry, J., Linford, M.R., Wigren, C., Cao, R., Pianetta, P., and Chidsey, C.E.D., *Appl. Phys. Lett.*, 1997, **71**, 1056.
60. Sieval, A.B., van den Hout, B., Zuilhof, H., and Sudhölter, E.J.R., *Langmuir*, 2001, **17**, 2172.
61. Boukherroub, R., Morin, S., Bensebaa, F., and Wayner, D.D.M., *Langmuir*, 1999, **15**, 3831.
62. Cicero, R.L., Chidsey, C.E.D., Lopinski, G.P., Wayner, D.D.M., and Wolkow, R.A., *Langmuir*, 2002, **18**, 305.
63. Boukherroub, R., Morin, S., Sharpe, P., Wayner, D.D.M., and Allongue, P., *Langmuir*, 2000, **16**, 7429.
64. Effenberger, F., Götz, G., Bidlingmaier, B., and Wezstein, M., *Angew. Chem. Int. Ed.*, 1998, **37**, 2462.
65. Wolkow, R., Tomietto, M., Boukherroub, R., and Wayner, D.D.M., *J. Am. Chem. Soc.*, 2001, **123**, 1535.
66. Sieval, A.B., Demirel, A.L., Nissink, J.W.M., Linford, M.R., van der Maas, J.H., de Jeu, W.H., Zuilhof, H., and Sudhölter, E.J.R., *Langmuir*, 1998, **14**, 1759.

67. Sieval, A.B., Vleeming, V., Zuilhof, H., and Sudhölter, E.J.R., *Langmuir*, 1999, **15**, 8288.
68. Lopinski, G.P., Wayner, D.D.M., and Wolkow, R.A., *Nature*, 2000, **406**, 48.
69. Furukawa, K., *Acc. Chem. Res.*, 2003, **36**, 102.
70. Fabre, B., Lopinski, G.P., and Wayner, D.D.M., *Chem. Commun.*, 2002, 2904.
71. Linford, M.R., and Chidsey, C.E.D., *Langmuir*, 2002, **18**, 6217.

LIST OF ABBREVIATIONS

AES	Auger electron spectroscopy
AFM	atomic force microscopy
AH_2	ascorbic acid
AIBN	2′,2′-azobisisobutyronitrile
AMVN	2′,2′-azobis-(4-methoxy)-3,4-dimethyl-valeronitrile
APPH	2′,2′-azobis-(2-methylpropionamide)
AO	atomic orbital
Boc	*tert*-butyloxycarbonyl
c-	cyclic
Cp	η^5-cyclopentadienyl
Cp^*	η^5-pentamethylcyclopentadienyl
de	diastereomeric excess
DFT	density functional theory
DMA	9,10-dimethoxyanthracene
DNA	deoxyribonucleic acid
ds	diastereomeric ratio
EA	electron affinity
ee	enantiomeric excess
EPR	electron paramagnetic resonance
FTIR	Fourier transform infrared
HALS	hindered amines light stabilizer
HMQC	heteronuclear multiple-quantum coherence
HOMO	highest occupied molecular orbital
HREELS	high resolution electron energy loss spectroscopy
hfs	hyperfine splitting
GC	gas chromatography
GPC	gel permeation chromatography
i-	iso
IP	ionization potential
INEPT	insensitive nuclei enhanced by polarization transfer
LFP	laser flash photolysis
LUMO	lowest unoccupied molecular orbital
MO	molecular orbital
n-	normal
NEXAFS	near-edge X-ray absorption fine structure
NMR	nuclear magnetic resonance

PED	photoelectron diffraction
PMB	*p*-methoxybenzyl
s-	secondary
SOMO	singly occupied molecular orbital
STM	scanning tunneling microscope (or microscopy)
t- or -*t*	tertiary
TEMPO	2,2,6,6-tetramethylpiperidine
TFA	trifluoroacetic acid
THF	tetrahydrofuran
TMS	trimethylsilyl
Ts	*p*-toluenesulfonyl
XPS	X-ray photoemission spectroscopy
UV	ultraviolet

SUBJECT INDEX